Shair Ahmad, Ivanka M. Stamova (Eds.)

Lotka–Volterra and Related Systems

De Gruyter Series in Mathematics and Life Sciences

Volume 2

Lotka–Volterra and Related Systems

Recent Developments in Population Dynamics

Edited by
Shair Ahmad,
Ivanka M. Stamova

DE GRUYTER

Mathematics Subject Classification 2010
34C60, 34K60, 34D20, 34K12, 34K20, 34A37, 37C25, 37D45, 37F15

Authors

Zhanyuan Hou
London Metropolitan University
Faculty of Life Sciences and Computing
School of Computing
166-220 Holloway Road
London N7 8DB
United Kingdom
z.hou@londonmet.ac.uk

Marina Pireddu
Università degli Studi di Milano-Bicocca
Dipartimento di Matematica e Applicazioni
Via R. Cozzi 53
20126 Milano
Italy
marina.pireddu@unimib.it

Benedetta Lisena
Università degli Studi di Bari
Dipartimento di Matematica
Via E. Orabona 4
70125 Bari
Italy
lisena@dm.uniba.it

Fabio Zamolin
Università degli Studi di Udine
Dipartimento di Matematica e Informatica
Via delle Scienze 206
33100 Udine
Italy
fabio.zamolin@uniud.it

ISBN 978-3-11-026951-2
e-ISBN 978-3-11-026984-0
ISSN 2195-5530

Library of Congress Cataloging-in-Publication Data
A CIP catalog record for this book has been applied for at the Library of Congress.

Bibliographic information published by the Deutsche Nationalbibliothek
The Deutsche Nationalbibliothek lists this publication in the Deutsche Nationalbibliografie; detailed bibliographic data are available in the Internet at http://dnb.dnb.de.

Typesetting: le-tex publishing services GmbH, Leipzig
Printing and binding: Hubert & Co. GmbH & Co. KG, Göttingen
♾ Printed on acid-free paper
Printed in Germany

www.degruyter.com

Preface

The purpose of this book is to facilitate research in the general area of population dynamics by presenting some of the recent developments involving theories, methods and application in this important area of mathematics. Recently, there has been a tremendous amount of research done in this area. It would be a monumental, if not impossible, task to address all of the various models and methods dealing with various types of problems. This book utilizes the expertise of four authors to address some of the recent developments involving certain models in the area of population dynamics. The underlying common feature of these studies is that they are related, either directly or indirectly, to the well-known Lotka–Volterra systems which offer a variety of mathematical concepts from both theoretical and application points of view.

In spite of the technological advances, many authors seem to be unaware of the bulk of work that has recently been done in this important area of research. We hope that this book will have a positive impact by providing relevant information on the theory, methods and applications of an important area of research.

This book consists of three chapters.

Chapter 1 is authored by Zhanyuan Hou (School of Computing, London Metropolitan University, United Kingdom). The focus is only on permanence and stability from many interesting and fast developing aspects of differential equations. Due to the availability of several good books such as [15, 36, 37, 44], in order to avoid unnecessary duplication and stay within a certain page limit, the author limits his presentation to an overview and an update of recent developments in the research front of the two topics mentioned above. For permanence, the author deals with nonautonomous Kolmogorov differential systems with delays and, as a special class, Lotka–Volterra differential systems with delays. For stability, the classic autonomous Lotka–Volterra differential systems without delays are considered.

Chapter 2 is authored by Benedetta Lisena (Dipartimento di Matematica, Università degli Studi di Bari, Italy). It is dedicated to some asymptotic stability questions concerning Lotka–Volterra competitive systems with periodic coefficients. The study of such models is relevant for many significant applications in which the interacting species are subject to the effects of a periodically varying environment. In such a situation, it is natural to suppose that the coefficients in the equations are periodic functions. The corresponding differential systems, in spite of their seemingly simple structure, sometimes require a delicate analysis. Interesting problems arise even in the lower dimensional case, namely, when only two or three species are considered. The results mainly concentrate on two-dimensional and three-dimensional models; competitive systems with impulses are also considered.

Finally, Chapter 3, which is the main part of this book, is coauthored by Marina Pireddu (Università degli Studi di Milano-Bicocca, Dipartimento di Matematica e

Applicazioni, Italy) and Fabio Zanolin (Università degli Studi di Udine, Dipartimento di Matematica e Informatica, Italy). They introduce a topological method for the search of fixed points and periodic points for continuous maps defined on rectangular domains in finite dimensional Euclidean spaces. Such a method can be applied to obtain the existence and multiplicity of periodic solutions as well as to detect (in a rigorous analytical manner) the presence of chaotic-like dynamics. Some nonlinear ODEs with periodic coefficients are analyzed, with specific emphasis on the applications in population dynamics. In more detail, a modified version of the Volterra predator–prey model is considered in which a periodic harvesting is included.

The book should be of particular importance to researchers in mathematics, biology, engineering and other scientists interested in these types of models.

San Antonio, TX, 2012 *The Editors*

Contents

Preface — **v**

Zhanyuan Hou
Permanence, global attraction and stability — 1
1 Introduction — **1**
2 Existence of a compact uniform attractor — **3**
3 Proof of Theorems 2.1, 2.2 and 2.3 — **8**
4 Partial permanence and permanence — **15**
5 Necessary conditions for permanence of Lotka–Volterra systems — **26**
6 Sufficient condition for permanence of Lotka–Volterra systems — **31**
7 Further notes — **39**
8 Global attraction and stability of Lotka–Volterra systems — **39**
9 Global stability by Lyapunov functions — **40**
10 Global stability by split Lyapunov functions — **42**
10.1 Checking the conditions (10.2) and (10.8) — **46**
10.2 Examples — **47**
11 Global stability of competitive Lotka–Volterra systems — **48**
12 Global attraction of competitive Lotka–Volterra systems — **55**
13 Some notes — **60**
Bibliography — **60**

Benedetta Lisena
Competitive Lotka–Volterra systems with periodic coefficients — 63
1 Introduction — **63**
2 The autonomous model. The logistic equation — **64**
3 Two species periodic models — **68**
4 Competitive exclusion — **76**
5 One species extinction in three-dimensional models — **82**
6 The impulsive logistic equation — **91**
7 Two species systems with impulsive effects. A look at the N-dimensional case — **95**
8 The influence of impulsive perturbations on extinction in three-species models — **109**
Bibliography — **121**

Marina Pireddu and Fabio Zanolin
Fixed points, periodic points and chaotic dynamics for continuous maps with applications to population dynamics — 123
1 Introduction — **123**
2 Notation — **125**
3 Search of fixed points for maps expansive along one direction — **127**
4 The planar case — **128**
4.1 Stretching along the paths and variants — **128**
4.2 The Crossing Lemma — **143**
5 The N-dimensional setting: Intersection Lemma — **152**
5.1 Zero-sets of maps depending on parameters — **157**
5.2 Stretching along the paths in the N-dimensional case — **163**
6 Chaotic dynamics for continuous maps — **168**
7 Definitions and main results — **172**
8 Symbolic dynamics — **181**
9 On various notions of chaos — **190**
10 Linked twist maps — **198**
11 Examples from the ODEs — **206**
12 Predator–prey model — **207**
12.1 The effects of a periodic harvesting — **207**
12.2 Technical details and proofs — **215**
Bibliography — **225**

Index — **235**

Zhanyuan Hou

Permanence, global attraction and stability

Zhanyuan Hou: School of Computing, London Metropolitan University, London, United Kingdom

1 Introduction

Kolmogorov systems of differential equations have been used to model many biological, ecological and other real world problems, and many variations of such systems have been extensively studied. Among the various investigations, permanence or uniform persistence is related to the problem of the coexistence of species and has received much attention in the last few decades.

In Sections 1–7, we consider the system

$$x_i'(t) = x_i(t) f_i(t, x_t)\,, \quad i \in I_N = \{1, \ldots, N\}\,, \tag{1.1}$$

where $f\colon \mathbb{R}_0 \times C \to \mathbb{R}^N$, $\mathbb{R}_0 = \mathbb{R}$ or $[\bar{t}, +\infty)$ or $(\bar{t}, +\infty)$ for some $\bar{t} \in \mathbb{R}$, $C = C([-\tau, 0], \mathbb{R}^N)$ for some $\tau > 0$ and $x_t(\theta) = x(t+\theta)$ for $\theta \in [-\tau, 0]$. For any integer $m \geq 1$, we denote the set $\{1, \ldots, m\}$ by I_m. The norm $|\cdot|$ for $\mathbb{R}^N$ is taken to be $|x| = \max\{|x_i|\colon i \in I_N\}$ and the norm $\|\cdot\|$ for C is taken to be $\|\varphi\| = \sup\{|\varphi(\theta)|\colon -\tau \leq \theta \leq 0\}$.

For (1.1), we assume that the $f_i(t, \varphi)$ are continuous on $\mathbb{R}_0 \times C$ and Lipschitzian in φ in each compact subset in $\mathbb{R}_0 \times C$. Then, each $\varphi_i(0) f_i(t, \varphi)$ is continuous on $\mathbb{R}_0 \times C$ and Lipschitzian in φ in each compact subset in $\mathbb{R}_0 \times C$. It is known from the general theory of functional differential equations (e. g. [12] or [27, Ch. 2]) that, for each $t_0 \in \mathbb{R}_0$ and every $\varphi \in C$, the solution $x(t, t_0, \varphi)$ of (1.1) with $x_{t_0} = \varphi$ exists for t in its maximum interval $[t_0 - \tau, \mu)$, and the solution is unique and depends on (t_0, φ, f) continuously. Moreover, due to the special form of (1.1), its solution also satisfies

$$\forall t \in [t_0, \mu)\,, \quad \forall i \in I_N\,, \quad x_i(t, t_0, \varphi) = \varphi_i(0) \exp\left(\int_{t_0}^{t} f_i(s, x_s(t_0, \varphi))\, ds \right). \tag{1.2}$$

Thus, $\varphi_i(0) > 0\,(= 0$ or $< 0)$ implies $x_i(t, t_0, \varphi) > 0\,(= 0$ or $< 0)$ for all $t \in [t_0, \mu)$.

Let $C^+ = C([-\tau, 0], \mathbb{R}_+^N)$. From (1.2), we see that $\varphi \in C^+$ implies $x_t(t_0, \varphi) \in C^+$ for all $t \in [t_0, \mu)$. Since $x_i(t)$ represents the population size or density of the ith species if (1.1) is viewed as a mathematical model of the populations in a biological or ecological community, the restriction of $x(t)$ to $\mathbb{R}_+^N$ and x_t to C^+ makes sense. From now on, we consider (1.1) only on $\mathbb{R}_0 \times C^+$. Viewing (1.1) as a population model, we are concerned with the problem of coexistence of the species and the long term future of the species. Mathematically, we are concerned with the global asymptotic behavior of the solutions in $\mathbb{R}_+^N$ in general, and with permanence and stability in particular.

Definition 1.1. We say that the solutions of (1.1) are uniformly bounded if there is a number $M > 0$ such that every solution of (1.1) in $\mathbb{R}_+^N$ exists on $[t_0, +\infty)$ and satisfies $|x(t)| < M$ for all large enough t. In this case, we simply say that (1.1) is uniformly bounded.

Definition 1.2. We say that the solutions of (1.1) in $\operatorname{int}\mathbb{R}_+^N$ are uniformly bounded away from the boundary $\partial\mathbb{R}_+^N$ if there is a $\delta > 0$ such that every solution in $\operatorname{int}\mathbb{R}_+^N$ satisfies

$$\forall i \in I_N\,, \quad \forall \text{ large enough } t\,, \quad x_i(t) > \delta\,.$$

Definition 1.3. For a nonempty subset $J \subset I_N$, (1.1) is said to be partially permanent with respect to J, or J-permanent in short, if there exist $\delta > 0$ and $M > \delta$ such that for each $(t_0, \varphi) \in \mathbb{R}_0 \times C^+$ with $\varphi_i(0) > 0$ for all $i \in J$, the solution $x(t, t_0, \varphi)$ satisfies

$$\forall i \in J\,, \quad \forall \text{ large enough } t\,, \quad \delta < x_i(t, t_0, \varphi) < M\,.$$

System (1.1) is said to be permanent if it is J-permanent with $J = I_N$.

From these definitions, it follows that (1.1) is permanent if it is uniformly bounded and its solutions in $\operatorname{int}\mathbb{R}_+^N$ are uniformly bounded away from $\partial\mathbb{R}_+^N$. Conversely, if (1.1) is permanent, then the solutions of (1.1) in $\operatorname{int}\mathbb{R}_+^N$ are uniformly bounded and uniformly bounded away from the boundary $\partial\mathbb{R}_+^N$. Before embarking on these properties in the following sections, we first make a few observations about the solution $x(t, t_0, \varphi)$ of (1.1) on its existing interval $[t_0, \mu)$.

(O1) In general, we may not have $\mu = +\infty$. For example, consider the equation

$$y'(t) = y(t)\left[1 + y(t) - y(t-1) + y(t-2)\right] \tag{1.3}$$

as a particular case of (1.1) with $N = 1$. The solution of (1.3) with $y_0(\theta) \equiv \alpha > 0$ for $\theta \in [-2, 0]$ satisfies

$$y'(t) = y(t)[1 + y(t)]\,, \quad t \in [0,1] \cap [0, \mu)\,.$$

Thus,

$$y(t) = \frac{e^t}{1 + \alpha^{-1} - e^t}\,, \quad t \in [0,1] \cap [0, \mu)\,.$$

From this, we obtain $\mu = \ln(1 + \alpha^{-1}) \le 1$ if $\alpha \ge (e-1)^{-1}$.

(O2) Even if $x(t, t_0, \varphi)$ exists on $[t_0, +\infty)$, it may not be bounded. For example, consider the system

$$x_1'(t) = x_1(t)\left[2 + \sin\left(x_2^2(t-1)\right)\right]\,, \quad x_2'(t) = x_2(t)\left[2 + \cos\left(x_1^2(t-1)\right)\right] \tag{1.4}$$

as a special case of (1.1) with $N = 2$. Then, the solution of (1.4) with $x_0 = \varphi \in C^+$ and $\varphi(0) \in \operatorname{int}\mathbb{R}^2_+$ exists on $[0, +\infty)$ and satisfies

$$\varphi_i(0)e^t \le x_i(t) \le \varphi_i(0)e^{3t}\,, \quad t \ge 0\,.$$

Clearly, $x_i(t) \to +\infty$ as $t \to +\infty$ so the solution is unbounded.

(O3) Even if every solution $x(t, t_0, \varphi)$ is bounded on $[t_0, +\infty)$, (1.1) may not be uniformly bounded. For example, the equation

$$x'(t) = x(t)\,[1 + \sin(x(t))] \tag{1.5}$$

has nonzero fixed points $x = (2k + \frac{3}{2})\pi\,, \quad k \in \mathbb{Z}$, and each positive solution is between two neighboring fixed points so it is bounded. Obviously, (1.5) is not uniformly bounded.

(O4) The solutions of (1.1) in $\operatorname{int}\mathbb{R}^N_+$ may not be uniformly bounded away from the boundary $\partial\mathbb{R}^N_+$ even if the system is uniformly bounded. For example, consider the two-dimensional competitive Lotka–Volterra system

$$x'(t) = x(t)\,[1 - x(t) - y(t)]\,, \quad y'(t) = y(t)\,[1 - 2x(t) - 3y(t)]\,. \tag{1.6}$$

Take $M > 1$. Then, $x'(t) \le 1 - M < 0$ if $x(t) \ge M$ and $y'(t) \le 1 - 3M < 0$ if $y(t) \ge M$. This shows that every solution of (1.6) in $\operatorname{int}\mathbb{R}^2_+$ satisfies $0 < x(t) < M$ and $0 < y(t) < M$ for all large enough t. Thus, (1.6) is uniformly bounded. Moreover, by a simple phase portrait, we see that the equilibrium $(1, 0)$ is globally asymptotically stable so that every solution in $\operatorname{int}\mathbb{R}^2_+$ satisfies $\lim_{t\to+\infty}(x(t), y(t)) = (1, 0)$. As $(1, 0)^T \in \partial\mathbb{R}^2_+$, the solutions of (1.6) in $\operatorname{int}\mathbb{R}^2_+$ are not uniformly bounded away from $\partial\mathbb{R}^2_+$. Indeed, (1.6) is partially permanent with respect to $J = \{1\}$.

(O5) A permanent system without delays may no longer be permanent after the replacement of some $x_i(t)$ by $x_i(t - \rho)$. For example, the cooperative system

$$x'(t) = x(t)\,[1 - ex(t) + y(t)]\,, \quad y'(t) = y(t)\,[1 + x(t) - ey(t)] \tag{1.7}$$

has a globally asymptotically stable fixed point $(\frac{1}{e-1}, \frac{1}{e-1})^T$. However, the delayed system

$$x'(t) = x(t)\,[1 - ex(t-1) + y(t)]\,, \quad y'(t) = y(t)\,[1 + x(t) - ey(t-1)] \tag{1.8}$$

has an unbounded solution $x(t) = y(t) = e^t$.

2 Existence of a compact uniform attractor

From Definitions 1.1–1.3, we see that if we manage to know the uniform boundedness of (1.1), then we need only know whether the solutions in $\operatorname{int}\mathbb{R}^N_+$ are uniformly bounded away from $\partial\mathbb{R}^N_+$ in order to determine whether the system is permanent. For the special case of (1.1) with $\tau = 0$, the initial value and state of the system is restricted to $\mathbb{R}_+^N$ instead of C^+. Since the closure of any bounded set in $\mathbb{R}_+^N$ is compact whereas the closure of any bounded set in C^+ may not be compact, any results or methods for (1.1) with $\tau = 0$ may no longer be valid for (1.1) with $\tau > 0$. However, in some cases, compactness of the closure of a bounded set in C^+ as a forward invariant set of (1.1) with $\tau > 0$ can be achieved.

In this section, we shall introduce a concept which describes a better property than uniform boundedness, and then find some classes of systems (1.1) that possess this property. For this purpose, we first recall the concept of a compact set.

Definition 2.1. A subset S of a normed vector space X is called a **compact set** if every open cover $\{U_x : x \in I\}$ of S (i. e. U_x is an open subset of X for each $x \in I$ such that $S \subset \cup_{x \in I} U_x$) contains a finite open cover $\{U_{x_i} : x_i \in I, i = 1, \dots k\}$ of S (i. e. $S \subset U_{x_1} \cup \cdots \cup U_{x_k}$). A set $S \subset X$ is said to be **relatively compact** if its closure is compact.

The following facts can be easily derived from the above definition or found in any textbook on real analysis (e. g. [29]).

(1) A nonempty subset of a normed vector space is compact if and only if it is closed and relatively compact.
(2) A nonempty subset S of a normed vector space is compact if every sequence in S has a subsequence that is convergent in S.
(3) A nonempty set $S \subset \mathbb{R}^N$ is compact if and only if S is closed and bounded.
(4) **Arzela–Ascoli Theorem** A nonempty set $S \subset C$ is relatively compact if S is bounded and the points in S as functions on $[-\tau, 0]$ are equicontinuous

$$\forall \varepsilon > 0, \quad \exists \delta > 0 \text{ such that } \forall x \in S, \quad \forall \theta_1, \theta_2 \in [-\tau, 0],$$
$$|\theta_1 - \theta_2| < \delta \text{ implies } |x(\theta_1) - x(\theta_2)| < \varepsilon.$$

Definition 2.2. System (1.1) is said to have a compact uniform attractor $\Omega \subset C^+$ if
(i) Ω is compact;
(ii) for each $\varphi \in C^+$, there is a $T(\varphi) > 0$ such that $x_t(t_0, \varphi) \in \Omega$ holds for all $t_0 \in \mathbb{R}_0$ and all $t \geq t_0 + T(\varphi)$;
(iii) $x_t(t_0, \varphi) \in \Omega$ holds for all $t_0 \in \mathbb{R}_0$, $\varphi \in \Omega$ and $t \geq t_0$;
(iv) for each $\varphi \in \Omega$ and every $i \in I_N$, $\varphi_i(\theta_0) = 0$ for some $\theta_0 \in [-\tau, 0]$ if and only if $\varphi_i(\theta) \equiv 0$ on $[-\tau, 0]$.

From this definition, we see that a compact uniform attractor is a compact global attractor in C^+ that is forward invariant and every solution $x_t(t_0, \varphi)$ will enter and stay in this attractor after a finite time (independent of $t_0 \in \mathbb{R}_0$). By (iv), φ is said to be in the boundary of this attractor if and only if $\varphi \in \Omega$ and $\varphi(\theta) \in \partial\mathbb{R}_+^N$ for all $\theta \in [-\tau, 0]$.

Clearly, if (1.1) has a compact uniform attractor, then it is uniformly bounded. Conversely, if (1.1) is uniformly bounded, does it have a compact uniform attractor?

We cannot answer this question in general, but the answer is positive if (1.1) is autonomous.

Proposition 2.1. *If (1.1) is autonomous and f is bounded on any bounded set $S \subset C^+$, then it is uniformly bounded if and only if it has a compact uniform attractor.*

Proof. We only need to prove the necessity. Suppose (1.1) is autonomous and uniformly bounded. Then,

$$\exists M > 0\,, \quad \forall \varphi \in C^+\,, \quad \exists T = T(\varphi) > 0 \text{ such that } \forall t \geq T\,, \quad |x(t,\varphi)| < M\,. \tag{2.1}$$

Thus, for each $\varphi \in C^+$ with $\|\varphi\| \leq M$, there is $t_1(\varphi) \in [0, T(\varphi) + \tau)$ such that $\|x_t(\varphi)\| < M$ for all $t > t_1$, but, if $t_1 > 0$, $\|x_{t_1}(\varphi)\| = M$. Let

$$S = \{x_t(\varphi) \colon \varphi \in C^+, \|\varphi\| \leq M, t \geq t_1(\varphi) + \tau\}\,.$$

Clearly, $S \neq \emptyset$. We show that $\overline{S}$, the closure of S, is a compact uniform attractor. Since $\overline{S}$ is bounded, by the assumption on f, there is a $\rho > 0$ such that $|f(\varphi)| \leq \rho$ for all $\varphi \in \overline{S}$. For each $\varphi \in C^+$ with $\|\varphi\| \leq M$ and for all $t \geq \tilde{t} \geq t_1(\varphi)$, we have

$$\forall i \in I_N\,, \quad x_i(t,\varphi) = x_i(\tilde{t},\varphi) \exp\left(\int_{\tilde{t}}^{t} f_i(s, x_s(\varphi))\,ds\right)$$

so

$$\forall i \in I_N\,, \quad x_i(\tilde{t},\varphi)e^{-\rho(t-\tilde{t})} \leq x_i(t,\varphi) \leq x_i(\tilde{t},\varphi)e^{\rho(t-\tilde{t})}\,.$$

This shows that every $\psi \in S$ satisfies $\|\psi\| \leq M$ and for all $i \in I_N$ and for all $\theta_1, \theta_2 \in [-\tau, 0]$ satisfying $-\tau \leq \theta_1 \leq \theta_2 \leq 0$,

$$\psi_i(\theta_1)e^{-\rho(\theta_2-\theta_1)} \leq \psi_i(\theta_2) \leq \psi_i(\theta_1)e^{\rho(\theta_2-\theta_1)}\,. \tag{2.2}$$

As these inequalities are retained by the limit of any convergent sequence in S, each $\psi \in \overline{S}$ also satisfies $\|\psi\| \leq M$ and (2.2). From (2.2), it follows that

$$\forall \theta_1, \theta_2 \in [-\tau, 0]\,, \quad |\psi(\theta_2) - \psi(\theta_1)| \leq M\rho e^{\rho\tau}|\theta_2 - \theta_1|\,.$$

This shows that the functions over $[-\tau, 0]$ in $\overline{S}$ are equicontinuous. Using the Arzela–Ascoli Theorem, $\overline{S}$ is relatively compact. As $\overline{S}$ is also closed, $\overline{S}$ is compact. The definition of S ensures its forward invariance. By continuous dependence, $\overline{S}$ is also forward invariant. From (2.1), we see that for each $\varphi \in C^+$, $x_t(\varphi) \in S$ for all $t \geq T(\varphi) + 2\tau$. From (2.2), we know that for any $\psi \in \overline{S}$, $\psi_i(\theta_0) = 0$ for some $\theta_0 \in [-\tau, 0]$ if and only if $\psi_i(\theta) \equiv 0$. Therefore, $\overline{S}$ is a compact uniform attractor. □

Since C^+ degenerates to $\mathbb{R}_+^N$ for $\tau = 0$, for autonomous Kolmogorov systems without delays as a special case of (1.1), from Proposition 2.1, we immediately obtain the following.

Corollary 2.1. *For every autonomous Kolmogorov system without delays and with a continuous f on $\mathbb{R}_+^N$, if it is uniformly bounded, then it has a forward invariant compact global attractor in $\mathbb{R}_+^N$.*

Open Problem 2.1. *If f is bounded on $\mathbb{R}_0 \times S$ for any bounded set $S \subset C^+$ and $f(t,\varphi)$ is* (i) *periodic in t or* (ii) *almost periodic in t or* (iii) *asymptotic to some continuous $g(\varphi)$, find an extra condition (if necessary) so that the uniform boundedness of* (1.1) *implies the existence of a compact uniform attractor.*

In the rest of this section, we give a few sufficient conditions for (1.1) to have a compact uniform attractor. For this purpose, we recall the concept of an M-matrix.

Definition 2.3 ([6]). A square matrix P with nonpositive off-diagonal entries is called an M-matrix if one of the following equivalent conditions is met:
(a) The leading principal minor determinants of P are all positive.
(b) There is a vector $x > 0$ (i. e. $x \in \operatorname{int} \mathbb{R}_+^N$) such that $Px > 0$.
(c) There is a vector $y > 0$ such that $P^T y > 0$.
(d) The matrix P is nonsingular and the entries of P^{-1} are all nonnegative.
(e) The real parts of the eigenvalues of P are all positive, i. e. the matrix $-P$ is stable.

For any vector $u \in \mathbb{R}^N$, let $D(u) = \operatorname{diag}[u_1, \ldots, u_N]$.

Theorem 2.1. *Assume that* (1.1) *meets the following requirements.*
(i) *The f_i are bounded on $\mathbb{R}_0 \times S$ for any bounded set $S \subset C^+$.*
(ii) *For all $(t,\varphi) \in \mathbb{R}_0 \times C^+$ and $i \in I_N$,*

$$f_i(t,\varphi) \le \beta_i + \sum_{j=1}^{N} a_{ij} \int_{-\tau}^{0} \varphi_j(\theta)\, d\xi_{ij}(t,\theta) - c_i \varphi_i(0)\,, \tag{2.3}$$

where $\beta_i > 0$, $a_{ij} \ge 0$, $c_i > 0$ and the $\xi_{ij}(t,\theta)$ are nondecreasing in $\theta \in [-\tau, 0]$ and continuous in $t \in \mathbb{R}_0$, satisfying

$$\forall i,j \in I_N\,, \quad \forall t \in \mathbb{R}_0\,, \quad \xi_{ij}(t,0) - \xi_{ij}(t,-\tau) = 1\,. \tag{2.4}$$

(iii) *The matrix $D(c) - A$ with $A = (a_{ij})$ is an M-matrix.*
Then, (1.1) *has a compact uniform attractor.*

Remark 2.1. Under the conditions (i) and (ii), the conclusion may not be true when (iii) is not met; even the boundedness of solutions may no longer hold. For example, the autonomous Lotka–Volterra system

$$x' = D(x)(r + Ax) \tag{2.5}$$

with $r_i > 0$, $a_{ij} \ge 0$ and $a_{ii} < 0$ for $i,j \in I_N$ $(i \ne j)$ satisfies (i) and (ii). It is shown that [15, Lemma 15.1.2] if A has an eigenvalue $\lambda > 0$ and a vector $v \in \mathbb{R}_+^N$ such that $v^T A = \lambda v^T$, and then (2.5) has an unbounded solution in $\operatorname{int} \mathbb{R}_+^N$. Indeed, by [21, Proof of Lemma 1], (2.5) has an unbounded solution in $\operatorname{int} \mathbb{R}_+^N$ if A is not stable, i. e. $-A$ is not an M-matrix.

Note that condition (ii) requires each $f_i(t,\varphi)$ to have a term $\varphi_i(0)$ with a negative coefficient. The next theorem gives an alternative condition when (ii) is not met.

Open Problem 2.2. *Suppose* (2.3) *is replaced by*

$$f_i(t,\varphi) \le \beta_i + \sum_{j=1}^{N} a_{ij} \int_{-\tau}^{0} \varphi_j(\theta)\, d\xi_{ij}(t,\theta) - c_i \int_{-\tau}^{0} \varphi_i(\theta)\, d\eta_i(\theta)\,,$$

where each η_i is nondecreasing, $\eta_i(0) - \eta(-\tau) = 1$ and $\eta_i(\theta) < \eta_i(0)$ for $\theta \in [-\tau, 0)$. If η_i is continuous at $\theta = 0$, then (2.3) *is not satisfied. Another case when* (2.3) *is not met is*

$$f_i(t,\varphi) \le \beta_i + \sum_{j=1}^{N} a_{ij} \int_{-\tau}^{-\delta} \varphi_j(\theta)\, d\xi_{ij}(t,\theta) - c_i\varphi_i(-\delta)\,,$$

where $\delta \in (0,\tau)$. In either case, find an essential condition to ensure the existence of a compact uniform attractor.

Theorem 2.2. *Assume that* (1.1) *meets the following requirements.*
(i) *The f_i are bounded on $\mathbb{R}_0 \times S$ for any bounded set $S \subset C^+$.*
(ii) *For all $(t,\varphi) \in \mathbb{R}_0 \times C^+$ and $i \in I_N$,*

$$f_i(t,\varphi) \le \beta_i - c_i \int_{-\tau}^{0} \varphi_i(\theta)\, d\xi_{ii}(t,\theta)\,, \tag{2.6}$$

where $\beta_i > 0$, $c_i > 0$ and the ξ_{ii} are the same as those in (2.3).
Then, (1.1) *has a compact uniform attractor.*

The next result is the combination of Theorems 2.1 and 2.2 when the system can be arranged into a triangular form of subsystems, of which each satisfies either Theorem 2.1 or Theorem 2.2.

Theorem 2.3. *Assume that I_N has a partition $\{I^1, \dots, I^m\}$ ($m > 1$) such that for $(t,\varphi) \in \mathbb{R}_0 \times C^+$ with $\varphi = (\varphi^1, \dots, \varphi^m)$,*

$$\forall i \in I^1\,, \quad f_i(t,\varphi) \le G_i^1\left(t,\varphi^1\right)\,, \tag{2.7}$$

$$\forall i \in I^k (k > 1)\,, \quad f_i(t,\varphi) \le F_i^k\left(t,\varphi^1,\dots,\varphi^{k-1}\right) + G^k\left(t,\varphi^k\right)\,, \tag{2.8}$$

where the F^k are bounded on $\mathbb{R}_0 \times S$ for any bounded set $S \subset C^+$ and each G_i^k has either the form

$$\forall i \in I^k\,, \quad G_i^k(t,\varphi^k) = \beta_i + \sum_{j\in I^k} a_{ij} \int_{-\tau}^{0} \varphi_j^k(\theta)\, d\xi_{ij}(t,\theta) - c_i\varphi_i^k(0)\,, \tag{2.9}$$

with $D(c)^k - A^k$ an M-matrix, or the form

$$\forall i \in I^k\,, \quad G_i^k(t,\varphi^k) = \beta_i - c_i \int_{-\tau}^{0} \varphi_i^k(\theta)\, d\xi_{ii}(t,\theta)\,, \tag{2.10}$$

where the β_i, a_{ij}, c_i and ξ_{ij} are the same as in (2.3), *and $D(c)^k - A^k$ is the corresponding $|I^k| \times |I^k|$ matrix. Then,* (1.1) *has a compact uniform attractor.*

3 Proof of Theorems 2.1, 2.2 and 2.3

The proof of Theorem 2.1 is lengthy. To avoid being bored by it, readers are suggested to skip it at the first reading.

Proof of Theorem 2.1. Since $D(c) - A$ is an M-matrix, there is a vector $d \in \mathrm{int}\mathbb{R}^N_+$ such that $(D(c) - A)d > 0$, i. e.

$$\forall i \in I_N\,, \quad c_i d_i - A_i d > 0\,. \tag{3.1}$$

Let $|v|_{d^{-1}} = \max\{|v_i| d_i^{-1} \colon i \in I_N\}$ for $v \in \mathbb{R}^N$ and

$$M_0 = \max\left\{\frac{\beta_i}{c_i d_i - A_i d} \colon i \in I_N\right\}\,, \tag{3.2}$$

$$\forall \varphi \in C^+\,, \quad \|\varphi\|_{d^{-1}} = \sup\left\{|\varphi(\theta)|_{d^{-1}} \colon \theta \in [-\tau, 0]\right\}\,. \tag{3.3}$$

We divide the rest of the proof into the following five steps.

Step 1. For any $\varphi \in C^+$ with $\varphi(0) \neq 0$ and $t_0 \in \mathbb{R}_0$, the solution $x_t(t_0, \varphi)$ exists on $[t_0, +\infty)$ and satisfies

$$\|x_t(t_0, \varphi)\|_{d^{-1}} \le \max\left\{\|\varphi\|_{d^{-1}}, M_0\right\}\,. \tag{3.4}$$

For if there is a $t_1 > t_0$ such that $\|x_{t_1}\|_{d^{-1}} > \max\{\|\varphi\|_{d^{-1}}, M_0\}$, then there are $t_2 \in (t_0, t_1]$ and $i \in I_N$ satisfying

$$x_i(t_2) d_i^{-1} = |x(t_2)|_{d^{-1}} = \max_{t_0 - \tau \le t \le t_1} |x(t)|_{d^{-1}} = \|x_{t_2}\|_{d^{-1}} \ge \|x_{t_1}\|_{d^{-1}} > M_0\,.$$

This means that $x_i(t) \le x_i(t_2)$ for all $t \in [t_0 - \tau, t_2]$. On the other hand, however, from (1.1), (2.3) and (3.1)–(3.3) we have

$$\begin{aligned} x_i'(t_2) &\le x_i(t_2)\left[\beta_i + A_i d\|x_{t_2}\|_{d^{-1}} - c_i d_i x_i(t_2) d_i^{-1}\right] \\ &= x_i(t_2)\left[\beta_i - (c_i d_i - A_i d)\|x_{t_2}\|_{d^{-1}}\right] \\ &\le x_i(t_2)(c_i d_i - A_i d)(M_0 - \|x_{t_2}\|_{d^{-1}}) < 0\,. \end{aligned}$$

From this, (2.3) and the continuity of x, there is a small open interval $\delta(t_2)$ centered at t_2 such that

$$\forall t \in \delta(t_2)\,, \quad x_i'(t) \le x_i(t)\left[\beta_i + A_i d\|x_t\|_{d^{-1}} - c_i d_i x_i(t) d_i^{-1}\right] < 0\,.$$

Thus, $x_i(t)$ is strictly decreasing on $\delta(t_2)$. This contradiction to $x_i(t) \le x_i(t_2)$ for $t \le t_2$ shows the truth of (3.4) on the existing interval of x and (3.4) ensures the extension of the solution to $[t_0, +\infty)$.

Step 2. We claim that if $\|\varphi\|_{d^{-1}} > M_0$, then $\|x_t\|_{d^{-1}}$ is nonincreasing as long as $\|x_t\|_{d^{-1}} \geq M_0$. For if there are $t_2 > t_1 \geq t_0$ such that

$$\|x_{t_2}\|_{d^{-1}} > \|x_{t_1}\|_{d^{-1}} > M_0 ,$$

then there are $t_3 \in (t_1, t_2]$ and $i \in I_N$ satisfying

$$x_i(t_3)d_i^{-1} = |x(t_3)|_{d^{-1}} = \max_{t_1-\tau\leq t\leq t_2} |x(t)|_{d^{-1}} = \|x_{t_3}\|_{d^{-1}} \geq \|x_{t_2}\|_{d^{-1}} .$$

Thus, $x_i(t) \leq x_i(t_3)$ for $t \in [t_3 - \tau, t_3]$. However, by the same technique as that used in the proof of (3.4), we derive $x_i'(t_3) < 0$ so $x_i'(t) < 0$ for t in a small open interval centered at t_3 and $x_i(t)$ is strictly decreasing in this interval, a contradiction to $x_i(t) \leq x_i(t_3)$ for $t \in [t_3 - \tau, t_3]$. This shows the truth of our claim.

Step 3. We show that for every $(t_0, \varphi) \in \mathbb{R}_0 \times C^+$,

$$\limsup_{t\to+\infty} \|x_t(t_0, \varphi)\|_{d^{-1}} \leq M_0 . \tag{3.5}$$

If this is not true, then some solution satisfies $\limsup_{t\to+\infty} \|x_t\|_{d^{-1}} > M_0$. From Step 2, we know that $\|x_t\|_{d^{-1}}$ is nonincreasing so $\lim_{t\to+\infty} \|x_t\|_{d^{-1}} = \bar{M}_0 > M_0$. Let

$$u_i = \limsup_{t\to+\infty} x_i(t)d_i^{-1} , \quad u_j = \max\{u_i \colon i \in I_N\}$$

for all i and some j in I_N. We look for an increasing sequence $\{t_k\}$ with $t_k \to \infty$ as $k \to \infty$ such that

$$\lim_{k\to\infty} x_j'(t_k) = 0 \quad \text{and} \quad \lim_{k\to\infty} x_j(t_k)d_j^{-1} = u_j . \tag{3.6}$$

If $x_j(t)$ is not monotone for large t, then we take a sequence $\{t_k\}$ so that each $x_j(t_k)$ is a local maximum of $x_j(t)$. This sequence certainly fulfills (3.6). If $x_j(t)$ is nondecreasing (nonincreasing), then $x_j'(t) \geq 0$ (≤ 0) and, by the boundedness of x, $\liminf_{t\to+\infty} x_j'(t) = 0$ ($\limsup_{t\to+\infty} x_j'(t) = 0$). Then, we can choose a sequence $\{t_k\}$ satisfying (3.6).

We check that $u_j = \bar{M}_0$. From (3.6), we have

$$u_j = \lim_{k\to\infty} x_j(t_k)d_j^{-1} \leq \lim_{k\to\infty} \|x_{t_k}\|_{d^{-1}} = \lim_{t\to+\infty} \|x_t\|_{d^{-1}} = \bar{M}_0 .$$

On the other hand, for any $\varepsilon > 0$, by the definition of u_j there exists $T > t_0$ such that

$$\forall i \in I_N , \quad \forall t \geq T , \quad x_i(t)d_i^{-1} < u_j + \varepsilon .$$

From this follows $\|x_t\|_{d^{-1}} < u_j + \varepsilon$ for $t \geq T + \tau$. Since $\|x_t\|_{d^{-1}}$ is nonincreasing, we have $\bar{M}_0 < u_j + \varepsilon$ so $\bar{M}_0 \leq u_j$ as $\varepsilon \to 0^+$. Therefore, $u_j = \bar{M}_0$. Then, it follows from this and (3.6) that

$$\lim_{k\to\infty} \varepsilon_k = \lim_{k\to\infty} \delta_k = 0 ,$$

where

$$\forall k \ge 1\,, \quad \varepsilon_k = x_j(t_k)d_j^{-1} - \bar{M}_0\,, \quad \delta_k = \|x_{t_k}\|_{d^{-1}} - \bar{M}_0\,.$$

Thus, from (1.1), (2.3), (3.1) and (3.2),

$$\begin{aligned} x_j'(t_k) &\le x_j(t_k)\left[\beta_j + A_j d\|x_{t_k}\|_{d^{-1}} - c_j d_j x_j(t_k)d_j^{-1}\right] \\ &= x_j(t_k)\left[\beta_j + A_j d(\bar{M}_0 + \delta_k) - c_j d_j(\bar{M}_0 + \varepsilon_k)\right] \\ &\le x_j(t_k)(c_j d_j - A_j d)\left[M_0 - \bar{M}_0 + \frac{A_j d\delta_k - c_j d_j \varepsilon_k}{c_j d_j - A_j d}\right] \\ &\to \bar{M}_0 d_j(c_j d_j - A_j d)(M_0 - \bar{M}_0) \quad (k \to \infty) \\ &< 0\,. \end{aligned}$$

This contradicts (3.6) and hence shows (3.5).

Step 4. We show that for each $M_1 > M_0$ and every $\varphi \in C^+$ with $\varphi(0) \ne 0$, there is a $T = T(\varphi, M_1) > 0$ such that

$$\forall t_0 \in \mathbb{R}_0\,, \quad \forall t \ge t_0 + T\,, \quad \|x_t(t_0, \varphi)\|_{d^{-1}} < M_1\,. \tag{3.7}$$

Suppose this is not true. Then, for some $\varphi \in C^+$ with $\varphi(0) \ne 0$ and $M_1 > M_0$, by (3.5) there are $\{t_k\} \subset \mathbb{R}_0$ and $\{T_k\} \subset [3\tau, +\infty)$, $T_k \uparrow +\infty$ as $k \to \infty$, such that

$$\forall k \ge 1\,, \quad \|x_{t_k+T_k}(t_k, \varphi)\|_{d^{-1}} = M_1\,, \quad \forall t > t_k + T_k\,, \quad \|x_t(t_k, \varphi)\|_{d^{-1}} < M_1\,. \tag{3.8}$$

From Step 2, $\|x_t(t_k, \varphi)\|_{d^{-1}}$ is nonincreasing for $t \in [t_k, t_k + T_k]$. Since the function $\|x_t(t_k, \varphi)\|_{d^{-1}}$ is continuous, with

$$g_k(t) = \|x_{t-2\tau}(t_k, \varphi)\|_{d^{-1}} - \|x_t(t_k, \varphi)\|_{d^{-1}}\,,$$

g_k is continuous and nonnegative for $t \in [t_k + 2\tau, t_k + T_k]$. Therefore, there is an $s_k \in [t_k + 2\tau, t_k + T_k]$ such that

$$g_k(s_k) = \min\{g_k(t) \colon t \in [t_k + 2\tau, t_k + T_k]\}\,.$$

(4a) We first show that

$$\lim_{k\to\infty} g_k(s_k) = 0\,. \tag{3.9}$$

Suppose (3.9) is not true. Then, $\limsup_{k\to\infty} g_k(s_k) > 0$. By choosing a subsequence of $\{s_k\}$ if necessary, we may assume the existence of $p_0 > 0$ and an integer K such that for all $k \ge K$, $g_k(s_k) \ge p_0$ so that

$$\forall t \in [t_k + 2\tau, t_k + T_k]\,, \quad \|x_{t-2\tau}(t_k, \varphi)\|_{d^{-1}} \ge \|x_t(t_k, \varphi)\|_{d^{-1}} + p_0\,. \tag{3.10}$$

Let m_k be the greatest integer part of $T_k/(2\tau)$. Then, $\{m_k\}$ is unbounded due to the unboundedness of $\{T_k\}$. However, by Step 2 and repeatedly using (3.10), we have

$$\begin{aligned}\|\varphi\|_{d^{-1}} &\geq \|x_{t_k+T_k-2\tau m_k}(t_k,\varphi)\|_{d^{-1}}\\ &\geq \|x_{t_k+T_k-2\tau(m_k-1)}(t_k,\varphi)\|_{d^{-1}} + p_0\\ &\geq m_k p_0 + \|x_{t_k+T_k}(t_k,\varphi)\|_{d^{-1}}\\ &= m_k p_0 + M_1\,.\end{aligned}$$

This contradiction to the unboundedness of $\{m_k\}$ shows the truth of (3.9).

(4b) We next show that for each $k \geq 1$, there is an $\ell_k \in [s_k - \tau, s_k]$ satisfying

$$\left(\tfrac{d}{dt}\|x_t(t_k,\varphi)\|_{d^{-1}}\right)_{t=\ell_k} \geq -\tfrac{2}{\tau}g_k(s_k)\,. \tag{3.11}$$

If $g_k(s_k) = 0$, then

$$\|x_t(t_k,\varphi)\|_{d^{-1}} \equiv \|x_{s_k}(t_k,\varphi)\|_{d^{-1}} \quad \text{for } t \in [s_k - 2\tau, s_k]$$

so $\frac{d}{dt}\|x_t(t_k,\varphi)\|_{d^{-1}} = 0$ for $t \in (s_k - 2\tau, s_k)$. In this case, $\ell_k = s_k - \tau$ meets the requirement of (3.11). Suppose $g_k(s_k) > 0$ and (3.11) does not hold for any $\ell_k \in [s_k - \tau, s_k]$. Then, $\frac{d}{dt}\|x_t(t_k,\varphi)\|_{d^{-1}} < -\frac{2}{\tau}g_k(s_k)$ for almost every $t \in [s_k - \tau, s_k]$ so

$$\begin{aligned}g_k(s_k) &= \|x_{s_k-2\tau}(t_k,\varphi)\|_{d^{-1}} - \|x_{s_k}(t_k,\varphi)\|_{d^{-1}}\\ &\geq \|x_{s_k-\tau}(t_k,\varphi)\|_{d^{-1}} - \|x_{s_k}(t_k,\varphi)\|_{d^{-1}}\\ &= -\int_{s_k-\tau}^{s_k}\left(\tfrac{d}{dt}\|x_t(t_k,\varphi)\|_{d^{-1}}\right)dt\\ &\geq 2g_k(s_k)\,.\end{aligned}$$

This contradiction to $0 < g_k(s_k) < 2g_k(s_k)$ shows the existence of ℓ_k satisfying (3.11).

(4c) We further show that for each $k \geq 1$, there are $w_k \in [\ell_k - \tau, \ell_k]$ and $i_k \in I_N$ such that

$$x'_{i_k}(w_k,t_k,\varphi) \geq -\frac{2d_{i_k}}{\tau}g_k(s_k)\,. \tag{3.12}$$

Indeed, for each $k \geq 1$, there are $w_k \in [\ell_k - \tau, \ell_k]$ and $i_k \in I_N$ such that

$$\|x_{\ell_k}(t_k,\varphi)\|_{d^{-1}} = |x(w_k,t_k,\varphi)|_{d^{-1}} = x_{i_k}(w_k,t_k,\varphi)d_{i_k}^{-1}\,.$$

For small $\delta > 0$, $w_k - \delta \in [\ell_k - \tau - \delta, \ell_k - \delta]$ so

$$\begin{aligned}&-\frac{1}{\delta}[x_{i_k}(w_k-\delta,t_k,\varphi) - x_{i_k}(w_k,t_k,\varphi)]d_{i_k}^{-1}\\ &\geq -\frac{1}{\delta}[|x(w_k-\delta,t_k,\varphi)|_{d^{-1}} - \|x_{\ell_k}(t_k,\varphi)\|_{d^{-1}}]\\ &\geq -\frac{1}{\delta}[\|x_{\ell_k-\delta}(t_k,\varphi)\|_{d^{-1}} - \|x_{\ell_k}(t_k,\varphi)\|_{d^{-1}}]\,.\end{aligned}$$

As $\delta \to 0+$, the above inequalities lead to

$$x'_{i_k}(w_k, t_k, \varphi)d_{i_k}^{-1} \geq \left(\tfrac{d}{dt}\|x_t(t_k, \varphi)\|_{d^{-1}}\right)_{t=\ell_k}.$$

Then, (3.12) follows from this and (3.11).

Now, armed with (4a)–(4c), we are able to construct a contradiction. By choosing a subsequence of $\{k\}$, if necessary, and without loss of generality, we may assume that $i_k = i_0 \in I_N$ for all $k \geq 1$ so that (3.12) becomes

$$x'_{i_0}(w_k, t_k, \varphi) \geq -\frac{2d_{i_0}}{\tau} g_k(s_k).$$

It then follows from this and (3.9) that

$$\liminf_{k\to\infty} x'_{i_0}(w_k, t_k, \varphi) \geq 0. \tag{3.13}$$

Nevertheless, from (1.1), (2.3), (3.1), (3.2), Step 2 and the equalities below (3.12),

$$\begin{aligned}
&x'_{i_0}(w_k, t_k, \varphi)/x_{i_0}(w_k, t_k, \varphi)\\
&\leq \beta_{i_0} + A_{i_0}d\|x_{w_k}(t_k, \varphi)\|_{d^{-1}} - c_{i_0}d_{i_0}x_{i_0}(w_k, t_k, \varphi)d_{i_0}^{-1}\\
&\leq \beta_{i_0} + A_{i_0}d\|x_{s_k-2\tau}(t_k, \varphi)\|_{d^{-1}} - c_{i_0}d_{i_0}\|x_{\ell_k}(t_k, \varphi)\|_{d^{-1}}\\
&\leq \beta_{i_0} + A_{i_0}d(\|x_{s_k}(t_k, \varphi)\|_{d^{-1}} + g_k(s_k)) - c_{i_0}d_{i_0}\|x_{s_k}(t_k, \varphi)\|_{d^{-1}}\\
&\leq \beta_{i_0} - (c_{i_0}d_{i_0} - A_{i_0}d)M_1 + A_{i_0}dg_k(s_k)\\
&\leq (c_{i_0}d_{i_0} - A_{i_0}d)(M_0 - M_1) + A_{i_0}dg_k(s_k).
\end{aligned}$$

From (3.8) and Step 2, we know that

$$x_{i_0}(w_k, t_k, \varphi)d_{i_0}^{-1} = \|x_{\ell_k}(t_k, \varphi)\|_{d^{-1}} \geq \|x_{s_k}(t_k, \varphi)\|_{d^{-1}} \geq M_1.$$

As $M_1 > M_0$, from (3.9) and the above inequalities, we obtain

$$\limsup_{k\to\infty} x'_{i_0}(w_k, t_k, \varphi) \leq d_{i_0}M_1(c_{i_0}d_{i_0} - A_{i_0}d)(M_0 - M_1) < 0.$$

This contradiction to (3.13) shows the existence of $T = T(\varphi, M_1) > 0$ satisfying (3.7).

Step 5. Construction of a compact uniform attractor $\Omega \subset C^+$. For this purpose, fix an $M_1 > M_0$ and let

$$S_0 = \{\varphi \in C^+ : \|\varphi\|_{d^{-1}} \leq M_1\}.$$

Then, by the assumption on f, there is a $\rho > 0$ such that

$$\forall i \in I_N, \quad \forall (t, \varphi) \in \mathbb{R}_0 \times S_0, \quad |f_i(t, \varphi)| \leq \rho.$$

Define

$$\Omega = \{\varphi \in S_0 : \forall i \in I_N, \forall \theta_1, \theta_2 \in [-\tau, 0] \text{ with } \theta_1 \leq \theta_2, \text{(3.15) holds}\}, \tag{3.14}$$

$$\varphi_i(\theta_1)e^{-\rho(\theta_2-\theta_1)} \leq \varphi_i(\theta_2) \leq \varphi_i(\theta_1)e^{\rho(\theta_2-\theta_1)}. \tag{3.15}$$

We check that this Ω is a compact uniform attractor.
(i) For any convergent sequence $\{\varphi^n\} \subset \Omega$ with a limit $\varphi^0 \in C^+$, since all the inequalities for each φ^n in the definition of S_0 and Ω are retained for φ^0, $\varphi^0 \in \Omega$ so Ω is closed. By (3.15), we have

$$\varphi_i(\theta_1)\left[e^{-\rho(\theta_2-\theta_1)}-1\right] \le \varphi_i(\theta_2)-\varphi_1(\theta_1) \le \varphi_i(\theta_1)\left[e^{\rho(\theta_2-\theta_1)}-1\right]$$

so

$$|\varphi_i(\theta_2)-\varphi_i(\theta_1)| \le |d|M_1\rho e^{\tau\rho}|\theta_2-\theta_1|\,.$$

This shows that the functions in Ω over $[-\tau, 0]$ are equicontinuous. Since Ω is bounded, by the Arzela–Ascoli Theorem, Ω is relatively compact. This, together with the closeness, shows that Ω is compact.
(ii) For each $\varphi \in \Omega$, from Step 2 we know that $x_t(t_0,\varphi) \in S_0$ for all $t_0 \in \mathbb{R}_0$ and $t \ge t_0$. For any $t_2 \ge t_1 \ge t_0$, integration of (1.1) gives

$$x_i(t_2,t_0,\varphi) = x_i(t_1,t_0,\varphi)\exp\left(\int_{t_1}^{t_2} f_i(s,x_s(t_0,\varphi))\,ds\right)$$

so

$$x_i(t_1,t_0,\varphi)e^{-\rho(t_2-t_1)} \le x_i(t_2,t_0,\varphi) \le x_i(t_1,t_0,\varphi)e^{\rho(t_2-t_1)}\,.$$

This shows that $x_t(t_0,\varphi) \in \Omega \neq \emptyset$ for $t \ge t_0+\tau$. For $t \in (t_0, t_0+\tau)$ and $-\tau \le \theta_1 < \theta_2 \le 0$, if $t+\theta_1 \ge t_0$, then

$$x_i(t+\theta_2,t_0,\varphi) = x_i(t+\theta_1,t_0,\varphi)\exp\left(\int_{\theta_1}^{\theta_2} f_i(t+s,x_{t+s}(t_0,\varphi))\,ds\right)$$

so

$$x_i(t+\theta_1,t_0,\varphi)e^{-\rho(\theta_2-\theta_1)} \le x_i(t+\theta_2,t_0,\varphi) \le x_i(t+\theta_1,t_0,\varphi)e^{\rho(\theta_2-\theta_1)}\,. \tag{3.16}$$

If $t+\theta_2 \le t_0$, then $x_t(\theta_j,t_0,\varphi) = \varphi(t-t_0+\theta_j)$ for $j = 1,2$ so (3.16) follows from (3.15). If $t+\theta_1 < t_0 < t+\theta_2$, then

$$\begin{aligned}
x_i(t+\theta_2,t_0,\varphi) &\le \varphi_i(0)e^{\rho(t+\theta_2-t_0)}\\
&\le \varphi_i(t-t_0+\theta_1)e^{\rho(t_0-t-\theta_1)}e^{\rho(t+\theta_2-t_0)}\\
&= x_i(t+\theta_1,t_0,\varphi)e^{\rho(\theta_2-\theta_1)},\\
x_i(t+\theta_2,t_0,\varphi) &\ge \varphi_i(0)e^{-\rho(t+\theta_2-t_0)}\\
&\ge \varphi_i(t-t_0+\theta_1)e^{-\rho(t_0-t-\theta_1)}e^{-\rho(t+\theta_2-t_0)}\\
&= x_i(t+\theta_1,t_0,\varphi)e^{-\rho(\theta_2-\theta_1)}\,.
\end{aligned}$$

Thus, (3.16) holds and $x_t(t_0,\varphi) \in \Omega$. Therefore, $\varphi \in \Omega$ implies $x_t(t_0,\varphi) \in \Omega$ for all $t_0 \in \mathbb{R}_0$ and all $t \ge t_0$.

(iii) For each $\varphi \in C^+$, from Step 4, we know the existence of $T = T(\varphi) > 0$ such that $x_t(t_0, \varphi) \in S_0$ holds for all $t_0 \in \mathbb{R}_0$ and $t \geq t_0 + T$. Then, from (ii), we obtain $x_t(t_0, \varphi) \in \Omega$ for $t \geq t_0 + T + \tau$.

(iv) For each $\varphi \in \Omega$, (3.15) implies that $\varphi_i(\theta_0) = 0$ for some $i \in I_N$ and some $\theta_0 \in [-\tau, 0]$ if and only if $\varphi_i(\theta) \equiv 0$ on $[-\tau, 0]$.

Therefore, Ω defined by (3.14) is a compact uniform attractor of (1.1). □

Proof of Theorem 2.2. We give a full outline following the proof of Theorem 2.1. For $(t_0, \varphi) \in \mathbb{R}_0 \times C^+$ with $\varphi_i(0) > 0$, we have

$$x_i'(t, t_0, \varphi) \leq \beta_i x_i(t, t_0, \varphi)$$

so for $t \geq t_0 + \tau$ and $\theta \in [-\tau, 0]$, $(x_i)_t(\theta) \geq x_i(t, t_0, \varphi)e^{\beta_i \theta}$. Hence, for $t \geq t_0 + \tau$ in its existing interval, the solution satisfies

$$x_i'(t, t_0, \varphi) \leq x_i(t, t_0, \varphi) \left[\beta_i - c_i \int_{-\tau}^{0} e^{\beta_i \theta} \, d\xi_{ii}(t, \theta) x_i(t, t_0, \varphi) \right] . \tag{3.17}$$

Let

$$\rho_i = \frac{\beta_i e^{\beta_i \tau}}{c_i}, \quad \rho^0 = \max\{\rho_i \colon i \in I_N\} . \tag{3.18}$$

Then, $x_i(t, t_0, \varphi)$ is decreasing as long as $x_i(t, t_0, \varphi) > \rho_i$. This shows the existence and boundedness of $x(t, t_0, \varphi)$ on $[t_0, \infty)$.

Now, multiplying (3.17) by $-x_i^{-2}(t, t_0, \varphi)e^{\beta_i(t-t_0-\tau)}$, changing it to

$$\left(e^{\beta_i(t-t_0-\tau)}/x_i(t, t_0, \varphi)\right)' \geq c_i e^{-\beta_i \tau} e^{\beta_i(t-t_0-\tau)},$$

and integrating from $t_0 + \tau$ to $t_0 + \tau + t$, we obtain

$$x_i(t_0 + \tau + t, t_0, \varphi) \leq \frac{e^{\beta_i t}}{x_i(t_0 + \tau, t_0, \varphi)^{-1} + \rho_i^{-1}(e^{\beta_i t} - 1)} . \tag{3.19}$$

From this,

$$\limsup_{t \to +\infty} x_i(t, t_0, \varphi) \leq \rho_i \leq \rho^0 .$$

Moreover, for any fixed $\rho > \rho^0$, as $x_i(t_0 + \tau, t_0, \varphi) \leq \varphi_i(0)e^{\beta_i \tau}$, (3.19) shows that for all $\varphi \in C^+$, there exists a $T = T(\varphi) > 0$ such that

$$\forall t_0 \in \mathbb{R}_0 , \quad \forall t \geq t_0 + T , \quad \|x_t(t_0, \varphi)\| < \rho . \tag{3.20}$$

Let

$$S_0 = \{\varphi \in C^+ \colon \|\varphi\| \leq \rho\} .$$

Then, by the assumption on f, there is an $\alpha < 0$ such that

$$\forall i \in I_N , \quad \forall (t, \varphi) \in \mathbb{R}_0 \times S_0 , \quad \alpha \leq f_i(t, \varphi) \leq \beta_i .$$

Define

$$\Omega = \{\varphi \in S_0\colon \forall i \in I_N, \forall \theta_1, \theta_2 \in [-\tau, 0] \text{ with } \theta_1 \le \theta_2, (3.22) \text{ holds}\}\ , \qquad (3.21)$$

$$\varphi_i(\theta_1)e^{\alpha(\theta_2-\theta_1)} \le \varphi_i(\theta_2) \le \varphi_i(\theta_1)e^{\beta_i(\theta_2-\theta_1)}\ . \qquad (3.22)$$

We check that this Ω is a compact uniform attractor.
(i) The compactness of Ω follows from the same proof as that of Theorem 2.1.
(ii) For each $\varphi \in \Omega$, we have $(x_i)_t(\theta) \ge x_i(t, t_0, \varphi)e^{\beta_i\theta}$ for all $t \ge t_0$. Therefore, (3.17) and (3.19) hold for all $t_0 \in \mathbb{R}_0$ and all $t \ge t_0$. Thus, $x_t(t_0, \varphi) \in S_0$ for all $t_0 \in \mathbb{R}_0$ and all $t \ge t_0$. Then, with minor modification to the proof of Theorem 2.1, we have $x_t(t_0, \varphi) \in \Omega$ for all $t_0 \in \mathbb{R}_0$ and all $t \ge t_0$.
(iii) From (3.20), it follows that

$$\forall \varphi \in C^+, \quad \exists T = T(\varphi) > 0 \text{ such that } \forall t_0 \in \mathbb{R}_0, \quad \forall t \ge t_0+T, \quad x_t(t_0, \varphi) \in \Omega.$$

(iv) For each $\varphi \in \Omega$, (3.22) implies that $\varphi_i(\theta_0) = 0$ for some $i \in I_N$ and some $\theta_0 \in [-\tau, 0]$ if and only if $\varphi_i(\theta) \equiv 0$ on $[-\tau, 0]$.
Therefore, Ω defined by (3.21) is a compact uniform attractor of (1.1). □

Proof of Theorem 2.3. Since $x_i'(t) \le x_i(t)G_i^1(t, x_t^1)$ for $i \in I^1$ with the assumption (2.9) or (2.10), by Theorem 2.1 or Theorem 2.2, this subsystem has a compact uniform attractor Ω^1. Since $F_i^2(t, \varphi^1)$ is bounded on $\mathbb{R}_0 \times \Omega^1$, there are $\beta_i' > 0$ for $i \in I^2$ such that $F_i^2(t, \varphi^1) \le \beta_i'$ on $\mathbb{R}_0 \times \Omega^1$. Then, $x_i'(t) \le x_i(t)(\beta_i' + G_i^2(t, x_t^2))$ for $i \in I^2$ and $(t_0, \varphi) \in \mathbb{R}_0 \times C^+$ with $\varphi^1 \in \Omega^1$. By the assumption (2.9) or (2.10) and Theorems 2.1 and 2.2, the subsystem for x^2 has a compact uniform attractor Ω^2. Repeating the above process, we obtain a compact uniform attractor Ω^k of the subsystem for x^k for each $k \in I_m$. Then, it can be verified that $\Omega^1 \times \cdots \times \Omega^m$ is a compact uniform attractor for (1.1). □

4 Partial permanence and permanence

In this section, we give some permanence criteria for systems having a compact uniform attractor. For any $i \in I_N$ and $\Omega \subset C^+$, let $\Omega_i = \{\varphi \in \Omega\colon \varphi_i = 0\}$ and denote the ith coordinate plane in $\mathbb{R}_+^N$ by $\pi_i = \{x \in \mathbb{R}_+^N\colon x_i = 0\}$.

Theorem 4.1. *Assume that* (1.1) *satisfies the following conditions.*
(i) *System* (1.1) *has a compact uniform attractor* $\Omega \subset C^+$.
(ii) *The function* f *is bounded on* $\mathbb{R}_0 \times \Omega$ *and uniformly Lipschitzian on* Ω, *i. e. there is a* $K > 0$ *such that*

$$\forall t \in \mathbb{R}_0\,, \quad \forall \varphi, \psi \in \Omega\,, \quad |f(t, \varphi) - f(t, \psi)| \le K\|\varphi - \psi\|\,.$$

(iii) *For a nonempty set* $J \subset I_N$ *and* $\varphi \in \cup_{i\in J}\Omega_i$, $f(t, \varphi)$ *is* T_0*-periodic.*

(iv) *There are* $q_i > 0$, $i \in J$, *for each* $(t_0, \varphi) \in \mathbb{R}_0 \times (\cup_{i\in J}\Omega_i)$, *there is a* $T(t_0, \varphi) > 0$ *such that*

$$\int_0^{T(t_0,\varphi)} \sum_{i\in J} q_i f_i(t_0 + s, x_{t_0+s}(t_0, \varphi))\, ds > 0\,. \tag{4.1}$$

Then, (1.1) *is* J*-permanent. In particular,* (1.1) *is permanent if* $J = I_N$.

Proof. By condition (i), (1.1) has a Ω. We first show the existence of $\rho > 0$ such that

$$\|x_t(t_0, \varphi) - x_t(t_0, \psi)\| \le \|\varphi - \psi\| e^{\rho(t-t_0)} \tag{4.2}$$

for all $t_0 \in \mathbb{R}_0$, $\varphi, \psi \in \Omega$ and $t \ge t_0$. From (1.1), we have

$$\begin{aligned} x_i(t, t_0, \varphi) - x_i(t, t_0, \psi) &= \varphi_i(0) - \psi_i(0) \\ &+ \int_{t_0}^{t} [x_i(s, t_0, \varphi) - x_i(s, t_0, \psi)] f_i(s, x_s(t_0, \varphi))\, ds \\ &+ \int_{t_0}^{t} x_i(s, t_0, \psi)[f_i(s, x_s(t_0, \varphi)) - f_i(s, x_s(t_0, \psi))]\, ds\,. \end{aligned}$$

By conditions (i) and (ii), there is a $\rho > 0$ independent of t_0, t, φ and ψ such that

$$|x_i(t, t_0, \varphi) - x_i(t, t_0, \psi)| \le |\varphi_i(0) - \psi_i(0)| + \rho \int_{t_0}^{t} \|x_s(t_0, \varphi) - x_s(t_0, \psi)\|\, ds$$

for all $t \ge t_0$ and all $i \in I_N$. Therefore,

$$\forall t \ge t_0\,, \quad \|x_t(t_0, \varphi) - x_t(t_0, \psi)\| \le \|\varphi - \psi\| + \rho \int_{t_0}^{t} \|x_s(t_0, \varphi) - x_s(t_0, \psi)\|\, ds\,.$$

Then, (4.2) follows from this and Gronwall's inequality.

By condition (iv), for each $(t_0, \varphi) \in \mathbb{R}_0 \times (\cup_{i\in J}\Omega_i)$, there is a $T(t_0, \varphi) > 0$ such that

$$\delta(t_0, \varphi) = \frac{1}{T(t_0, \varphi)} \int_0^{T(t_0,\varphi)} \sum_{i\in J} q_i f_i(t_0 + s, x_{t_0+s}(t_0, \varphi))\, ds > 0\,. \tag{4.3}$$

Then, by (4.2) and the continuous dependence of $x_t(t_0, \varphi)$ on (t_0, φ), there is an open interval $I(t_0, \varphi) \subset \mathbb{R}$ centered at t_0 and an open ball $B(t_0, \varphi)$ of Ω centered at φ such that

$$\frac{1}{T(t_0, \varphi)} \int_0^{T(t_0,\varphi)} \sum_{i\in J} q_i f_i(t_1 + s, x_{t_1+s}(t_1, \psi))\, ds \ge \tfrac{1}{2}\delta(t_0, \varphi) \tag{4.4}$$

for all $t_1 \in I(t_0, \varphi) \cap \mathbb{R}_0$ and $\psi \in B(t_0, \varphi)$. Since $f(t, \varphi)$ is T_0-periodic by condition (iii), we may assume that, for any integer k satisfying $t_0 + kT_0 \in \mathbb{R}_0$, $T(t_0 + kT_0, \varphi) = T(t_0, \varphi)$ so $\delta(t_0 + kT_0, \varphi) = \delta(t_0, \varphi)$. By (4.2), (ii) and (iii), we may also assume that $I(t_0 + kT_0, \varphi)$ centered at $t_0 + kT_0$ has the same length as $I(t_0, \varphi)$ and

$$B(t_0 + kT_0, \varphi) = B(t_0, \varphi) . \tag{4.5}$$

Then, for any fixed $\ell \in \mathbb{R}_0$, $[\ell, \ell + T_0] \times \{\varphi\}$ is a compact set of $\mathbb{R}_0 \times \Omega$ and

$$\{I(t_0, \varphi) \times B(t_0, \varphi) \colon t_0 \in [\ell, \ell + T_0]\}$$

is an open cover of $[\ell, \ell + T_0] \times \{\varphi\}$. Thus, there is a finite open cover of $[\ell, \ell + T_0] \times \{\varphi\}$. Combining (4.4) and (4.5) with this finite open cover, we obtain an open ball $B(\varphi)$ of Ω, positive numbers $\delta_1(\varphi), \dots, \delta_m(\varphi), T_1(\varphi), \dots, T_m(\varphi)$, and a finite open cover $\{I^1, \dots, I^m\}$ of $\mathbb{R}_0$ such that for each $k \in I_m$ and for all $(t_1, \psi) \in I^k \times B(\varphi)$,

$$\frac{1}{T_k(\varphi)} \int_0^{T_k(\varphi)} \sum_{i \in J} q_i f_i(t_1 + s, x_{t_1+s}(t_1, \psi))\, ds \ge \delta_k(\varphi) . \tag{4.6}$$

Since $\cup_{i \in J} \Omega_i$ is compact and $\{B(\varphi) \colon \varphi \in \cup_{i \in J} \Omega_i\}$ is an open cover of $\cup_{i \in J} \Omega_i$, there are $\varphi^1, \dots, \varphi^p \in \cup_{i \in J} \Omega_i$ such that $\{B(\varphi^j) \colon j \in I_p\}$ is a finite open cover of $\cup_{i \in J} \Omega_i$. Then, for each $j \in I_p$, there is an integer $m_j > 0$ such that $\{I^{j1}, \dots, I^{jm_j}\}$ is a finite open cover of $\mathbb{R}_0$ and for all $(t_1, \psi) \in (I^{jk} \cap \mathbb{R}_0) \times B(\varphi^j)$, (4.6) holds after the replacement of φ by φ^j. Now, put

$$\begin{aligned} T^0 &= \min\left\{T_k(\varphi^j) \colon j \in I_p, k \in I_{m_j}\right\} , \\ T^1 &= \max\left\{T_k(\varphi^j) \colon j \in I_p, k \in I_{m_j}\right\} , \\ \delta_0 &= \min\left\{\delta_k(\varphi^j) \colon j \in I_p, k \in I_{m_j}\right\} , \end{aligned}$$

and define the function $V \colon \mathbb{R}_+^N \to \mathbb{R}_+$ by

$$V(x) = \prod_{i \in J} x_i^{q_i} .$$

Then, V is continuous on $\mathbb{R}_+^N$ and $V(x) = 0$ if and only if $x \in \cup_{i \in J} \pi_i$. Thus, $x \in \mathbb{R}_+^N$ is close to $\cup_{i \in J} \pi_i$ if and only if $V(x)$ is small. By the properties of Ω, $\varphi(0)$ is close to $\cup_{i \in J} \pi_i$ if and only if $\varphi \in \Omega$ is close to $\cup_{i \in J} \Omega_i$. Then, we can choose $\mu > 0$ sufficiently small so that the set

$$S_1 = \{\varphi \in \Omega \colon 0 \le V(\varphi(0)) \le \mu\}$$

is contained in $\cup_{j \in I_p} B(\varphi^j)$.

We claim that for each $\varphi \in S_1$ with $\varphi_i(0) > 0$ for all $i \in J$ and every $t_0 \in \mathbb{R}_0$, there is a $T > t_0$ such that $V(x(T, t_0, \varphi)) > \mu$. Indeed, if $V(x(t, t_0, \varphi)) \le \mu$ for all

$t \geq t_0$, then $x_t(t_0, \varphi) \in S_1 \subset \cup_{j \in I_p} B(\varphi^j)$ for all $t \geq t_0$. Since $\varphi \in S_1$, we have $\varphi \in B(\varphi^j)$ for some $j \in I_p$. As $t_0 \in I^{jk}$ for some $k \in I_{m_j}$, by (4.6) and the definition of T^0 and δ_0, we have

$$\int_0^{T_k(\varphi^j)} \sum_{i \in J} q_i f_i(t_0 + s, x_{t_0+s}(t_0, \varphi)) ds \geq \delta_k \left(\varphi^j\right) T_k \left(\varphi^j\right) \geq \delta_0 T^0 .$$

Differentiation of $V(x(t, t_0, \varphi))$ gives

$$V(x(t, t_0, \varphi))' = V(x(t, t_0, \varphi)) \sum_{i \in J} q_i f_i(t, x_t(t_0, \varphi)) .$$

Then, with $t_1 = t_0 + T_k(\varphi^j)$, we obtain

$$\begin{aligned} V(t_1, t_0, \varphi)) &= V(\varphi(0)) \exp\left(\int_0^{T_k(\varphi^j)} \sum_{i \in J} q_i f(i(t_0 + s, x_{t_0+s}(t_0, \varphi)) \, ds \right) \\ &\geq V(\varphi(0)) e^{\delta_0 T^0} . \end{aligned}$$

Since $\psi = x_{t_1}(t_0, \varphi) \in B(\varphi^n)$ for some $n \in I_p$ and $t_1 \in I^{nk}$ for some $k \in I_{m_n}$, by the same procedure as above and with $t_2 = t_1 + T_k(\varphi^n)$, we obtain

$$V(x(t_2, t_0, \varphi) = V(x(t_2, t_1, \psi)) \geq V(\psi(0)) e^{\delta_0 T^0} \geq V(\varphi(0)) e^{2\delta_0 T^0} .$$

Repetition of the above process infinitely many times leads to the unboundedness of $V(x(t, t_0, \varphi))$ for $t \geq t_0$. This contradiction to our assumption $V(x(t, t_0, \varphi)) \leq \mu$ for $t \geq t_0$ shows our claim.

Let $\alpha = \inf\{\sum_{i \in J} q_i f_i(t, \varphi) \colon (t, \varphi) \in \mathbb{R}_0 \times \Omega\}$. The boundedness of f on $\mathbb{R}_0 \times \Omega$ implies $\alpha \in \mathbb{R}$. If $\alpha \geq 0$, then, for any $\varphi \in \Omega$ with $V(\varphi(0)) > \mu$,

$$\begin{aligned} V(x(t, t_0, \varphi)) &= V(\varphi(0)) \exp\left(\int_{t_0}^{t} \sum_{i \in J} q_i f_i(s, x_s(t_0, \varphi)) \, ds \right) \\ &\geq V(\varphi(0)) e^{\alpha(t - t_0)} > \mu \end{aligned}$$

for all $t_0 \in \mathbb{R}_0$ and $t \geq t_0$.

If $\alpha < 0$, then $\rho = \mu e^{\alpha T^1} \in (0, \mu)$. We show that $V(x(t, t_0, \varphi)) > \rho$ for all $\varphi \in \Omega$ with $V(\varphi(0)) > \mu$ and all $t_0 \in \mathbb{R}_0$ and $t \geq t_0$. In fact, for fixed φ and t_0, we have either $V(x(t, t_0, \varphi)) > \mu$ for all $t \geq t_0$ or $V(x(t_1, t_0, \varphi)) = \mu$ for some $t_1 > t_0$, but $V(x(t, t_0, \varphi)) > \mu$ for all $t \in [t_0, t_1)$. In the latter case, as $\psi = x_{t_1}(t_0, \varphi) \in S_1$, for some $j \in I_p$ and $k \in I_{m_j}$, we have

$$\int_0^{T_k(\varphi^j)} \sum_{i \in J} q_i f_i(t_1 + s, x_{t_1+s}(t_1, \psi)) ds > \delta_0 T^0 .$$

Thus, with $t_2 = t_1 + T_k(\varphi^j)$,

$$V(x(t_2, t_1, \psi)) \geq V(\psi(0))e^{\delta_0 T^0} > V(\psi(0)) = V(x(t_1, t_0, \varphi)) = \mu > \rho .$$

For $t \in [t_1, t_2)$,

$$\begin{aligned} V(x(t, t_1, \psi)) &= V(\psi(0)) \exp\left(\int_{t_1}^{t} \sum_{i\in J} q_i f_i(s, x_s(t_1, \psi))\, ds\right) \\ &\geq \mu e^{\alpha(t-t_1)} > \mu e^{\alpha T^1} = \rho . \end{aligned}$$

Thus, we have $V(x(t_2, t_0, \varphi)) > \mu$ and $V(x(t, t_0, \varphi)) > \rho$ for all $t \in [t_0, t_2]$. If $V(x(t, t_0, \varphi)) > \mu$ for all $t \geq t_2$, then $V(x(t, t_0, \varphi)) > \rho$ for all $t \geq t_0$. Otherwise, there is $t_3 > t_2$ such that $V(x(t, t_0, \varphi)) > \mu$ for all $t \in [t_2, t_3)$ and $V(x(t_3, t_0, \varphi)) = \mu$. Then, repeating the above process with t_1 replaced by t_3, we obtain $V(x(t, t_0, \varphi)) > \rho$ for all $t \geq t_0$.

This shows that for each $\varphi \in \Omega$ with $\varphi_i(0) > 0$ for all $i \in J$ and for every $t_0 \in \mathbb{R}_0$, there is a $T > t_0$ such that $V(x(t, t_0, \varphi)) > \rho\ (\mu)$, if $\alpha < 0\ (\geq 0)$, for all $t \geq T$. Let

$$\delta = \inf\{\varphi_i(0) \colon V(\varphi(0)) = \rho, \varphi \in \Omega, i \in J\}\ , \tag{4.7}$$

$$M = \sup\{\|\varphi\| \colon \varphi \in \Omega\} \tag{4.8}$$

if $\alpha < 0$ and replace ρ by μ in (4.7) if $\alpha \geq 0$. Then,

$$\forall i \in J\,, \quad \forall t \geq T\,, \quad \delta \leq x_i(t, t_0, \varphi) \leq M\,.$$

Therefore, (1.1) is partially permanent with respect to J. □

Remark 4.1. (a) If $f(t, \varphi) \equiv g(\varphi)$ for some g and all $(t, \varphi) \in \mathbb{R}_0 \times (\bigcup_{i\in J} \Omega_i)$, then condition (iii) of Theorem 4.1 is met for any $T_0 > 0$.

(b) This theorem, when $J = I_N$, can be viewed as an extension of [15, Theorem 12.2.1] from a system without delays on a closed set $S_n \subset \mathbb{R}_+^N$ to (1.1) on $\mathbb{R}_0 \times C^+$.

(c) Unfortunately, it is not easily applicable to any concrete system as condition (iv) is hardly checkable. However, we can develop an easily checkable condition for a class of systems based on this.

A particular case of (1.1) is that

$$f_i(t, \varphi) = r_i(t) + L_i(\varphi) - F_i(t, \varphi)\,, \tag{4.9}$$

$$L_i(\varphi) = \sum_{j=1}^{N} a_{ij} \int_{-\tau}^{0} \varphi_j(\theta)\, d\xi_{ij}(\theta) - \sum_{j=1}^{N} b_{ij} \int_{-\tau}^{0} \varphi_j(\theta)\, d\eta_{ij}(\theta)\,, \tag{4.10}$$

$$\forall i, j \in I_N\,, \quad a_{ij} \geq 0, b_{ij} \geq 0, b_{ii} > 0\,, \tag{4.11}$$

where the r_i are continuous T_0-periodic with

$$\forall i \in I_N\,, \quad \forall t_0 \in \mathbb{R}_0\,, \quad \frac{1}{T_0}\int_0^{T_0} r_i(t_0+s)ds = \bar{r}_i > 0\,, \tag{4.12}$$

and the ξ_{ij} and η_{ij} are nondecreasing with

$$\forall i,j \in I_N\,, \quad \xi_{ij}(0)-\xi_{ij}(-\tau) = 1, \eta_{ij}(0)-\eta_{ij}(-\tau) = 1\,. \tag{4.13}$$

We assume that the F_i are nonnegative, bounded on $\mathbb{R}_0 \times S$ and uniformly Lipschitzian in $\varphi \in S$ for any bounded set $S \subset C^+$. We also assume that (1.1) with (4.9)–(4.13) meets the requirement of one of Theorems 2.1–2.3 or Proposition 2.1. Then, (1.1) with (4.9)–(4.13) has a compact uniform attractor $\Omega \subset C^+$ and f is bounded on $\mathbb{R}_0\times\Omega$ and uniformly Lipschitzian on Ω. From now on, (1.1) with (4.9)–(4.13) is always assumed to have these properties.

In addition to (1.1) with (4.9)–(4.13), also consider the autonomous Lotka–Volterra system

$$x_i' = x_i(\bar{r}_i + (A-B)_i x)\,, \quad i \in I_N\,, \tag{4.14}$$

where $A = (a_{ij})$ and $B = (b_{ij})$.

Theorem 4.2. *For* (1.1) *with* (4.9)–(4.13) *and a nonempty set $J \subset I_N$, also assume that*

$$\forall j \in I_N\,, \quad \forall t \in \mathbb{R}_0\,, \quad \forall \varphi \in \cup_{i\in J}\Omega_i\,, \quad F_j(t,\varphi) \equiv 0\,, \tag{4.15}$$

$$\exists q_i > 0 \text{ for } i \in J \text{ such that } \sum_{i\in J} q_i(\bar{r}_i + (A-B)_i\hat{x}) > 0 \tag{4.16}$$

for all fixed point $\hat{x}$ of (4.14) *in $\cup_{i\in J}\pi_i$. Then,* (1.1) *with* (4.9)–(4.13) *is partially permanent with respect to J. In particular,* (1.1) *with* (4.9)–(4.13) *is permanent if $J = I_N$.*

Remark 4.2. (a) Theorem 4.2 includes the case of $r_i(t) \equiv \bar{r}_i$.
(b) This theorem, when $J = I_N$, is the extension of Jansen's result [26] from autonomous replicator and Lotka–Volterra systems to (1.1) with (4.9)–(4.13) (see also [15, Theorem 13.6.1 and Exercise 13.6.3]).

Note that permanence of (1.1) with respect to $\{i\}$ for every $i \in J$ implies permanence with respect to J. Then, applying Theorems 4.1 and 4.2 to each $i \in J$, we obtain the following corollaries.

Corollary 4.1. *Assume that* (1.1) *satisfies the conditions* (i)–(iii) *of Theorem* 4.1. *Moreover, for each $i \in J$ and every $(t_0,\varphi) \in \mathbb{R}_0 \times \Omega_i$, there is a $T(t_0,\varphi) > 0$ such that*

$$\int_0^{T(t_0,\varphi)} f_i(t_0+s, x_{t_0+s}(t_0,\varphi))\,ds > 0\,.$$

Then, (1.1) *is partially permanent with respect to J. If also $J = I_N$, then* (1.1) *is permanent.*

Corollary 4.2. *Assume that* (1.1) *with* (4.9)–(4.13) *satisfies* (4.15). *Moreover, for each* $i \in J$ *and every fixed point* $\hat{x}$ *of* (4.14) *in* π_i, *we have* $\bar{r}_i + (A-B)_i\hat{x} > 0$. *Then,* (1.1) *with* (4.9)–(4.13) *is partially permanent with respect to* J. *If also* $J = I_N$, *then* (1.1) *with* (4.9)–(4.13) *is permanent.*

When $J = I_N$, if we apply Corollary 4.2 to every subsystem of (1.1) with (4.9)–(4.13) and (4.15), and we then obtain the following.

Corollary 4.3. *Assume that* (1.1) *with* (4.9)–(4.13) *and* (4.15) *satisfies*

$$\forall i \in I_N\,, \quad \forall \hat{x} \in \pi_i \text{ (fixed points of (4.14))}\,, \quad \bar{r}_i + (A-B)_i\hat{x} > 0\,.$$

Then, (1.1) *and all of its subsystems are permanent.*

Example 4.1. Consider the following system, that is,

$$\begin{aligned} x_1'(t) &= x_1(t)\Bigg[b_1 + \sin t - a_{11}x_1(t) \\ &\quad + a_{12}\int_{-1}^{0} x_2(t+\theta)\,d\theta - e^{-t}x_1(t-2)x_2(t)\Bigg] \\ x_2'(t) &= x_2(t)\left[b_2 - \cos t + a_{21}x_1(t-1) + a_{22}'x_2(t-2) - a_{22}x_2(t)\right]\,. \end{aligned} \tag{4.17}$$

Assume that for all $i, j \in \{1, 2\}$,

$$b_i > 0\,, \quad a_{ii} > 0\,, \quad a_{ij} \geq 0\,, \quad a_{22}' \geq 0\,, \quad a_{11}(a_{22} - a_{22}') > a_{12}a_{21}\,.$$

Then, the matrix

$$\begin{pmatrix} a_{11} & -a_{12} \\ -a_{21} & a_{22} - a_{22}' \end{pmatrix}$$

is an M-matrix. By Theorem 2.1, (4.17) has a compact uniform attractor. The corresponding Lotka–Volterra system (4.14) is

$$x_1' = x_1\left(b_1 - a_{11}x_1 + a_{12}x_2\right)\,, \quad x_2' = x_2\left(b_2 + a_{21}x_1 + \left(a_{22}' - a_{22}\right)x_2\right)\,. \tag{4.18}$$

System (4.18) has three fixed points on $\partial\mathbb{R}^2_+$: $p^1 = (0,0)^T$, $p^2 = (b_1/a_{11}, 0)^T$, $p^3 = (0, b_2/[a_{22} - a_{22}'])^T$. Taking $q_1 = q_2 = 1$, we have

$$q_1\left(b_1 + (-a_{11}, a_{12})\,p^j\right) + q_2\left(b_2 + (a_{21}, a_{22}' - a_{22})\,p^j\right) > 0\,, \quad j = 1, 2, 3\,.$$

By Theorem 4.2, (4.17) is permanent. Note that Corollary 4.2 instead of Theorem 4.2 is also applicable here.

To prove Theorem 4.2, we need a lemma about the solution of the system

$$x_i'(t) = x_i(t)[e_i(t) + L_i(x_t)]\,, \quad i \in I_N\,. \tag{4.19}$$

For any function $g\colon [a,b] \to \mathbb{R}$, denote the average of g over $[a,b]$ by

$$m(g,a,b) = \frac{1}{b-a}\int_a^b g(s)\,ds\,.$$

Assume that the $e_i(t)$ are bounded continuous with $|e_i(t)-u_i(t)| \leq \varepsilon_0$ on $\mathbb{R}_0$, the u_i satisfy

$$\lim_{T\to+\infty} m(u_i, t_0, t_0+T) = \bar{u}_i > \varepsilon_0 \quad \text{uniformly for} \quad t_0 \in \mathbb{R}_0\,, \tag{4.20}$$

and the $L_i(\varphi)$ are defined by (4.10) and (4.11). Then, under the assumption that (1.1) with (4.9)–(4.13) and (4.15) has a compact uniform attractor by one of the Theorems 2.1, 2.2 and 2.3, (4.19) also has a compact uniform attractor.

Lemma 4.1. *For each $\varphi \in C^+$ with $\varphi(0) \neq 0$ and every $t_0 \in \mathbb{R}_0$, the solution of* (4.19) *with $x_{t_0} = \varphi$ satisfies*

$$\liminf_{t\to+\infty} \max\{x_i(t)\colon i \in I_N\} > 0\,.$$

Proof. Let Ω be a compact uniform attractor and let

$$\rho = \sup\{|e(t) + L(\psi)|\colon t \in \mathbb{R}_0, \psi \in \Omega\} + 1\,.$$

Suppose the conclusion is not true. Then, for some $t_0 \in \mathbb{R}_0$ and $\varphi \in \Omega$ with $\varphi(0) \neq 0$, we have

$$\liminf_{t\to+\infty} \max\{x_i(t)\colon i \in I_N\} = 0\,.$$

Take $\alpha_0 > 0$ such that for each $i \in I_N$, if $\varphi_i(0) > 0$, and then $\varphi_i(0) \geq 2\alpha_0$. Then, there is a sequence $\{t_n\} \subset (t_0, \infty)$ such that

$$\forall n \geq 1\,, \quad \forall i \in I_N\,, \quad x_i(t_n) < \frac{\alpha_0}{n} e^{-n\rho}\,. \tag{4.21}$$

As $\max\{x_i(t_0)\colon i \in I_N\} = \max\{\varphi_i(0)\colon i \in I_N\} \geq 2\alpha_0$ and $\max\{x_i(t)\colon i \in I_N\}$ is continuous in t, there are $s_n \in (t_0, t_n)$ and $i_n \in I_N$ such that

$$\forall n \geq 1\,, \quad \max\{x_i(s_n)\colon i \in I_N\} = x_{i_n}(s_n) = \frac{\alpha_0}{n}\,, \tag{4.22}$$

$$\forall n \geq 1\,, \ \forall t \in (s_n, t_n]\,, \quad \max\{x_i(t)\colon i \in I_N\} < \frac{\alpha_0}{n}\,. \tag{4.23}$$

Then, considering the integration of (4.19) over $[s_n, t_n]$ and by (4.22), we have

$$\begin{aligned} x_{i_n}(t_n) &\geq x_{i_n}(s_n) \exp\left(-\int_{s_n}^{t_n} |e(t) + L(x_t)|\,dt\right) \\ &\geq \frac{\alpha_0}{n} e^{-\rho(t_n - s_n)}\,, \quad \forall n \geq 1\,. \end{aligned}$$

This, together with (4.21), gives

$$\forall n \geq 1\,, \quad t_n - s_n \geq n\,. \tag{4.24}$$

Since I_N is finite, by choosing a subsequence if necessary, we may assume that $i_n = i_0$ for all $n \geq 1$. Now, dividing the i_0th component equation of (4.19) by $x_{i_0}(t)$ and integrating over $[s_n, t_n]$, we can check that

$$\frac{1}{t_n - s_n} \ln \frac{x_{i_0}(t_n)}{x_{i_0}(s_n)} = m(e_{i_0}, s_n, t_n) + \sum_{j=1}^{N} (a_{i_0 j} - b_{i_0 j}) m(x_j, s_n, t_n) + o(1)$$

as $n \to \infty$, where the $o(1)$ term has the precise expression

$$\frac{1}{t_n - s_n} \sum_{j=1}^{N} \int_{-\tau}^{0} \left[\int_{s_n+\theta}^{s_n} x_j(s)ds - \int_{t_n+\theta}^{t_n} x_j(s)ds \right] d[a_{i_0 j}\xi_{i_0 j}(\theta) - b_{i_0 j}\eta_{i_0 j}(\theta)]\,.$$

From (4.22) and (4.23), we see that the left-hand side of the above equality is negative and

$$\forall j \in I_N\,, \quad m(x_j, s_n, t_n) = o(1) \quad (n \to \infty)\,.$$

Thus, $\limsup_{n\to\infty} m(e_{i_0}, s_n, t_n) \leq 0$. This is against the assumption that

$$\limsup_{n\to\infty} m(e_{i_0}, s_n, t_n) \geq \liminf_{n\to\infty} m(e_{i_0}, s_n, t_n) \geq \bar{u}_{i_0} - \varepsilon_0 > 0\,.$$

This contradiction shows the conclusion of the lemma. □

Proof of Theorem 4.2. By Theorem 4.1, we need only prove that for all (t_0, φ) in $\mathbb{R}_0 \times (\cup_{i\in J}\Omega_i)$, there exists a $T = T(t_0, \varphi) > 0$ such that

$$\frac{1}{T} \int_{0}^{T} \sum_{i\in J} q_i \left[r_i(t_0 + s) + L_i(x_{t_0+s}(t_0, \varphi)) \right] ds > 0\,. \tag{4.25}$$

We proceed by induction on the number m of positive components of φ.

When $m = 1$, we have $\varphi_k(\theta) > 0$ for some $k \in I_N$ and all $\theta \in [-\tau, 0]$, and $\varphi_j(\theta) \equiv 0$ for $j \in I_N \setminus \{k\}$. Then, $x_j(t, t_0, \varphi) \equiv 0$ for $t \geq t_0$ and $j \neq k$, and $x_k(t, t_0, \varphi)$ satisfies

$$x_k'(t) = x_k(t) \left[r_k(t) + a_{kk} \int_{-\tau}^{0} x_k(t+\theta)\, d\xi_{kk}(\theta) - b_{kk} \int_{-\tau}^{0} x_k(t+\theta)\, d\eta_{kk}(\theta) \right].$$

By Lemma 4.1, $\liminf_{t\to+\infty} x_k(t) > 0$. Then, $\ln x_k(t)$ is bounded. Now, integration of the above equation gives

$$\frac{\ln x_k(t) - \ln x_k(t_0)}{t - t_0} = m(r_k, t_0, t) + (a_{kk} - b_{kk}) m(x_k, t_0, t) + o(1)$$

as $t \to +\infty$. As the left-hand side vanishes and $m(r_k, t_0, t) \to \bar{r}_k$ when $t \to +\infty$, we must have

$$\lim_{t\to+\infty} m(x_k, t_0, t) = \frac{\bar{r}_k}{b_{kk} - a_{kk}} > 0 .$$

Then, $\lim_{t\to+\infty} m(x, t_0, t) = \hat{x}$ with $\hat{x}_k = \bar{r}_k/(b_{kk} - a_{kk})$ and $\hat{x}_j = 0$ for all $j \in I_N \setminus \{k\}$ and

$$\begin{aligned}\lim_{t\to+\infty} \frac{1}{t - t_0} \int_{t_0}^{t} \sum_{i\in J} q_i \left[r_i(s) + L_i(x_s(t_0, \varphi))\right] ds &= \lim_{t\to+\infty} \sum_{i\in J} q_i \left[m(r_i, t_0, t) + (A - B)_i m(x, t_0, t)\right] \\ &= \sum_{i\in J} q_i \left[\bar{r}_i + (A - B)_i \hat{x}\right] .\end{aligned}$$

As $\hat{x} \in \cup_{i\in J}\pi_i$ is a fixed point of (4.14), by (4.16), we have $\sum_{i\in J} q_i(\bar{r}_i + (A - B)_i\hat{x}) > 0$. Then, (4.25) holds for large enough $T > 0$ when $m = 1$.

Assume that (4.25) holds for some $m \geq 1$ and all $\varphi \in \cup_{i\in J}\Omega_i$ with at most m positive components. Now, suppose $\varphi^0 \in \cup_{i\in J}\Omega_i$ has $m + 1$ positive components and we show that (4.25) also holds. Let $J_1 = \{j \in I_N : \varphi_j^0(0) > 0\}$ with $|J_1| = m + 1$ and let

$$\Omega^{J_1} = \cap_{j\in I_N\setminus J_1}\Omega_j .$$

Note that $\Omega^{J_1} \subset \cup_{i\in J}\Omega_i$. Since Ω is a compact uniform attractor of (1.1), for each $j \in I_N$, Ω_j is a compact uniform attractor of the $(N - 1)$-dimensional subsystem of (1.1) with $x_j \equiv 0$, and Ω^{J_1} is a compact uniform attractor of the corresponding $(m + 1)$-dimensional subsystem. Since $\varphi^0 \in \operatorname{int}\Omega^{J_1}$, we have $x_t(t_0, \varphi^0) \in \operatorname{int}\Omega^{J_1}$ for all $t \geq t_0$. There are two possible cases for the limit set $\omega(t_0, \varphi^0)$ of $x_t(t_0, \varphi^0)$ as $t \to +\infty$: (a) $\omega(t_0, \varphi^0) \not\subset \partial\Omega^{J_1}$ and (b) $\omega(t_0, \varphi^0) \subset \partial\Omega^{J_1}$.

(a) In this case, there is a $\psi \in \operatorname{int}\Omega^{J_1}$ and a sequence $\{t_n\}$ with $t_n \to +\infty$ as $n \to \infty$ such that $\lim_{n\to\infty} x_{t_n}(t_0, \varphi^0) = \psi$. Then, $\lim_{n\to\infty} x(t_n, t_0, \varphi^0) = \psi(0)$, and thus the set $\{\ln x_j(t_n) : j \in J_1, n \geq 1\}$ is bounded and

$$\forall j \in J_1 , \quad \lim_{n\to\infty} \frac{\ln x_j(t_n) - \ln \varphi_j^0(0)}{t_n - t_0} = 0 . \tag{4.26}$$

Integrating the jth component equation of (1.1) with (4.9)–(4.13) as we did in the base case $m = 1$, we obtain

$$\frac{\ln x_j(t_n) - \ln \varphi_j^0(0)}{t_n - t_0} = m(r_j, t_0, t_n) + (A - B)_j m(x, t_0, t_n) + o(1)$$

as $n \to \infty$. This, together with (4.26) and $\lim_{n\to\infty} m(r_j, t_0, t_n) = \bar{r}_j$, gives

$$\forall j \in J_1 , \quad \lim_{n\to\infty} (A - B)_j m(x, t_0, t_n) = -\bar{r}_j .$$

By choosing a subsequence of $\{t_n\}$ if necessary, we may assume that $m(x, t_0, t_n)$ tends to $\bar{x}$ as $n \to \infty$. Then, $\bar{x} \in \cap_{j \in I_N \setminus J_1} \pi_j \subset \cup_{i \in J} \pi_i$ and $(A - B)_j \bar{x} = -\bar{r}_j$ for all $j \in J_1$. Thus, $\bar{x}$ is a fixed point of (4.14) in $\cup_{i \in J} \pi_i$ and, by (4.16), $\sum_{i \in J} q_i[\bar{r}_i + (A - B)_i \bar{x}] > 0$. It then follows that

$$\begin{aligned}
&\lim_{n\to\infty} \frac{1}{t_n - t_0} \int_{t_0}^{t_n} \sum_{i\in J} q_i \left[r_i(s) + L_i\left(x_s\left(t_0, \varphi^0\right)\right)\right] ds \\
&= \lim_{n\to\infty} \sum_{i\in J} q_i \left[m(r_i, t_0, t_n) + (A - B)_i m(x, t_0, t_n)\right] \\
&= \sum_{i\in J} q_i \left[\bar{r}_i + (A - B)_i \bar{x}\right] > 0 .
\end{aligned}$$

Therefore, for n large enough, (4.25) holds with $T = t_n - t_0$.

(b) For each $(t_1, \varphi) \in \mathbb{R}_0 \times \partial\Omega^{J_1}$, since φ has at most m positive components, by the inductive hypothesis, there is a $T(t_1, \varphi) > 0$ such that

$$\int_0^{T(t_1,\varphi)} \sum_{i\in J} q_i \left[r_i(t_1 + s) + L_i(x_{t_1+s}(t_1, \varphi))\right] ds > 0 .$$

For this fixed (t_1, φ), by continuous dependence, there is an open interval $I(t_1, \varphi)$ of $\mathbb{R}$ centered at t_1 and an open ball $B(t_1, \varphi)$ of Ω^{J_1} centered at φ such that for all $(\sigma, \psi) \in (I(t_1, \varphi) \cap \mathbb{R}_0) \times B(t_1, \varphi)$,

$$\frac{1}{T(t_1,\varphi)} \int_0^{T(t_1,\varphi)} \sum_{i\in J} q_i \left[r_i(\sigma + s) + L_i(x_{\sigma+s}(\sigma, \psi))\right] ds > \delta(t_1, \varphi) ,$$

$$\delta(t_1, \varphi) = \frac{1}{2T(t_1,\varphi)} \int_0^{T(t_1,\varphi)} \sum_{i\in J} q_i \left[r_i(t_1 + s) + L_i(x_{t_1+s}(t_1, \varphi))\right] ds > 0 .$$

Since $r(t)$ is T_0-periodic and Ω^{J_1} and $\partial\Omega^{J_1}$ are compact, by the same technique as that used in the proof of Theorem 4.1, we obtain an open set S_0 of Ω^{J_1} with $\partial\Omega^{J_1} \subset S_0$ and numbers $\delta_0 > 0$, $T^0 > 0$ and $T^1 > T^0$, satisfying, for all $(\sigma, \psi) \in \mathbb{R}_0 \times S_0$,

$$\exists T \in \left[T^0, T^1\right] \text{ such that } \frac{1}{T} \int_0^T \sum_{i\in J} q_i \left[r_i(\sigma + s) + L_i(x_{\sigma+s}(\sigma, \psi))\right] ds > \delta_0 . \tag{4.27}$$

From (4.27), we see that for any $(\sigma, \psi) \in \mathbb{R}_0 \times S_0$, if $x_t(\sigma, \psi) \in S_0$ for all $t \geq \sigma$, then there are $T_n \geq nT^0$ such that

$$\int_0^{T_n} \sum_{i\in J} q_i \left[r_i(\sigma + s) + L_i(x_{\sigma+s}(\sigma, \psi))\right] ds > nT^0\delta_0 \to +\infty \quad (n \to \infty) . \tag{4.28}$$

Now that $\omega(t_0, \varphi^0) \subset \partial\Omega^{J_1} \subset S_0$, there is a $\sigma > t_0$ such that $x_t(t_0, \varphi^0) \in S_0$ for all $t \geq \sigma$. Then, there is a $T > 0$ for (t_0, φ^0) such that (4.25) follows from (4.28).

By induction, (4.25) holds for all $(t_0, \varphi) \in \mathbb{R}_0 \times (\cup_{i \in J} \Omega_i)$. Therefore, (1.1) with (4.9)–(4.13) is partially permanent with respect to J. □

5 Necessary conditions for permanence of Lotka–Volterra systems

In this section, we consider delayed Lotka–Volterra systems of the form

$$x_i'(t) = x_i[r_i(t) + L_i(x_t)], \quad i \in I_N, \tag{5.1}$$

where the r_i are continuous on $\mathbb{R}_0$, satisfying, for all $i \in I_N$,

$$\forall t \in \mathbb{R}_0, -\infty < r_{iL} \leq r_i(t) \leq r_{iM} < \infty, \quad r_{iM} > 0, \tag{5.2}$$

$$\lim_{T \to +\infty} m(r_i, t_0, t_0 + T) = \bar{r}_i > 0 \ \text{uniformly for} \ t_0 \in \mathbb{R}_0. \tag{5.3}$$

The L_i can be written as $L_i(\varphi) = \sum_{j=1}^{N} L_{ij}(\varphi_j)$ with

$$L_{ij}(\varphi_j) = a_{ij} \int_{-\tau}^{0} \varphi_j(\theta) d_\theta \xi_{ij}(\theta) - b_{ij} \int_{-\tau}^{0} \varphi_j(\theta) d_\theta \eta_{ij}(\theta), \tag{5.4}$$

$$a_{ij} \geq 0, \quad b_{ij} \geq 0, \quad b_{ii} > 0 \tag{5.5}$$

for all $i, j \in I_N$, and the ξ_{ij} and η_{ij} are nondecreasing, satisfying

$$\forall i, j \in I_N, \quad \xi_{ij}(0) - \xi_{ij}(-\tau) = 1, \quad \eta_{ij}(0) - \eta_{ij}(-\tau) = 1. \tag{5.6}$$

We assume that (5.1) with (5.2)–(5.6) satisfies the conditions of Theorem 2.1, Theorem 2.2 or Theorem 2.3 so that it has a compact uniform attractor $\Omega \subset C^+$. The corresponding autonomous system

$$x_i'(t) = x_i[\bar{r}_i + L_i(x_t)], \quad i \in I_N \tag{5.7}$$

also has a compact uniform attractor. When $\tau = 0$, (5.7) degenerates to system

$$x_i'(t) = x_i[\bar{r}_i + (A - B)_i x], \quad i \in I_N. \tag{5.8}$$

For convenience, we adopt the following notation: $E = \mathbb{R}_+^N$, $E^0 = \operatorname{int} \mathbb{R}_+^N$ and for $I \subset I_N$,

$$E_I = \{x \in E \colon \forall i \in I, x_i = 0\},$$

$$E_I^0 = \left\{x \in E \colon x_j > 0 \ \text{if and only if} \ j \in I_N \setminus I\right\}.$$

Theorem 5.1. *Assume that* (5.1) *with* (5.2)–(5.6) *is permanent. Then,* (5.8) *has a fixed point* $x^* \in E^0$. *If* x^* *is the unique fixed point of* (5.8) *in* E^0, *then for each* $t_0 \in \mathbb{R}_0$, *every solution of* (5.1) *in* E^0 *on* $[t_0, +\infty)$ *satisfies*

$$\forall i \in I_N\,, \quad \lim_{t\to+\infty} m(x_i, t_0, t) = x_i^*\,. \tag{5.9}$$

Proof. Note that every solution of (5.1) in E^0 on $[t_0, +\infty)$ satisfies

$$\forall i \in I_N\,, \quad \frac{1}{t-t_0}\ln\frac{x_i(t)}{x_i(t_0)} = m(r_i, t_0, t) + (A-B)_i m(x, t_0, t) + o(1)$$

as $t \to +\infty$. Thus, the permanence assumption of (5.1) ensures that the left-hand side vanishes so

$$\forall i \in I_N\,, \quad \lim_{t\to+\infty} (B-A)_i m(x, t_0, t) = \bar{r}_i\,. \tag{5.10}$$

The permanence assumption of (5.1) also implies the existence of $\delta_1 > 0$ and $\delta_2 > \delta_1$ such that

$$\forall j \in I_N\,, \quad \delta_1 \le \liminf_{t\to+\infty} m(x_j, t_0, t) \le \limsup_{t\to+\infty} m(x_j, t_0, t) \le \delta_2\,.$$

Then, there exist $x_1^* \in [\delta_1, \delta_2]$ and an increasing sequence $\{t_n\}$ with $t_n \to +\infty$ as $n \to \infty$ such that

$$\lim_{n\to\infty} m(x_1, t_0, t_n) = \liminf_{t\to+\infty} m(x_1, t_0, t) = x_1^*\,.$$

By choosing subsequences of $\{t_n\}$ if necessary, without loss of generality, we may assume that

$$\forall j \in I_N\,, \quad \lim_{n\to\infty} m(x_j, t_0, t_n) = x_j^* \in [\delta_1, \delta_2]\,. \tag{5.11}$$

Then, after the replacement of t by t_n in (5.10), substitution of (5.11) into (5.10) gives $(B-A)x^* = \bar{r}$. This shows that x^* is a fixed point of (5.8) in E^0.

If $\lim_{t\to+\infty} m(x_i, t_0, t) \ne x_i^*$ for some $i \in I_N$, by the same reasoning as above, we can find a sequence $\{t'_n\}$ such that $\lim_{n\to\infty} m(x, t_0, t'_n)$ exists and is another fixed point of (5.8) in E^0. Hence, if x^* is the unique fixed point of (5.8) in E^0, then we must have (5.9). □

Theorem 5.2. *If* (5.7) *is permanent, then it has a unique fixed point* x^* *in* E^0, *and* (5.9) *holds for all* $t_0 \in \mathbb{R}_0$ *and every solution of* (5.7) *in* E^0 *on* $[t_0, +\infty)$.

Proof. Note that (5.7) and (5.8) have the same fixed points. By Theorem 5.1, (5.7) has at least one fixed point x^* in E^0. If (5.7) has more than one fixed point in E^0, then the set of infinitely many solutions of $(B-A)x = \bar{r}$ contains at least a line segment connecting x^* to the boundary ∂E. Since this line segment consists of constant solutions of (5.7), this contradicts the permanence of (5.7). Thus, x^* is the unique equilibrium of (5.7) in E^0 and (5.9) holds by Theorem 5.1. □

Definition 5.1. A vector $v \in E^0$ and a matrix $U = (u_{ij}) \in \mathbb{R}^{N\times N}$ are said to satisfy the $(I - J)$-condition if for each $I \subset I_N$ with $J = I_N \setminus I \neq \emptyset$, there exists a unique $x_I^* \in E_I^0$ such that

$$\forall j \in J, \quad U_j x_I^* = v_j \tag{5.12}$$

and, if $I \neq \emptyset$, this x_I^* also satisfies

$$\forall i \in I, \quad U_i x_I^* < v_i . \tag{5.13}$$

We give a geometric interpretation of the $(I - J)$-condition below. Let $P_i = \{x \in E\colon U_i x = v_i\}$. Then, P_i can be viewed as a hyperplane in E. A point $p \in E$ is said to be **below** (**on** or **above**) P_i if $U_i p < v_i$ ($U_i p = v_i$ or $U_i p > v_i$). Then, v and U satisfy the $(I - J)$-condition if and only if for each $I \subset I_N$ with $J = I_N \setminus I \neq \emptyset$, the planes P_j $(j \in J)$ have a unique intersection point $x_I^* \in E_I^0$ and, if $I \neq \emptyset$, x_I^* is below all the planes P_i $(i \in I)$.

We give another interpretation of the $(I - J)$-condition in terms of fixed points of the system (5.8). Let $\gamma_i = \{x \in E\colon (B - A)_i x = \bar{r}_i\}$ be the ith nullcline plane of (5.8). Then, $\bar{r}$ and $B - A$ satisfy the $(I - J)$-condition if and only if each subsystem of (5.8), including (5.8) itself, has a unique fixed point with positive components corresponding to the subsystem and this fixed point is below all the other nullcline planes.

Theorem 5.3. *Assume that all small perturbations of* (5.1) *on B and A and their subsystems are permanent. Then, $\bar{r}$ and $B - A$ satisfy the $(I - J)$-condition.*

Proof. Since all small perturbations of (5.1) on A and B and their subsystems are permanent, for each $I \subset I_N$ with $I \neq I_N$, application of Theorem 5.1 to the corresponding subsystem ($x_i = 0$ for $i \in I$) shows the existence of a fixed point $x_I^* \in E_I^0$ of (5.8). Obviously, for each $i \in I_N$, the fixed point $x^*_{(I_N\setminus\{i\})}$ is uniquely given by $(x^*_{(I_N\setminus\{i\})})_i = \bar{r}_i/(b_{ii} - a_{ii})$ and $(x^*_{(I_N\setminus\{i\})})_k = 0$ for $k \in I_N \setminus \{i\}$. If, for some $I \subset I_N$, the fixed point x_I^* of (5.8) is not unique in E_I^0, then the set of solutions of the linear algebraic system $(B-A)_j x = \bar{r}_j$ for $j \in I_N \setminus I$ contains a line segment connecting x^* to the boundary ∂E_I^0. Thus, we can always find a unique fixed point x_J^* of (5.8) in E_J^0 for some $J \subset I_N$ with $I \subset J$ and $I \neq J$ satisfying $(B - A)_j x_J^* = \bar{r}_j$ for some $j \in J \setminus I$.

Now, suppose $\bar{r}$ and $B - A$ do not satisfy the $(I - J)$-condition. From the above paragraph, there is a unique fixed point x_I^* of (5.8) in E_I^0 for some $I \subset I_N$ with $I \neq I_N$ such that $(B - A)_j x_I^* = \bar{r}_j$ for all $j \in J = I_N \setminus I$, but $(B - A)_{i_0} x_I^* \geq \bar{r}_{i_0}$ for some $i_0 \in I$. If $(B - A)_{i_0} x_I^* = \bar{r}_{i_0}$, then we can choose a small perturbation $\tilde{B}$ of B such that $(\tilde{B} - A)_j x_I^* = \bar{r}_j$ for all $j \in J$ but $(\tilde{B} - A)_{i_0} x_I^* > \bar{r}_{i_0}$. Without loss of generality, we assume that

$$\forall j \in I_N \setminus I, \quad (B - A)_j x_I^* = \bar{r}_j; \quad \exists i_0 \in I, \quad (B - A)_{i_0} x_I^* > \bar{r}_{i_0} \tag{5.14}$$

and let

$$y_0 = (B - A)_{i_0} x_I^* - \bar{r}_{i_0} > 0 . \tag{5.15}$$

Let $y(t)$ be a solution of (5.1) on $[t_0, +\infty)$ with

$$\forall i \in I\,, \quad y_i(t) = 0\,; \quad \forall j \in I_N \setminus I, y_j(t) > 0 \tag{5.16}$$

for all $t \geq t_0 - \tau$. Then, applying Theorem 5.1 to the subsystem of (5.1) corresponding to this I, we have $\lim_{t\to+\infty} m(y, t_0, t) = x_I^*$ so that

$$\forall i \in I_N\,, \quad \lim_{T\to+\infty} \frac{1}{T} \int_{t_0}^{t_0+T} L_i(y_t)dt = (A - B)_i x_I^*\,. \tag{5.17}$$

By (5.3), (5.15) and (5.17), there is a $T_1 > t_0$ such that

$$-\frac{1}{t - t_0} \int_{t_0}^{t} L_{i_0}(y_s)ds - m(r_{i_0}, t_0, t) \geq \frac{1}{2}\gamma_0 \tag{5.18}$$

for all $t \geq T_1$. Since all subsystems of (5.1) are permanent, there are $\delta_1 > 0$ and $\delta_2 > \delta_1$ such that for each $t_0 \in \mathbb{R}_0$ and $i \in I_N$, every solution of (5.1) on $[t_0, +\infty)$ satisfies

$$x_i(t_0) > 0 \Longrightarrow \delta_1 \leq x_i(t) \leq \delta_2 \text{ for large enough } t \geq t_0\,. \tag{5.19}$$

Put

$$\Phi = \{\varphi \in C^+ \colon \forall i \in I_N \setminus \{i_0\}, \varphi_i = (y_{t_0})_i, \forall h \in [0, \delta_1/2], \varphi_{i_0}(\theta) \equiv h\} \tag{5.20}$$

and consider the set U of solutions of (5.1) on $[t_0, +\infty)$ defined by

$$U = \{x(t, t_0, \varphi) \colon \varphi \in \Phi\}\,. \tag{5.21}$$

Since $x(t, t_0, \varphi)$ is continuous in (t, t_0, φ), Φ defined by (5.20) is compact and $y_i(t_0) > 0$ for $i \in I_N \setminus I$, by (5.19) there are $u > 0$ and $v > u$, satisfying

$$\forall i \in I_N \setminus I\,, \quad \forall \varphi \in \Phi\,, \quad \forall t \geq t_0\,, \quad u \leq x_i(t, t_0, \varphi) \leq v\,. \tag{5.22}$$

For each integer $k \geq 2$ and $T_k = T_1 + k$, by continuous dependence of the solution on φ, there is an $h_k \in (0, \delta_1/2)$ such that every $x(t, t_0, \varphi)$ in U with $\varphi_{i_0}(0) \leq h_k$ satisfies

$$\forall t \in [t_0, T_k]\,, \quad \|x_t(t_0, \varphi) - y_t\| < \frac{\gamma_0}{4ka^0}\,, \tag{5.23}$$

where $a^0 = \max\{\sum_{j=1}^{N}(a_{ij} + b_{ij}) \colon i \in I_N\}$. Take $\varphi^k \in \Phi$ with $\varphi_{i_0}^k(\theta) \equiv h_k$. Then, from (5.1), (5.18) and (5.23), we have

$$
\begin{aligned}
\frac{1}{T_k - t_0} \ln \frac{x_{i_0}\left(T_k, t_0, \varphi^k\right)}{h_k} &= m(r_{i_0}, t_0, T_k) + \frac{1}{T_k - t_0} \int_{t_0}^{T_k} L_{i_0}\left(x_t\left(t_0, \varphi^k\right)\right) dt \\
&\le m(r_{i_0}, t_0, T_k) + \frac{1}{T_k - t_0} \int_{t_0}^{T_k} L_{i_0}(y_t)\, dt \\
&\quad + \frac{1}{T_k - t_0} \int_{t_0}^{T_k} \left| L_{i_0}\left(x_t\left(t_0, \varphi^k\right) - y_t\right)\right| dt \\
&\le -\tfrac{1}{2}\gamma_0 + \tfrac{1}{4k}\gamma_0 \\
&< -\tfrac{1}{4}\gamma_0 .
\end{aligned}
$$

Thus, $x_{i_0}(T_k, t_0, \varphi^k) < h_k$. From (5.19) and the continuity of $x(t, t_0, \varphi)$, there is a $t_k > T_k$ satisfying

$$
\forall t \in [T_k, t_k), \quad x_{i_0}(t, t_0, \varphi^k) < h_k = x_{i_0}(t_k, t_0, \varphi^k). \tag{5.24}
$$

Note that (5.23) implies $h_k < \gamma_0/(4ka^0)$. Then, (5.24) together with (5.23) and $y_{i_0}(t) \equiv 0$ gives

$$
\forall t \in [t_0 - \tau, t_k], \quad 0 < x_{i_0}(t, t_0, \varphi^k) < \frac{\gamma_0}{4ka^0}. \tag{5.25}
$$

From (5.1), we see that each solution $x(t, t_0, \varphi^k)$ on $[t_0, t_k]$ satisfies

$$
\begin{aligned}
\frac{1}{t_k - t_0} \ln \frac{x_i(t_k, t_0, \varphi^k)}{x_i(t_0)} &= (A - B)_i m\left(x\left(\cdot, t_0, \varphi^k\right), t_0, t_k\right) \\
&\quad + m(r_i, t_0, t_k) + o(1) \quad (k \to \infty)
\end{aligned} \tag{5.26}
$$

for $i \in (I_N \setminus I) \cup \{i_0\}$. Since $t_k - t_0 > T_1 + k - t_0 \to \infty$ as $k \to \infty$ and $\ln x_i(t_k, t_0, \varphi^k) - \ln x_i(t_0)$ is bounded by (5.22) and (5.24), the left-hand side of (5.26) vanishes as $k \to \infty$, i. e.

$$
\forall i \in (I_N \setminus I) \cup \{i_0\}, \quad \lim_{k \to \infty} \frac{1}{t_k - t_0} \ln \frac{x_i(t_k, t_0, \varphi^k)}{x_i(t_0)} = 0. \tag{5.27}
$$

Then, by (5.3), we can rewrite (5.26) as

$$
\forall i \in (I_N \setminus I) \cup \{i_0\}, \quad \lim_{k \to \infty} (B - A)_i m(x(\cdot, t_0, \varphi^k), t_0, t_k) = \bar{r}_i. \tag{5.28}
$$

By (5.25), we have $\lim_{k \to \infty} m(x_{i_0}(\cdot, t_0, \varphi^k), t_0, t_k) = 0$. From (5.22), we know that for each $j \in I_N \setminus I$, the sequence $\{m(x_j(\cdot, t_0, \varphi^k), t_0, t_k)\}$ has convergent subsequences. By choosing convergent subsequences one after another $N - |I|$ times, we find a subsequence $\{n_k\} \subset \{k\}$ with $n_k \to \infty$ as $k \to \infty$ such that all of the above sequences for $j \in I_N \setminus I$ after the replacement of k by n_k are convergent. Without loss of generality, we assume that

$$
\forall j \in I_N \setminus I, \quad \lim_{k \to \infty} m(x_j(\cdot, t_0, \varphi^k), t_0, t_k) = z_j \in [u, v]. \tag{5.29}
$$

Then, with $z_i = 0$ for $i \in I$, it follows from (5.28) and (5.29) that

$$\forall i \in (I_N \setminus I) \cup \{i_0\}, \quad (B - A)_i z = \bar{r}_i . \tag{5.30}$$

As x_I^* is the unique fixed point of (5.8) in E_I^0, from (5.29) and (5.30), we obtain $z_j = (x_I^*)_j$ for all $j \in I_N \setminus I$. Thus, (5.30) can be written as

$$\forall i \in (I_N \setminus I) \cup \{i_0\}, \quad (B - A)_i x_I^* = \bar{r}_i ,$$

a contradiction to (5.14). This contradiction shows that $\bar{r}$ and $B - A$ satisfy the $(I - J)$-condition. □

Theorem 5.4. *Assume that* (5.7) *and its subsystems are permanent. Then, $\bar{r}$ and $B - A$ satisfy the $(I - J)$-condition.*

Proof. By Theorem 5.2, every x_I^* is the unique fixed point of (5.7) in E_I^0. If $\bar{r}$ and $B - A$ do not satisfy the $(I - J)$-condition, then there is a proper $I \subset I_N$ such that $(B - A)_j x_I^* = \bar{r}_j$ for all $j \in J = I_N \setminus I$, but $(B - A)_{i_0} x_I^* \geq \bar{r}_{i_0}$ for some $i_0 \in I$. If $(B - A)_{i_0} x_I^* = \bar{r}_{i_0}$, then, with $J = I \setminus \{i_0\}$, x_J^* is the unique fixed point of (5.7) in E_J^0. However, since both x_I^* and x_J^* satisfy the algebraic system $(B - A)_i x = \bar{r}_i$ for all $i \in I_N \setminus J$, every point on the line segment joining x_I^* to x_J^* also satisfies the system. This shows that (5.8) has infinitely many fixed points in E_J^0, a contradiction to the uniqueness of x_J^* in E_J^0. Therefore, we must have (5.14). The rest of the proof for Theorem 5.1 is still valid here. □

6 Sufficient condition for permanence of Lotka–Volterra systems

With a perturbation in mind, we also consider systems of the form

$$y_i'(t) = y_i(t)[\tilde{r}_i(t) + \tilde{L}_i(t, y_t)] , \quad i \in I_N , \tag{6.1}$$

where $\tilde{L}_i(t, \varphi) = \tilde{L}_{i1}(t, \varphi_1) + \cdots + \tilde{L}_{iN}(t, \varphi_N)$ with

$$\tilde{L}_{ij}(t, \varphi_j) = \tilde{a}_{ij}(t) \int_{-\tau}^{0} \varphi_j(\theta)\, d_\theta \tilde{\xi}_{ij}(t, \theta) - \tilde{b}_{ij}(t) \int_{-\tau}^{0} \varphi_j(\theta)\, d_\theta \tilde{\eta}_{ij}(t, \theta) . \tag{6.2}$$

We assume that the $\tilde{\xi}_{ij}(t, \theta)$ and $\tilde{\eta}_{ij}(t, \theta)$ are bounded, continuous in t and nondecreasing in θ, and satisfy

$$\tilde{\xi}_{ij}(t, 0) - \tilde{\xi}_{ij}(t, -\tau) = 1 , \quad \tilde{\eta}_{ij}(t, 0) - \tilde{\eta}_{ij}(t, -\tau) = 1 . \tag{6.3}$$

Let

$$c_i = b_{ii}[\eta_{ii}(0) - \eta_{ii}(0^-)] , \quad \tilde{c}_i(t) = \tilde{b}_{ii}[\tilde{\eta}_{ii}(t, 0) - \tilde{\eta}_{ii}(t, 0^-)] .$$

Assume also that

$$|\tilde{r}_i(t) - r_i(t)|\,, \quad |\tilde{a}_{ij}(t) - a_{ij}|\,, \quad |\tilde{b}_{ij}(t) - b_{ij}|\,, \quad |\tilde{c}_i(t) - c_i| \tag{6.4}$$

are sufficiently small for all $t \in \mathbb{R}_0$ and (6.1) with (6.2), and (6.3) also satisfies the requirements of one of the theorems given in Section 2 so that it has a compact uniform attractor $\Omega' \subset C^+$. Let $M > 0$ and $\rho > 0$ such that $\|\varphi\| \le M$ and $|\varphi_i(0)[\tilde{r}_i(t) + \tilde{L}_i(t,\varphi)]| \le \rho$ for all $t \in \mathbb{R}_0$ and $\varphi \in \Omega'$. Since Ω' is forward invariant, the solutions in Ω' satisfies $|x(t)| \le M$ and $|x'(t)| \le \rho$ so there is a subset $S_0 \subset \Omega'$ such that each $\varphi \in S_0$ is differentiable with $\|\varphi'\| \le \rho$.

Definition 6.1. For any given $\varepsilon > 0$, (6.1) with (6.2) and (6.3) is called an ε-perturbation of (5.1) with (5.2)–(5.6) if

$$\forall (t,\varphi) \in \mathbb{R}_0 \times S_0\,, \quad |\tilde{r}(t) + \tilde{L}(t,\varphi) - r_i(t) - L_i(\varphi)| < \varepsilon\,. \tag{6.5}$$

Note that this definition not only includes a small perturbation on the coefficients listed in (6.4), but also covers small perturbations on delays such as $\varphi_i(-\tilde{\tau}_{ij}(t))$ where $\tilde{\tau}_{ij}(t) - \tau_{ij}$ is small. In the case of $\tilde{\tau}_{ij}(t) - \tau_{ij}$ not necessarily small for all $t \in \mathbb{R}_0$ but $\lim_{t\to+\infty} \tilde{\tau}_{ij}(t) = \tau_{ij}$, we may define an ε-perturbation of (5.1) for large t only with the replacement of (6.5) by

$$\limsup_{t\to+\infty} |\tilde{r}(t) + \tilde{L}(t,\varphi) - r_i(t) - L_i(\varphi)| < \varepsilon\,. \tag{6.6}$$

Theorem 6.1. *Assume that* (5.1) *with* (5.2)–(5.6) *has a compact uniform attractor and that $\bar{r}$ and $B - A$ satisfy the $(I - J)$-condition. Then, there is an $\varepsilon > 0$ such that if* (6.1) *with* (6.2) *and* (6.3) *as an ε-perturbation of* (5.1) *with* (5.2)–(5.6) *has a compact uniform attractor, then* (6.1) *and all of its subsystems are permanent.*

If (6.1) is an ε-perturbation of (5.1) and $x(t)$ is a solution of (6.1) on $[t_0, +\infty)$, then $x(t)$ is also a solution of the system

$$x_i'(t) = x_i(t)[e_i(t) + L_i(x_t)]\,, \quad i \in I_N \tag{6.7}$$

with $e(t) = \tilde{r}(t) + \tilde{L}(t,x_t) - L(x_t)$ so $|e(t) - r(t)| < \varepsilon$ for all $t \ge t_0$. Thus, to prove Theorem 6.1, we need only prove the existence of $\varepsilon > 0$, $\delta > 0$ and $M > 0$ such that for all continuous $e(t)$ satisfying $|e(t) - r(t)| < \varepsilon$ on $\mathbb{R}_0$ and for all $(t_0, \varphi) \in \mathbb{R}_0 \times C^+$, the solution of (6.7) satisfies

$$\forall i \in I_N\,, \quad \varphi_i(0) > 0 \Longrightarrow \delta < x_i(t,t_0,\varphi) < M \;\text{ for all large } t\,. \tag{6.8}$$

To this end, we start with the $(I - J)$-condition satisfied by $\bar{r}$ and $B - A$. Since each $x_I^* \in E_I^0$ in the definition of $(I - J)$-condition depends on $\bar{r}$ linearly, there is a small ε_0 with $0 < \varepsilon_0 < \min\{\bar{r}_1, \dots, \bar{r}_N\}$ such that $B - A$ and each $\hat{r}$ in

$$B(\bar{r}, \varepsilon_0) = \left\{x \in E^0 \colon \forall i \in I_N, |x_i - \bar{r}_i| \le \varepsilon_0\right\}$$

also satisfy the $(I-J)$-condition. Hence, $(B-A)^{-1}B(\bar{r},\varepsilon_0)\subset E^0$. For this ε_0, take $\varepsilon_1\in(0,\varepsilon_0)$ and let

$$E(r,\varepsilon_1)=\left\{e\in C\left(\mathbb{R}_0,\mathbb{R}^N\right):\forall t\in\mathbb{R}_0,|e(t)-r(t)|\le\varepsilon_1\right\}. \tag{6.9}$$

Then, from (5.2) and (5.3), each $e\in E(\bar{r},\varepsilon_1)$ satisfies

$$\forall i\in I_N\,,\quad -\infty<r_{iL}-\varepsilon_1\le e_i(t)\le r_{iM}+\varepsilon_1<+\infty\,, \tag{6.10}$$

$$\forall i\in I_N\,,\quad 0<\bar{r}_i-\varepsilon_1\le\liminf_{t\to\infty}m(e_i,t_0,t)\le\limsup_{t\to\infty}m(e_i,t_0,t)\le\bar{r}_i+\varepsilon_1<+\infty\,. \tag{6.11}$$

Lemma 6.1. *Assume that* (5.1) *with* (5.2)–(5.6) *has a compact uniform attractor by one of the theorems given in Section 2 and that $\bar{r}$ and $B-A$ satisfy the $(I-J)$-condition. Then,*

(i) (6.7) *for all $e\in E(r,\varepsilon_1)$ also has a compact uniform attractor $\Omega\subset C^+$ independent of e, and thus all solutions of* (6.7) *for all $e\in E(r,\varepsilon_1)$ have a uniform upper bound $M>0$;*
(ii) *for each $e\in E(r,\varepsilon_1)$, every solution of* (6.7) *in E satisfies*

$$\forall i\in I_N\,,\quad\forall t_0\in\mathbb{R}_0\,,\quad x_i(t_0)>0\ \textit{implies}\ \liminf_{t\to+\infty}x_i(t)>0\,.$$

Proof. (i) Since (5.1) has a compact uniform attractor by the theorems given in Section 2, from (6.9)–(6.11), we see that $e\in E(r,\varepsilon_1)$ implies that the corresponding theorem in Section 2 also holds for (6.7) so it has a compact uniform attractor Ω. From Section 2, we know that this Ω is determined by ε_1 rather than by $e\in E(r,\varepsilon_1)$. Thus, Ω is independent of e. This shows the existence of $M>0$ such that all solutions of (6.7) for all $e\in E(r,\varepsilon_1)$ satisfy $|x(t)|<M$ for large enough t. Moreover, there is a $\rho>0$ such that $|e(t)+L(\varphi)|<\rho$ for all $(t,\varphi)\in\mathbb{R}_0\times\Omega$.

(ii) Suppose the conclusion is not true. Then, there exist an $e\in E(r,\varepsilon_1)$, a $t_0\in\mathbb{R}_0$, an $i_0\in I_N$ and a $\varphi\in\Omega$ with $\varphi_{i_0}(0)>0$ such that the solution $x(t)$ of (6.7) with (6.10) and (6.11) satisfies $\liminf_{t\to+\infty}x_{i_0}(t)=0$. Let $J_0\subset I_N$ such that $\varphi_i(0)>0$ if and only if $i\in J_0$. Then, for all $t\ge t_0$, $x_i(t)>0$ for $i\in J_0$ and $x_j(t)\equiv0$ for $j\in I_N\setminus J_0$. By Lemma 4.1, $x_1(t)+\cdots+x_N(t)\ge\sigma(>0)$ holds for sufficiently large t. Hence, there is a proper subset $J\subset I_N$ with $i_0\in J$ such that

$$\liminf_{t\to+\infty}\max\{x_j(t)\colon j\in J\}=0\,, \tag{6.12}$$

$$\forall i\in I_N\setminus J\,,\quad\liminf_{t\to+\infty}\max\{x_j(t)\colon j\in J\cup\{i\}\}>0\,. \tag{6.13}$$

From (6.13) and the definition of J_0, we have

$$\rho_1=\inf\{\max\{x_j(t)\colon j\in J\cup\{i\}\}\colon t\ge t_0,i\in I_N\setminus J\}>0\,, \tag{6.14}$$

$$\rho_2=\tfrac{1}{2}\min\{\rho_1,x_i(t_0)\colon i\in J_0\}>0\,. \tag{6.15}$$

By (6.12), there is an increasing sequence $\{t_n\}$ satisfying

$$\forall n \geq 1\,, \quad \max\{x_j(t_n)\colon j \in J\} < \frac{\rho_2}{n} e^{-n\rho}\,. \tag{6.16}$$

Then, as (6.15) implies $\max\{x_j(t_0)\colon j \in J\} \geq 2\rho_2$, by continuity, there are $s_n \in (t_0, t_n)$ and $j_n \in J$ such that

$$\forall n \geq 1\,, \quad \max\{x_j(s_n)\colon j \in J\} = x_{j_n}(s_n) = \frac{\rho_2}{n}\,, \tag{6.17}$$

$$\forall n \geq 1\,, \quad \forall t \in (s_n, t_n], \max\{x_j(t)\colon j \in J\} < \frac{\rho_2}{n}\,. \tag{6.18}$$

From (6.14), (6.15) and (6.18), we deduce that

$$\forall i \in I_N \setminus J\,, \quad \forall t \in [s_n, t_n]\,, \quad \rho_1 \leq x_i(t) \leq M\,. \tag{6.19}$$

Since integration of (6.7) gives

$$x_{j_n}(t_n) = x_{j_n}(s_n) \exp\left(\int_{s_n}^{t_n} [e_{j_n}(t) + L_{j_n}(x_t)]\, dt\right) > x_{j_n}(s_n) e^{-\rho(t_n - s_n)}\,, \quad \forall n \geq 1\,,$$

this together with (6.16) and (6.17) leads to

$$\forall n \geq 1\,, \quad \frac{\rho_2}{n} e^{-n\rho} > \frac{\rho_2}{n} e^{-\rho(t_n - s_n)}\,,$$

which is equivalent to

$$\forall n \geq 1\,, \quad t_n - s_n > n\,. \tag{6.20}$$

For $i \in I_N \setminus J$, from (6.7), we have

$$\frac{1}{t_n - s_n} \ln \frac{x_i(t_n)}{x_i(s_n)} = m(e_i, s_n, t_n) + \sum_{j=1}^{N} (a_{ij} - b_{ij}) m(x_j, s_n, t_n) + o(1) \tag{6.21}$$

as $n \to \infty$. Since (6.17) and (6.18) imply $\lim_{n\to\infty} m(x_j, s_n, t_n) = 0$ for all $j \in J$ and an immediate consequence of (6.19) and (6.20) is

$$\forall i \in I_N \setminus J\,, \quad \lim_{n\to\infty} \frac{1}{t_n - s_n} \ln \frac{x_i(t_n)}{x_i(s_n)} = 0\,,$$

we can rewrite (6.21) as, for all $i \in I_N \setminus J$,

$$\sum_{j \in I_N \setminus J} (b_{ij} - a_{ij}) m(x_j, s_n, t_n) = m(e_i, s_n, t_n) + o(1)\ (n \to \infty)\,. \tag{6.22}$$

By (6.19), (6.10) and (6.11), there is a subsequence $\{n_\ell\} \subset \{n\}$ such that for all $\ell \geq 1$, $j_{n_\ell} = j_0 \in J$ and

$$\forall i \in I_N \setminus J\,, \quad \lim_{\ell\to\infty} m(x_i, s_{n_\ell}, t_{n_\ell}) = x_i^* \in [\rho_1, M],$$

$$\forall i \in I_n\,, \quad \lim_{\ell\to\infty} m(e_i, s_{n_\ell}, t_{n_\ell}) = e_i^* \in [\bar{r}_i - \varepsilon_1, \bar{r}_i + \varepsilon_1]\,.$$

Replacing n by n_ℓ in (6.22) and letting $\ell \to \infty$, we obtain

$$\forall i \in I_N \setminus J, \sum_{j \in I_N \setminus J} (b_{ij} - a_{ij}) x_j^* = e_j^* . \tag{6.23}$$

Since (6.16) and (6.17) imply $x_{j_n}(t_n) < x_{j_n}(s_n)$, from this and (6.21) with the replacement of i by j_n and n by n_ℓ and letting $\ell \to \infty$, we obtain

$$\sum_{j \in I_N \setminus J} (b_{j_0 j} - a_{j_0 j}) x_j^* \geq e_{j_0}^* . \tag{6.24}$$

By the choice of ε_1, $e^* \in B(\bar{r}, \varepsilon_0)$ so e^* and $B - A$ satisfy the $(I - J)$-condition. Then, (6.23) should imply

$$\forall i \in J, \quad \sum_{j \in I_N \setminus J} (b_{ij} - a_{ij}) x_j^* < e_i^* ,$$

which contradicts (6.24). This contradiction shows the truth of the conclusion. □

Lemma 6.2. *Under the assumptions of Lemma 6.1, there is a $\mu_0 > 0$ such that for all $e \in E(r, \varepsilon_1)$, every solution of (6.7) in E^0 satisfies*

$$\forall i \in I_N , \quad \limsup_{t \to +\infty} x_i(t) \geq \mu_0 .$$

Proof. For each $e \in E(r, \varepsilon_1)$, $t_0 \in \mathbb{R}_0$ and every solution of (6.7) on $[t_0, +\infty)$, let

$$F_i(t) = (B - A)_i m(x, t_0, t) , \quad t \geq t_0, i \in I_N .$$

Then, integration of (6.7) gives

$$\forall i \in I_N , \quad F_i(t) = m(e_i, t_0, t) - \frac{1}{t - t_0} \ln \frac{x_i(t)}{x_i(t_0)} + o(1) \quad (t \to +\infty) .$$

By Lemma 6.1, we have

$$\forall i \in I_N , \quad \lim_{t \to +\infty} [F_i(t) - m(e_i, t_0, t)] = 0 . \tag{6.25}$$

Take $\varepsilon' = (\varepsilon_0 + \varepsilon_1)/2$. Then, $\varepsilon_1 < \varepsilon' < \varepsilon_0$ so $B(\bar{r}, \varepsilon_1) \subset B(\bar{r}, \varepsilon') \subset B(\bar{r}, \varepsilon_0)$. By (6.11) and (6.25), there is a $T > t_0$ such that $F(t) \in B(\bar{r}, \varepsilon')$ for all $t \geq T$. Hence, from the definition of F, we have $m(x, t_0, t) = (B - A)^{-1} F(t) \in (B - A)^{-1} B(\bar{r}, \varepsilon')$ for all $t \geq T$. Since $(B - A)^{-1} B(\bar{r}, \varepsilon') \subset (B - A)^{-1} B(\bar{r}, \varepsilon_0) \subset E^0$ and both $B(\bar{r}, \varepsilon')$ and $(B - A)^{-1} B(\bar{r}, \varepsilon')$ are compact, with

$$\mu_0 = \min \left\{ z_i \colon \forall z \in (B - A)^{-1} B(\bar{r}, \varepsilon'), \forall i \in I_N \right\} ,$$

we have

$$\forall i \in I_N , \quad \forall t \geq T , \quad m(x_i, t_0, t) \geq \mu_0 > 0 .$$

Then, the conclusion follows from this and

$$\limsup_{t \to +\infty} x_i(t) \geq \limsup_{t \to +\infty} m(x_i, t_0, t) . \qquad \square$$

Lemma 6.3. *Under the assumptions of Lemma 6.1, there is a $\sigma > 0$ such that for each $e \in E(r, \varepsilon_1)$, $t_0 \in \mathbb{R}_0$ and every $\varphi \in C^+$ with $\varphi(0) \neq 0$, the solution of (6.7) with $x_{t_0} = \varphi$ satisfies*

$$\liminf_{t\to+\infty} \max\{x_i(t) : i \in I_N\} \geq \sigma .$$

Proof. Suppose the conclusion is not true. Then, there are $\{e^n\} \subset E(\bar{r}, \varepsilon_1)$, $\{T_n\} \subset \mathbb{R}_0$, $\{\varphi^n\} \subset \Omega$ and $\{\alpha_n\} \subset \mathbb{R}_+$ such that the solution $y^n(t) = x(t, e^n, T_n, \varphi^n)$ of (6.7) with e replaced by e^n and $(y^n)_{T_n} = \varphi^n$ satisfies

$$\liminf_{t\to+\infty} \max\left\{y_i^n(t) : i \in I_N\right\} = \alpha_n \downarrow 0 \quad (n \to \infty) . \tag{6.26}$$

Applying Lemma 6.2 to all subsystems of (6.7), we have

$$\forall n \geq 1, \quad \limsup_{t\to\infty} \max\left\{y_i^n(t) : i \in I_N\right\} \geq \mu_0 > 0 .$$

Thus,

$$\forall n \geq 1, \quad \exists T_n' \geq T_n \text{ such that } \max\{y_i^n(T_n') : i \in I_N\} \geq \mu_0/2 . \tag{6.27}$$

By (6.26), there is a subsequence $\{n_k\}$ of $\{n\}$ such that

$$\forall k \geq 1, \quad \liminf_{t\to+\infty} \max\left\{y_i^{n_k}(t) : i \in I_N\right\} < \frac{\mu_0}{2k} e^{-k\rho} .$$

Then,

$$\forall k \geq 1, \quad \exists t_k > T_{n_k}' \text{ such that } \max\left\{y_i^{n_k}(t_k) : i \in I_N\right\} < \frac{\mu_0}{2k} e^{-k\rho} .$$

By (6.27) and the continuity of $\max\{y_i^n(t) : i \in I_N\}$, there are $s_k \in (T_{n_k}', t_k)$ and $i_k \in I_N$, satisfying

$$\forall k \geq 1 : \quad \max\left\{y_i^{n_k}(s_k) : i \in I_N\right\} = y_{i_k}^{n_k}(s_k) = \frac{\mu_0}{2k} , \tag{6.28}$$

$$\forall k \geq 1, \ \forall t \in (s_k, t_k], \quad \max\left\{y_i^{n_k}(t) : i \in I_N\right\} < \frac{\mu_0}{2k} . \tag{6.29}$$

Then, following the same technique as that used in the proof of Lemma 4.1, we have

$$\forall k \geq 1, \quad t_k - s_k \geq k .$$

By assuming $i_k = i_0$ for all $k \geq 1$ and by the same reasoning as that in the proof of Lemma 4.1, we obtain $\limsup_{k\to\infty} m(e_{i_0}^{n_k}, s_k, t_k) \leq 0$ as well as by (6.11), $\limsup_{k\to\infty} m(e_{i_0}^{n_k}, s_k, t_k) > 0$. This contradiction shows the conclusion of the lemma. □

Lemma 6.4. *Under the assumptions of Lemma 6.1, there is a $\mu_1 > 0$ such that for all $e \in E(r, \varepsilon_1)$, every solution of (6.7) in E^0 satisfies*

$$\forall i \in I_N , \quad \liminf_{t\to\infty} x_i(t) \geq \mu_1 . \tag{6.30}$$

Proof. Suppose the conclusion of the lemma is not true and we aim to derive a contradiction. By Lemma 6.1, for each $e \in E(r, \varepsilon_1)$, each positive component of any solution of (6.7) in E has a positive lower limit as $t \to +\infty$. Then, the nonexistence of $\mu_1 > 0$ for (6.30) means the existence of $\{e^n\} \subset E(r, \varepsilon_1)$, $\{T_n\} \subset \mathbb{R}_0$, $\{\varphi^n\} \subset \Omega$ and $\{\alpha_n\} \subset \mathbb{R}_+$ such that for each $n \geq 1$, $y^n(t) = x(t, e^n, T_n, \varphi^n)$ is the solution of (6.7) with $(y^n)_{T_n} = \varphi^n$ after the substitution $e(t) = e^n(t)$ and satisfies

$$\min\left\{\liminf_{t\to+\infty} y_i^n(t) \colon i \in I_N\right\} = \alpha_n \downarrow 0 \quad (n \to \infty)\,. \tag{6.31}$$

By (6.31), for each $n \geq 1$, there is an $i_0 \in I_N$ such that

$$\min\left\{\liminf_{t\to+\infty} y_i^n(t) \colon i \in I_N\right\} = \liminf_{t\to+\infty} y_{i_0}^n(t) = \alpha_n\,. \tag{6.32}$$

As I_N is a finite set, by choosing a subsequence of $\{n\}$ if necessary, without loss of generality, we assume that this fixed $i_0 \in I_N$ in (6.32) suits all $n \geq 1$. By Lemma 6.3,

$$\inf\left\{\liminf_{t\to+\infty} \max\left\{y_i^n(t) \colon i \in I_N\right\} : n \geq 1\right\} > 0\,.$$

This, along with (6.31) and (6.32), justifies the existence of a proper subset $J \subset I_N$ with $i_0 \in J$, fulfilling

$$\inf\left\{\liminf_{t\to+\infty} \max\left\{y_j^n(t) \colon j \in J\right\} : n \geq 1\right\} = 0\,, \tag{6.33}$$

$$\forall i \in I_N \setminus J\,, \quad \inf\left\{\liminf_{t\to\infty} \max\left\{y_j^n(t) \colon j \in J \cup \{i\}\right\} : n \geq 1\right\} > 0\,. \tag{6.34}$$

By (6.34), there exist a $\sigma_0 > 0$ and a sequence $\{T_n'\}$ with $T_n' \geq T_n$ such that

$$\forall n \geq 1\,, \quad \forall i \in I_N \setminus J\,, \quad \forall t \geq T_n'\,, \quad \max\left\{y_j^n(t) \colon j \in J \cup \{i\}\right\} \geq \sigma_0\,. \tag{6.35}$$

By Lemma 6.2, there is a sequence $\{T_n''\}$ with $T_n'' \geq T_n'$ that satisfies

$$\forall n \geq 1\,, \quad \max\left\{y_j^n(T_n'') \colon j \in J\right\} \geq \mu_0/2\,. \tag{6.36}$$

Note that (6.33) implies the existence of a subsequence $\{n_k\} \subset \{n\}$, meeting the requirement of

$$\forall k \geq 1\,, \quad \liminf_{t\to+\infty} \max\left\{y_j^{n_k}(t) \colon j \in J\right\} < \frac{\min\{\mu_0, \sigma_0\}}{4k} e^{-k\rho}\,. \tag{6.37}$$

By (6.37), we can choose a sequence $\{t_k\}$ with $t_k > T_{n_k}''$ to satisfy

$$\forall k \geq 1\,, \quad \max\left\{y_j^{n_k}(t_k) \colon j \in J\right\} < \frac{\min\{\mu_0, \sigma_0\}}{4k} e^{-k\rho}\,. \tag{6.38}$$

Then, by (6.36), (6.38) and continuity, we can choose $\{s_k\}$ with $T''_{n_k} < s_k < t_k$ and $\{j_k\}$ with $j_k \in J$ such that

$$\forall k \geq 1, \quad \max\left\{y_j^{n_k}(s_k) : j \in J\right\} = y_{j_k}^{n_k}(s_k) = \frac{\min\{\mu_0, \sigma_0\}}{4k}, \tag{6.39}$$

$$\forall k \geq 1, \ \forall t \in (s_k, t_k], \quad \max\left\{y_j^{n_k}(t) : j \in J\right\} < \frac{\min\{\mu_0, \sigma_0\}}{4k}. \tag{6.40}$$

From (6.7) with the replacements of $e(t)$ by $e^{n_k}(t)$ and $x(t)$ by $y^{n_k}(t)$, we have

$$\forall k \geq 1, \quad \forall j \in J, \quad y_j^{n_k}(t_k) \geq y_j^{n_k}(s_k) e^{-\rho(t_k - s_k)}.$$

Combining this with (6.38) and (6.39), we obtain

$$\forall k \geq 1, \quad t_k - s_k > k. \tag{6.41}$$

As $j_k \in J$ for all $k \geq 1$, there exist a subsequence $\{k_m\} \subset \{k\}$ and a $j_0 \in J$ such that $j_{k_m} = j_0$ for all $m \geq 1$. Without loss of generality, we view $\{k\}$ as $\{k_m\}$ so that $j_k = j_0$ for all $k \geq 1$ and (6.39) becomes

$$\forall k \geq 1, \quad \max\left\{y_j^{n_k}(s_k) : j \in J\right\} = y_{j_0}^{n_k}(s_k) = \frac{\min\{\mu_0, \sigma_0\}}{4k}. \tag{6.42}$$

As $y_{j_0}^{n_k}(t_k) < y_{j_0}^{n_k}(s_k)$ by (6.40) and (6.42), using (6.40)–(6.42) and the same technique as that used in the proof of Lemma 6.1, we deduce that

$$\forall i \in I_N \setminus J, \quad \sum_{j \in I_N \setminus J} (b_{ij} - a_{ij}) x_j^* = r_i^*, \quad \sum_{j \in I_N \setminus J} (b_{j_0 j} - a_{j_0 j}) x_j^* \geq r_{j_0}^*,$$

where $r^* \in B(\bar{r}, \varepsilon_1)$ and $x^* \in E_J^0$. This shows that r^* and $B - A$ do not satisfy the $(I - J)$-condition, a contradiction to the fact that every vector in $B(\bar{r}, \varepsilon_1)$ and $B - A$ still satisfy the $(I - J)$-condition. Therefore, the conclusion of the lemma must be true. □

Proof of Theorem 6.1. The conclusion follows from applying Lemma 6.4 to all subsystems of (6.7). □

Now, combining Proposition 2.1, Theorems 5.3 and 5.4 with Theorem 6.1, we obtain the following results.

Corollary 6.1. *Assume that* (5.8) *is uniformly bounded. Then,* (5.8) *and all of its subsystems are permanent if and only if $\bar{r}$ and $B - A$ satisfy the $(I - J)$-condition $(I - J)$-condition.*

Theorem 6.2. *Assume that* (5.1) *with* (5.2)–(5.6) *has a compact uniform attractor. Then, all small perturbations of* (5.1) *on B and A and their subsystems are permanent if and only if $\bar{r}$ and $B - A$ satisfy the $(I - J)$-condition.*

Theorem 6.3. *Assume that* (5.1) *with* (5.2)–(5.6) *has a compact uniform attractor. Then,* (5.7) *and all of its subsystems are permanent if and only if $\bar{r}$ and $B - A$ satisfy the $(I - J)$-condition.*

7 Further notes

Sections 2–4 are based on the author's recent work [24] and Sections 5–6 are based on [20, 21]. The $(I-J)$-condition was first proposed by Ahmad and Lazer [4] for Lotka–Volterra systems without delays. The techniques used in the proof of Theorem 6.1 are similar to those of [4] and [5]. Further applications of the $(I-J)$-condition can be found in [22, 23]. Some related publications are mentioned here. Mierczyński and Schreiber [35] dealt with autonomous Kolmogorov systems with robustly permanent subsystems. Battauz and Zanolin [8] investigated coexistence states for periodic competitive Kolmogorov systems. Kuang [28] and Tang [38] investigated delayed nonautonomous Kolmogorov systems and obtained permanence criteria which depend on successful construction of Lyapunov functions or functionals. Yang [40] studied the persistence of a single-species Kolmogorov equation with delays. Examples of permanence of special classes of Kolmogorov systems without delays, including Lotka–Volterra differential systems, are given by Ahmad and Lazer [4], Ahmad and Stamova [5], Baigent and Hou [7], Hofbauer and Schreiber [14], and Zhao and Jiang [45]. Examples of permanence for special delayed Kolmogorov systems are given by Chen, Lu and Wang [10], Li and Teng [30], Liu and Chen [31], Lu, Lu and Enatsu [32], Mukherjee [34], Teng [39], and the references therein. In particular, for autonomous Lotka–Volterra differential systems with multiple delays, sufficient conditions for permanence, which are easily checkable inequalities involving the constant coefficients of the system, were obtained in [32].

8 Global attraction and stability of Lotka–Volterra systems

In Sections 8–13, we turn our attention to the typical autonomous Lotka–Volterra system

$$x_i' = x_i(r_i + A_i x), \quad i \in I_N, \tag{8.1}$$

where $A_i = (a_{i1}, \ldots, a_{iN})$ is the ith row of a matrix A and the r_i and a_{ij} are real numbers. System (8.1) can be written in the concise form $x' = D(x)(r + Ax)$ where $D(x) = \mathrm{diag}[x_1, \ldots, x_N]$. Suppose $x^* \in E$ is a nontrivial fixed point of (8.1), x^* is either in the interior E^0 or boundary ∂E of E. Our main concern is the global attractivity and global asymptotic stability of x^*.

Definition 8.1. Let $I \subset I_N$ such that $x_i^* > 0$ if and only if $i \in I$. We say that x^* is globally attractive in E if for all $x^0 \in E$ with $x_i^0 > 0$ for all $i \in I$, the solution $x(t, x^0)$ of (8.1) exists on $[0, +\infty)$ and satisfies $\lim_{t\to+\infty} x(t, x^0) = x^*$. In addition, if x^* is locally stable, that is,

$$\forall \varepsilon > 0\,, \ \exists \delta > 0 \text{ such that } \forall x^0 \in E\,, \forall t \geq 0\,, \\ |x^0 - x^*| < \delta \Longrightarrow |x(t, x^0) - x^*| < \varepsilon\,,$$

we say that x^* (or (8.1)) is globally asymptotically stable in E (at x^*).

If $r_i + A_i x^* > 0$ for some $i \in I_N$, then, since x^* is a fixed point, we must have $x_i^* = 0$. Thus, $r_i + A_i x^*$ is a positive eigenvalue of the Jacobian matrix of (8.1) at x^* and x^* is unstable. Therefore, for x^* to be globally asymptotically stable, x^* is required to be *saturated*, i. e. for any $i \in I_N$, if $x_i^* = 0$, then $r_i + A_i x^* \leq 0$.

Recall that a typical and most effective method for stability is the use of Lyapunov functions. For any $x^0 \in E$, if $x(t, x^0)$ exists and is bounded on $[0, +\infty)$, then $\omega(x^0)$, the set of $y \in E$ such that $\lim_{k\to\infty} x(t_k, x^0) = y$ for some sequence with $t_k \to +\infty$ as $k \to \infty$, is nonempty, closed, connected and invariant in E.

Theorem 8.1 ([15, Theorem 2.6.1]). *Consider $x' = f(x)$ for $x \in G \subset \mathbb{R}^N$. Let $V\colon G \to \mathbb{R}$ be continuously differentiable. If for some solution $x(t, x^0)$, we have $dV(x(t, x^0))/dt \geq 0$ (or ≤ 0), then $\omega(x^0) \cap G$ is contained in the set $\{x \in G\colon dV(x)/dt = 0\}$.*

Theorem 8.2. *Let $x^* \in G$ be a fixed point of $x' = f(x)$. Assume that $V\colon G \to R$ satisfies*
(i) $V(x^*) = 0$ *and* $V(x) > 0$ *for* $x \in G \setminus \{x^*\}$;
(ii) $V(x) \to \infty$ *as* $|x| \to \infty$ *or* $x \to \overline{G} \setminus G$;
(iii) $\dot{V}(x)|_{(8.1)} = \frac{\partial V}{\partial x} f(x) < 0$ *for* $x \in G \setminus \{x^*\}$.
Then, x^* is globally asymptotically stable in G.

9 Global stability by Lyapunov functions

If the matrix A in (8.1) is *Volterra–Lyapunov stable* (VL-stable), i. e. there exists $D = \operatorname{diag}[d_1, \ldots, d_N]$ with $d_i > 0$ for all $i \in I_N$ such that $DA + A^T D$ is negative definite, then the trivial solution of the linear system $x' = Ax$ is globally asymptotically stable in $\mathbb{R}^N$ as $V(x) = d_1 x_1^2 + \cdots + d_N x_N^2$ is a Lyapunov function satisfying the conditions of Theorem 8.2. It is interesting to see that if A is VL-stable, then (8.1) always has a globally asymptotically stable fixed point.

Theorem 9.1 ([37, Theorem 3.2.1]). *Assume that A is VL-stable. Then, for each $r \in \mathbb{R}^N$, (8.1) has a saturated fixed point $x^* \in E$ and x^* is globally asymptotically stable in E.*

Proof. Since A is VL-stable, for each $x \in \mathbb{R}^N$ with $x \neq 0$, we have

$$0 > x^T(DA + A^T D)x = 2x^T DAx = 2(d_1 x_1 A_1 x + \cdots + d_N x_N A_N x)\,.$$

In particular, this holds for all $x \in E$ with $x \neq 0$. Thus, A is a B-matrix, that is,

$$\forall x \in E(x \neq 0)\,, \quad \exists i \in I_N \text{ such that } x_i > 0, A_i x < 0\,. \tag{9.1}$$

Then, by [15, Exercise 15.2.10], for each $r \in \mathbb{R}^N$, (8.1) has a saturated fixed point $x^* \in E$.

Let $I \subset I_N$ such that $x_i^* > 0$ if and only if $i \in I$. Then, for all $i \in I_N \setminus I$, we have $r_i + A_i x^* \leq 0$. Let $G = \{x \in E\colon \forall i \in I, x_i > 0\}$ and define a Lyapunov function $V\colon G \to \mathbb{R}$ by

$$V(x) = \sum_{i\in I} d_i \left[x_i - x_i^* - x_i^* \ln \frac{x_i}{x_i^*}\right] + \sum_{i \in I_N \setminus I} d_i x_i\,. \tag{9.2}$$

Then, $V(x^*) = 0$ and $V(x) > 0$ for $x \in G \setminus \{x^*\}$. Clearly, $V(x) \to \infty$ if $x_i \to 0^+$ for $i \in I$ or $|x| \to \infty$. Moreover,

$$\begin{aligned}
\dot V(x)|_{(8.1)} &= \sum_{i\in I} d_i (x_i - x_i^*) x_i'/x_i + \sum_{i \in I_N\setminus I} d_i x_i' \\
&= \sum_{i\in I} d_i (x_i - x_i^*) A_i (x - x^*) + \sum_{i\in I_N \setminus I} d_i x_i A_i (x - x_i^*) \\
&\quad + \sum_{i \in I_N\setminus I} d_i x_i (r_i + A_i x^*) \\
&\leq (x - x^*)^T DA(x - x^*)\,.
\end{aligned}$$

By (9.1), we have $\dot V(x) < 0$ for $x \in G \setminus \{x^*\}$. Then, by Theorem 8.2, x^* is globally asymptotically stable. □

The VL-stability of A can be guaranteed for some particular cases. One particular case is when A is real symmetric and stable, i. e. every eigenvalue of A has a negative real part. In this case, A is negative definite so $DA + A^T D = 2A$ is negative definite with D the identity matrix. Another case is that A has a negative diagonal dominant, i. e. there exists a $d \in E^0$ such that $-d_i a_{ii} > \sum_{j \in I_N \setminus\{i\}} d_j |a_{ij}|$ for all $i \in I_N$. In this case, it is shown [9] that A is VL-stable. If (8.1) is cooperative, i. e. $a_{ij} \geq 0$ for all $i \neq j$, and $-A$ is not an M-matrix, we know from Chapter 1 that (8.1) may have unbounded solutions. However, if $-A$ is an M-matrix, then (8.1) has a global uniform attractor. Moreover, by [6], A is VL-stable. Then, the corollary below follows directly from Theorem 9.1.

Corollary 9.1. *System* (8.1) *has a globally asymptotically stable fixed point* $x^* \in E$ *if one of the conditions holds:*
(i) *A is a real symmetric stable matrix;*
(ii) *A has a negative diagonal dominant;*
(iii) *$-A$ is an M-matrix.*

Note that A is a B-matrix if and only if for each $r \in \mathbb{R}^N$, (8.1) is uniformly bounded. Many results relating properties of solutions of (8.1) to A being a particular type of matrices (B-matrix, M-matrix, P-matrix, etc.) can be found in [15] and [37].

10 Global stability by split Lyapunov functions

A function $V: E \to \mathbb{R}$ is called a *split Lyapunov function* for (8.1) if there is an $(N-1)$-dimensional hyperplane $\pi \subset \mathbb{R}^N$ passing through a fixed point $x^* \in E$ such that for $x \in E$, $\dot{V}(x)|_{(8.1)} = 0$ on $\pi \cap E$, < 0 on one side of π and > 0 on the other side of π. Then, along any solution trajectory, the function $V(x(t))$ is monotone on either side of π; but if the trajectory passes through a point $p \in \pi$ from one side to the other side, then $V(x(t))$ has a local maximum or minimum $V(p)$. If $x^* \in E^0$, then, for any $\alpha \in E^0$, we check that $V(x) = \prod_{i=1}^N x_i^{\alpha_i}$ is a split Lyapunov function. Indeed,

$$\dot{V}|_{(8.1)} = \frac{\partial V}{\partial x} D(x)(r + Ax) = V\alpha^T A(x - x^*) .$$

So V is a split Lyapunov function with $\pi = \{x \in \mathbb{R}^N : \alpha^T A(x - x^*) = 0\}$. If $x^* \in \partial E$ with $x_i^* > 0$, if and only if $i \in I \subset I_N$, for $J = I_N \setminus I$ and any $\alpha \in E_J^0$, the function $V(x) = \prod_{i \in I} x_i^{\alpha_i}$ satisfies

$$\dot{V}|_{(8.1)} = \sum_{i \in I} \frac{\partial V}{\partial x_i} x_i' = V\alpha^T A(x - x^*),$$

and thus V is a split Lyapunov function. In this section, we use an appropriate split Lyapunov function to obtain a sufficient condition for x^* to be globally asymptotically stable.

In the following, we set $B = \frac{\partial F}{\partial x}(x^*)$ where $F_i(x) = x_i(r_i + A_i x)$ for $i \in I_N$. Explicitly, for $i, j \in I_N$,

$$\left.\frac{\partial F_i}{\partial x_j}\right|_{x=x^*} = \begin{cases} r_i + A_i x^* + x_i^* a_{ii}, & i = j, \\ x_i^* a_{ij}, & i \neq j . \end{cases}$$

Since $x_i^* > 0$ implies $r_i + A_i x^* = 0$, we have, for the ith row of B,

$$B_i = \begin{cases} x_i^* A_i , & \text{if } x_i^* > 0 , \\ (0 \cdots 0\, r_i + A_i x^*\, 0 \cdots 0) , & \text{if } x_i^* = 0 . \end{cases} \tag{10.1}$$

When x^* is an interior fixed point, we obtain $B = D(x^*)A$.

Theorem 10.1. *Suppose that for the system (8.1), the following conditions hold:*
(i) *The system is permanent, so that there is a unique interior fixed point x^*.*
(ii) *There is a left eigenvector $\alpha \in \mathbb{R}^N$ of $D(x^*)A$ associated with an eigenvalue $\lambda < 0$ and $\alpha_i \neq 0$ for all $i \in I_N$.*
(iii) *The following holds:*

$$y^T D(\alpha) A y < 0 \text{ for all nonzero } y \in \mathbb{R}^N \text{ such that } \alpha^T y = 0 . \tag{10.2}$$

Then, x^* is globally asymptotically stable in E.

Proof. Our assumption means that the unique interior fixed point x^* satisfies $Ax^* = -r$ and the Jacobian matrix at x^* is $B = D(x^*)A$. Notice that (8.1) can be rewritten as

$$x' = D(x)(r + Ax) = B(x - x^*) + D(x - x^*)A(x - x^*)\,.$$

By assumption, $\alpha^T B = \lambda\alpha^T$. Now, let $y = x - x^*$ and $\psi = \alpha^T(x - x^*)$. Then,

$$\psi' = \alpha^T x' = \alpha^T By + \alpha^T D(y)Ay = \lambda\psi + y^T D(\alpha)Ay\,. \tag{10.3}$$

Let us also define $V(x) = \prod_{i=1}^{N} x_i^{x_i^* \alpha_i}$. Then,

$$\begin{aligned}\dot V|_{(8.1)} &= V\sum_{i=1}^{N} x_i^*\alpha_i \frac{x_i'}{x_i}\\ &= V\sum_{i=1}^{N} x_i^*\alpha_i(r_i + A_i x) = V\sum_{i=1}^{N} x_i^*\alpha_i A_i(x - x^*)\\ &= V\alpha^T B(x - x^*) = \lambda V\alpha^T(x - x^*) = \lambda\psi V\,.\end{aligned}$$

In summary, we find that

$$\psi' = \lambda\psi + y^T D(\alpha)A\,y, \tag{10.4}$$

$$V' = \lambda\psi V\,. \tag{10.5}$$

Then, the assumption (10.2) becomes

$$y^T D(\alpha)A\,y < 0 \;\text{ for }\; y \in \psi^{-1}(0)\setminus\{0\}\,. \tag{10.6}$$

Now, suppose that $x^0 \in E^0 \setminus \{x^*\}$ is given with $y^0 = x^0 - x^*$ and $y(t, y^0) = x(t, x^0) - x^*$. We consider the two cases $\psi(y^0) \le 0$ and $\psi(y^0) > 0$ separately. Let $\mathcal{Y} = \{x \in \mathbb{R}^N : \psi(x - x^*) = 0\}$.

First, suppose that $\psi(y^0) \le 0$. If $\psi(y^0) = 0$, then, since $y^0 \ne 0$, by (10.4) and (10.6), we have $\psi' < 0$ at $t = 0$ and hence, by continuity, $\psi' < 0$ so $\psi(y(t, y^0)) < 0$ is strictly decreasing for sufficiently small $t \ge 0$. This shows that $\psi(y(t, y^0))$ will never come back to 0 once it is below 0. Therefore, $\psi(y^0) \le 0$ implies $\psi(y(t, y^0)) < 0$ for all $t > 0$. Thus, from (10.5), we have $V' > 0$ and hence $V(x(t, x^0))$ is increasing for all $t > 0$. Since (8.1) is assumed permanent, $V(x(t, x^0))$ is bounded and $V(x(t, x^0)) \ge \varepsilon$ for some $\varepsilon > 0$ and all $t \ge 0$. Moreover, $\omega(x^0) \subset E^0$. By Theorem 8.1, we have $\omega(x^0) \subset \{x \in E^0 : V' = 0\} = V^{-1}(0) \cup \mathcal{Y}$. However, $V(u) \ge \varepsilon$ for all $u \in \omega(x^0)$ so $\omega(x^0) \cap V^{-1}(0) = \emptyset$. Thus, we must have $\omega(x^0) \subset \mathcal{Y}$. From (10.4) and (10.6), we see that the only invariant set in $\mathcal{Y}$ is $\{x^*\}$. Hence, $\omega(x^0) = \{x^*\}$ and $\lim_{t\to+\infty} x(t, x^0) = x^*$.

Next, we suppose $\psi(y^0) > 0$. If there is a $T > 0$ such that $\psi(y(t, y^0)) > 0$ for $0 \le t < T$ but $\psi(y(T, y^0)) = 0$, then $\lim_{t\to+\infty} x(t, x^0) = x^*$ from the first case. Otherwise, $\psi(y(t,y^0)) > 0$, so $V' < 0$ by (10.5) for all $t > 0$. By Theorem 8.1 and the permanence assumption, the same reasoning as above shows that $\lim_{t\to+\infty} x(t,x^0) = x^*$.

Finally, we prove the stability of x^*. Denote the open ball centered at x^* with a radius ρ by $\mathcal{B}_\rho(x^*)$. We need to show that for any given $\varepsilon > 0$, there is a $\delta > 0$ such that $x^0 \in \mathcal{B}_\delta(x^*)$ implies $x(t,x^0) \in \mathcal{B}_\varepsilon(x^*)$ for all $t \geq 0$.

Let W be an $N \times N$ matrix such that the columns of W, W_{cj} for $j \in I_N$ satisfy

$$W_{c1} = (\alpha_1^{-1}, 0, \ldots, 0)^T, \quad \psi^{-1}(0) = \text{span}\{W_{c2}, \ldots, W_{cN}\}.$$

Since $\alpha^T W_{c1} = \alpha_1 \alpha_1^{-1} = 1 \neq 0$, $W_{c1} \notin \psi^{-1}(0)$. Then, W is invertible. Let $z^T = (\psi, \tilde{z}^T)$ with $\tilde{z} \in \mathbb{R}^{N-1}$ and $y = x - x^* = Wz$. We have $\psi = (1, 0, \ldots, 0)z = \alpha^T(x - x^*)$ and, by (10.4),

$$\psi' = (\lambda + \Theta(z))\psi + \tilde{z}^T M_0 \tilde{z}, \tag{10.7}$$

where Θ is linear with $\Theta(0) = 0$ and M_0 is an $(N-1)\times(N-1)$ real symmetric matrix. Then, condition (10.6) implies the negative definiteness of M_0. From (10.7), we see the existence of $\varepsilon_0 > 0$ such that $x^0 \in \mathcal{B}_{\varepsilon_0}(x^*)$ and $\psi(y^0) > 0$ imply $\psi' < 0$ at $t = 0$.

For each $l > 0$, the set $S_l = \{x \in E \colon V(x) = l\}$ defines an $(N-1)$-dimensional surface and by the definition of V, $l^* = V(x^*) > 0$. For each $x^0 \in E$ with $y^0 \in \psi^{-1}(0) \setminus \{0\}$, since $\psi(y(t,y^0)) < 0$ and $V(x(t,x^0))' > 0$ for all $t > 0$ so that $V(x(t,x^0)) \uparrow l^*$ as $t \to +\infty$, we have $V(x^0) < l^*$ so $S_{l^*} \cap \gamma = \{x^*\}$.

Then, for any fixed $\varepsilon \in (0, \varepsilon_0)$, there is an $l \in (0, l^*)$ such that $V_l = \{x \in E \colon V(x) \geq l, \psi(x - x^*) \leq 0\} \subset \mathcal{B}_{\varepsilon/2}(x^*)$. The set V_l is bounded by S_l and γ with an $(N-2)$-dimensional edge $S_l \cap \gamma$. Since $x^* \notin S_l \cap \gamma$, for each $x^0 \in S_l \cap \gamma$, there is a $t_1(x^0) < 0$ such that $\psi(y(t,y^0)) > 0$ and $x(t,x^0) \in \mathcal{B}_{\varepsilon/2}(x^*)$ for $t \in (t_1(x^0), 0)$, but $x(t_1(x^0), x^0) \notin \mathcal{B}_{\varepsilon/2}(x^*)$. Let

$$\Gamma = \left\{x\left(t, x^0\right) : x^0 \in S_l \cap \gamma, t \in \left[t_1\left(x^0\right), 0\right]\right\}.$$

Then, the open subset of $\mathcal{B}_{\varepsilon/2}(x^*)$ bounded "below" by S_l, "surrounding" by Γ and "above" by the upper boundary of $\mathcal{B}_{\varepsilon/2}(x^*)$ satisfying $\psi > 0$ contains x^* and is positively invariant. Therefore, it contains $\mathcal{B}_\delta(x^*)$ for a small $\delta \in (0, \varepsilon/2)$ and $x^0 \in \mathcal{B}_\delta(x^*)$ implies $x(t,x^0) \in \mathcal{B}_\varepsilon(x^*)$ for all $t \geq 0$. □

For a boundary fixed point x^* of (8.1) to be stable in E, it is necessary that it is saturated.

Theorem 10.2. *Suppose* (8.1) *satisfies the following conditions:*
(a) *The solutions of* (8.1) *are uniformly bounded.*
(b) *For a proper subset $J \subset I_N$ with $I = I_N \setminus J$,* (8.1) *is J-permanent and the unique fixed point x^* in E_I^0 is saturated.*
(c) *The matrix $D(x^*)A$ has an eigenvalue $\lambda < 0$ and an associated left eigenvector $\alpha \in \mathbb{R}^N$ such that $\alpha_j \neq 0$ for all $j \in J$ and $\alpha_i > 0$ for all $i \in I$.*

(d) *The following holds:*

$$y^T D(\alpha) A\, y < 0 \ \textit{for all nonzero} \ y \in \mathbb{R}^N \ \textit{such that} \ \alpha^T y = 0\,. \tag{10.8}$$

Then, x^* is globally asymptotically stable in E.

Proof. We first rewrite (8.1) as

$$x' = B(x - x^*) + D(x - x^*)A(x - x^*)\,,$$

where B is given by (10.1). Now, let $\psi = \alpha^T(x - x^*)$ and $y = x - x^*$. Then, $\psi' = \alpha^T x' = \alpha^T B y + \alpha^T D(y) A\, y$. As $y_i = x_i$ for $i \in I$, by condition (c), we can further write

$$\psi' = \lambda\psi + \sum_{i\in I} \alpha_i(r_i + A_i x^*)x_i + y^T D(\alpha) A\, y\,. \tag{10.9}$$

With $V(x) = \prod_{i\in J} x_i^{x_i^* \alpha_i}$, we have

$$\begin{aligned} \dot V|_{(8.1)} &= V \sum_{i\in J} x_i^* \alpha_i \frac{x_i'}{x_i} \\ &= V \sum_{i\in J} x_i^* \alpha_i (r_i + A_i x) = V \sum_{i\in J} x_i^* \alpha_i A_i (x - x^*) \\ &= \lambda V \alpha^T (x - x^*) = \lambda \psi V\,. \end{aligned}$$

Thus, instead of (10.4) and (10.5) in the proof of Theorem 10.1, we obtain

$$\psi' = \lambda\psi + \sum_{i\in I} \alpha_i(r_i + A_i x^*)x_i + y^T D(\alpha) A\, y, \tag{10.10}$$

$$V' = \lambda\psi V\,. \tag{10.11}$$

Since x^* is saturated by condition (b), and $\alpha_i > 0$ for all $i \in I$ by condition (c), for $x \in E$, the term $\sum_{i\in I} \alpha_i(r_i + A_i x^*)x_i$ is always nonpositive. Then, by replacing permanence by J-permanence in the proof of Theorem 10.1, we obtain the global attraction of x^*.

To show the stability of (8.1) at x^* in E, since $x^* \in \partial E$, the balls centered at x^* are restricted to E only. Referring to the proof of Theorem 10.1, we note that (10.10) implies

$$\psi' \le (\lambda + \Theta(z))\psi + \tilde z^T M_0 \tilde z \tag{10.12}$$

instead of (10.7), the set V_l is bounded by S_l, γ and $\cup_{i\in I}\pi_i$, and the open subset of $\mathcal{B}_{\varepsilon/2}(x^*) \cap E$ relative to E bounded "below" by S_l, "surrounding" by Γ and $\cup_{i\in I}\pi_i$, and "above" by the upper boundary of $\mathcal{B}_{\varepsilon/2}(x^*)$ satisfying $\psi > 0$ contains x^* and is positively invariant. Then, the stability follows. □

For a real symmetric matrix M, it is known that M is positive definite if and only if its leading principal minor determinants are positive, and that M is negative definite if and only if $-M$ is positive definite. For effective application of Theorems 10.1 and 10.2, we need to convert (10.2) and (10.8) into a negative definite condition of a real symmetric matrix.

10.1 Checking the conditions (10.2) and (10.8)

In what follows, for a given square matrix M, set $M^S = M + M^T$. Let $W_{c2}, \dots, W_{cN}$ be $N-1$ linearly independent column vectors that span $\psi^{-1}(0)$. For any $\alpha \in \mathbb{R}^N$ with $\alpha_i \neq 0$ for all $i \in I_N$, from (10.7) or (10.12), we see that $y^T(D(\alpha)A)^S y < 0$ for all $y \in \psi^{-1}(0) \setminus \{0\}$ if and only if the $(N-1) \times (N-1)$ real symmetric matrix $M_0 = \frac{1}{2}U$ is negative definite where

$$U = [W_{c2}, \dots, W_{cN}]^T (D(\alpha)A)^S [W_{c2}, \dots, W_{cN}]\,. \tag{10.13}$$

Since the negative definiteness of U does not depend on the choice of the vectors $W_{c2}, \dots, W_{cN}$, we are at liberty to choose the basis so that U has a simple explicit expression in terms of $D(\alpha)$ and A. For example, if we choose

$$[W_{c2}, \dots, W_{cN}] = D(\alpha)^{-1} \begin{pmatrix} 1 & & & 0 \\ -1 & & \ddots & \\ & \ddots & & 1 \\ 0 & & & -1 \end{pmatrix} \tag{10.14}$$

and denote by M_{ij} the submatrix of M obtained by deleting the ith row and the jth column of M for any square matrix M, then

$$U = (AD(\alpha)^{-1})^S_{11} + (AD(\alpha)^{-1})^S_{NN} - (AD(\alpha)^{-1})^S_{1N} - (AD(\alpha)^{-1})^S_{N1}\,. \tag{10.15}$$

Note that U given by (10.15) is obtained from $(AD(\alpha)^{-1})^S$ as follows: the ith column minus the $(i+1)$th column for $i = 1, \dots, N-1$ gives a matrix U^0, the ith row of U^0 minus its $(i+1)$th row for $i = 1, \dots, N-1$ results in U^1, and then deletion of the last row and the last column of U^1 produces U.

Alternatively, if we choose

$$[W_{c2}, \dots, W_{cN}] = D(\alpha)^{-1} \begin{pmatrix} 1 & & 0 \\ & \ddots & \\ 0 & & 1 \\ -1 & \cdots & -1 \end{pmatrix} \tag{10.16}$$

and write

$$(AD(\alpha)^{-1})^S = (d_{ij})_{N \times N}\,, \tag{10.17}$$

then

$$U = (d_{ij} + d_{NN} - d_{iN} - d_{Nj})_{(N-1)\times(N-1)}\,. \tag{10.18}$$

Note that U given by (10.18) is obtained from $(AD(\tilde{\alpha})^{-1})^S$ as follows: the ith column minus the Nth column of $(AD(\tilde{\alpha})^{-1})^S$ for $i = 1, \dots, N-1$ gives a matrix U^0, the ith row of U^0 minus its Nth row for $i = 1, \dots, N-1$ gives U^1, and then deletion of the Nth row and the Nth column of U^1 gives U.

In addition to (10.15) and (10.18), there are many other ways to form the matrix U. Thus, in checking the conditions (10.2) and (10.8), we simply check that U given by (10.15) or (10.18) is negative definite.

10.2 Examples

Example 10.1. Consider system (8.1) with

$$A = -\begin{pmatrix} 5 & -1 & 1 \\ 4 & 3 & 1 \\ -1 & -\frac{1}{2} & 2 \end{pmatrix}, \quad r = \frac{1}{3}\begin{pmatrix} 25 \\ 46 \\ 26 \end{pmatrix}. \tag{10.19}$$

For each $x \in \mathbb{R}^3_+$ ($x \neq 0$), if $x_2 > 0$, then $A_2x < 0$; if $x_2 = 0$ but $x_1 > 0$, then $A_1x < 0$; if $x_1 = x_2 = 0$ but $x_3 > 0$, then $A_3x < 0$. Hence, by (9.1), A is a B-matrix so (8.1) with (10.19) is uniformly bounded. The boundary fixed points $\hat{x}^T$ are: $(0,0,0)$, $(\frac{5}{3},0,0)$, $(0,\frac{46}{9},0)$, $(0,0,\frac{13}{3})$, $(\frac{121}{57},\frac{130}{57},0)$, $(\frac{8}{11},0,\frac{155}{33})$, and $(0,\frac{132}{39},\frac{202}{39})$. If $q = (1,1,1)^T \in E^0$, we can check that $q^T(r + A\hat{x}) > 0$ for all $\hat{x}$. Thus, by Theorem 5.3, (8.1) is permanent. The system has a unique interior fixed point $x^* = (1,2,\frac{16}{3})^T$. The matrix $D(x^*)A$ has a negative eigenvalue $\lambda = -8$ and an associated left eigenvector $\alpha^T = (48,4,-21)$. Then, from (10.15), we have

$$\left(AD(\alpha)^{-1}\right)^S = -\begin{pmatrix} \frac{5}{24} & -\frac{1}{6} & -\frac{23}{336} \\ -\frac{1}{6} & \frac{3}{2} & -\frac{29}{168} \\ -\frac{23}{336} & -\frac{29}{168} & -\frac{4}{21} \end{pmatrix}, \quad U = -\begin{pmatrix} \frac{49}{24} & -\frac{85}{48} \\ -\frac{85}{48} & \frac{139}{84} \end{pmatrix}.$$

Since $-U$ is positive definite, by Theorem 10.1, (8.1) with (10.19) at x^* is globally asymptotically stable in $\mathbb{R}^3_+$.

Example 10.2. Consider system (8.1) with

$$A = -\begin{pmatrix} 2 & 0 & 1 \\ 1 & 3 & 2 \\ 3 & -\frac{1}{2} & 4 \end{pmatrix}, \quad r = \begin{pmatrix} 2 \\ 7 \\ 1 \end{pmatrix}. \tag{10.20}$$

Clearly, A is a B-matrix so (8.1) with (10.20) is uniformly bounded. There are five fixed points with the first or second component being zero: $p_0^T = (0,0,0)$, $p_1^T = (1,0,0)$, $p_2^T = (0,\frac{7}{3},0)$, $p_3^T = (0,0,\frac{1}{4})$, and $p_4^T = (0,2,\frac{1}{2})$, and they satisfy $r_1 + A_1p_0 > 0$, $r_1 + A_1p_2 > 0$, $r_1 + A_1p_3 > 0$, $r_+A_1p_4 > 0$, $r_2 + A_2p_0 > 0$, $r_2 + A_2p_1 > 0$ and $r_2 + A_2p_3 > 0$. Then, by Corollary 4.2, (8.1) with (10.20) is partially permanent with respect to $J = \{1,2\}$. Thus, it has a unique fixed point $p = (1,2,0)^T$ in $E^0_{\{3\}}$. As $r_3 + A_3p < 0$, p is saturated. The matrix $D(p)A$ has a negative eigenvalue $\lambda = -6$

and an associated left eigenvector $\alpha^T = (2,4,3)$. By (10.18), we have

$$\left(AD(\alpha)^{-1}\right)^S = -\begin{pmatrix} 2 & \frac{1}{2} & \frac{11}{6} \\ \frac{1}{2} & \frac{3}{2} & \frac{13}{24} \\ \frac{11}{6} & \frac{13}{24} & \frac{8}{3} \end{pmatrix}, \quad U = -\begin{pmatrix} 1 & \frac{19}{24} \\ \frac{19}{24} & \frac{37}{12} \end{pmatrix}.$$

It can be checked that $-U$ is positive definite. By Theorem 10.2, (8.1) with (10.20) at p is globally asymptotically stable in $\mathbb{R}^3_+$.

Example 10.3. Consider system (8.1) with

$$A = -\begin{pmatrix} 2 & 0 & 1 & 1 \\ 1 & 3 & 2 & 4 \\ 3 & -\frac{1}{2} & 4 & 1 \\ 2 & \frac{1}{2} & \frac{1}{2} & 2 \end{pmatrix}, \quad r = \begin{pmatrix} 2 \\ 7 \\ 1 \\ 4 \end{pmatrix}. \tag{10.21}$$

Again, A is a B-matrix so (8.1) with (10.21) is uniformly bounded. The subsystem of (8.1) with (10.21) and $x_4 = 0$ is the system (8.1) with (10.20). Thus, from the above example, we know that by adding 0 as the fourth component to each of the fixed points $p_i (0 \le i \le 4)$ and p, they are the fixed points of (8.1) with (10.21) and $x_4 = 0$. Since $r_4 + A_4 p > 0$ and $r_4 + A_4 p_i > 0$ for all $0 \le i \le 4$, by Corollary 4.2, (8.1) with (10.21) is partially permanent with respect to $J = \{4\}$. The unique fixed point in E_I^0 ($I = \{1,2,3\}$) is $x^* = (0,0,0,2)^T$. As $r_i + A_i x^* \le 0$ for all $i \in I$, x^* is saturated. The matrix $D(x^*)A$ has a negative eigenvalue $\lambda = -4$ and an associated left eigenvector $\alpha^T = (2, \frac{1}{2}, \frac{1}{2}, 2)$. By (10.15), we have

$$\left(AD(\alpha)^{-1}\right)^S = -\begin{pmatrix} 2 & \frac{1}{2} & \frac{7}{2} & \frac{3}{2} \\ \frac{1}{2} & 12 & 3 & 3 \\ \frac{7}{2} & 3 & 16 & \frac{3}{2} \\ \frac{3}{2} & 3 & \frac{3}{2} & 2 \end{pmatrix}, \quad U = -\begin{pmatrix} 13 & -12 & 2 \\ -12 & 22 & -\frac{29}{2} \\ 2 & -\frac{29}{2} & 15 \end{pmatrix}.$$

Since

$$\begin{vmatrix} 13 & -12 \\ -12 & 22 \end{vmatrix} > 0$$

and $\det(-U) = 4.75 > 0$, $-U$ is positive definite. By Theorem 10.2, (8.1) with (10.21) at x^* is globally asymptotically stable in $\mathbb{R}^4_+$.

11 Global stability of competitive Lotka–Volterra systems

From the last section, we see that one of the advantages of Theorems 10.1 and 10.2 is that they do not have any restriction on the sign of each entry of A. However, the disadvantage is the requirement of permanence or J-permanence and uniform boundedness of (8.1). In some cases, these requirements are unnecessary and can be omitted. One such case is when (8.1) is *competitive*: $a_{ij} \leq 0$ for all $i, j \in I_N (i \neq j)$. For convenience, we rewrite the competitive system (8.1) as

$$x_i' = x_i(b_i - A_i x)\,, \quad i \in I_N \quad \text{or} \quad x' = D(x)(b - Ax)\,, \tag{11.1}$$

where $b_i > 0$, $a_{ii} > 0$ and $a_{ij} \geq 0$ for all $i, j \in I_N$. System (11.1) is called *strongly competitive* if $a_{ij} > 0$ for all $i, j \in I_N$.

For (11.1), the origin O repels trajectories and its basin of repulsion in E, $B(O)$, is open and bounded, and thus we have the following famous result regarding the so-called carrying simplex $\Sigma = \overline{B(O)} \setminus B(O)$ ([41, 43] or [13]).

Theorem 11.1 (Hirsch). *If system* (11.1) *is strongly competitive, then every trajectory in* $E \setminus \{O\}$ *is asymptotic to one in* Σ*;* Σ *is a balanced Lipschitz submanifold homeomorphic to the closed unit simplex in E via radial projection, and* $\operatorname{int}\Sigma$ *is strongly balanced.*

(A set S is said to be *balanced* if both $u - v \notin E^0$ and $v - u \notin E^0$ hold, and *strongly balanced* if both $u - v \notin E$ and $v - u \notin E$ hold for any distinct $u, v \in S$.)

From Theorem 11.1, it is clear that the global dynamics of (11.1) is completely determined by the dynamics on Σ. All limit sets, and in particular fixed points, belong to Σ. We shall say fixed point $p \in \Sigma$ is a *global attractor* (*repellor*) relative to Σ if for any $x^0 \in \Sigma$ such that $x_i^0 > 0$ whenever $p_i > 0$, for any $i \in I_N$, then $\lim_{t\to+\infty} x(t, x^0) = p$ ($\lim_{t\to-\infty} x(t, x^0) = p$).

For any hyperplane P in E not containing the origin, the side containing the origin is said to be *below P* and the other side *above P*. In the following, for a fixed point $p \in \Sigma$, if we can find a hyperplane P with $p \in P$ such that the trajectories of (11.1) pass through P from above (below) P to below (above) P, then $\Sigma \setminus \{p\}$ ie below (above) P. We can show that p is globally asymptotically stable (a global repellor relative to Σ).

Assume that (11.1) is strongly competitive and $p \in \Sigma_J^0 = \{x \in \Sigma\colon x_i > 0 \iff i \in I_N \setminus J\}$ for some $J \subsetneq I_N$ is an interior or boundary fixed point of (11.1). Then, every entry of the submatrix $(p_i a_{ij})$ of $D(p)A$ for all $i, j \in I_N \setminus J$ is positive. By the Perron–Frobenius Theorem [16, Theorem 8.44], this matrix has a positive eigenvalue and a positive left eigenvector, i. e. there are $\alpha_i > 0$ for all $i \in I_N \setminus J$ and $\lambda_0 > 0$ such that

$$\forall j \in I_N \setminus J\,, \quad \sum_{i \in I_N \setminus J} \alpha_i p_i a_{ij} = \lambda_0 \alpha_j\,. \tag{11.2}$$

For $j \in J$ such that $A_j p - b_j < \lambda_0$, we define

$$\alpha_j = \frac{1}{\lambda_0 - (A_j p - b_j)} \sum_{i \in I_N \setminus J} \alpha_i p_i a_{ij}\,. \tag{11.3}$$

Let

$$\hat{\alpha}_j = \frac{1}{\lambda_0} \sum_{i \in I_N \setminus J} \alpha_i p_i a_{ij}\,. \tag{11.4}$$

Then, we define a normal vector $\tilde{\alpha} \in E^0$ for the required plane via

$$\tilde{\alpha}_j = \begin{cases} \alpha_j & \text{if } j \in I_N \setminus J, \\ \hat{\alpha}_j & \text{if } j \in J \text{ and } A_j p - b_j \geq \lambda_0, \end{cases} \tag{11.5}$$

$$\tilde{\alpha}_j \in [\alpha_j, \hat{\alpha}_j] \text{ or } [\hat{\alpha}_j, \alpha_j] \text{ if } j \in J \text{ and } A_j p - b_j < \lambda_0. \tag{11.6}$$

The required plane is then defined by

$$T_p = \left\{ x \in \mathbb{R}^N \colon \tilde{\alpha}^T (x - p) = 0 \right\}. \tag{11.7}$$

Theorem 11.2. *Assume that* (11.1) *is strongly competitive and* $p \in \Sigma_J^0$ *for some* $J \subsetneq I_N$ *is an isolated saturated fixed point of* (11.1). *If* $\tilde{\alpha}$ *is the positive vector defined by* (11.5) *and* (11.6), *and*

$$\forall x \in T_p \setminus \{p\}, \quad (x - p)^T D(\tilde{\alpha}) A (x - p) > 0, \tag{11.8}$$

then p *is globally asymptotically stable in* E.

Proof. Let W be an $N \times N$ matrix such that the columns of W, W_{cj} $(1 \leq j \leq N)$, satisfy

$$W_{c1} = (\tilde{\alpha}_1^{-1} 0 \cdots 0)^T, \quad T_p = \operatorname{span}\{W_{c2}, \ldots, W_{cN}\}. \tag{11.9}$$

Then, $\tilde{\alpha}^T W = (1\, 0 \cdots 0)$ so $W_{c1}, \ldots, W_{cN}$ are linearly independent and W is invertible. Let $x - p = Wz$. Then, $\tilde{\alpha}^T (x - p) = \tilde{\alpha}^T W z = z_1$ and

$$\begin{aligned} z_1'|_{(11.1)} = \tilde{\alpha}^T x'|_{(11.1)} &= \sum_{i=1}^N \tilde{\alpha}_i x_i (b_i - A_i x) \\ &= -\tilde{\alpha}^T B (x - p) - (x - p)^T D(\tilde{\alpha}) A (x - p), \end{aligned}$$

where $B_i = p_i A_i$ for $i \in I_N \setminus J$ and for $j \in J$, the jth component of B_j is $A_j p - b_j$ and the rest are 0. When p is an interior fixed point, $\tilde{\alpha} = \alpha$ is a positive left eigenvector of B and the linear term $\tilde{\alpha}^T B(x - p) = \lambda_0 \alpha^T (x - p) = 0$ on $z_1 = 0$ and $z_1' < 0$ by virtue of (11.8). When p belongs to the boundary of Σ, we use the definition of $\tilde{\alpha}$ as follows. For $j \in I_N \setminus J$, from (11.2) and (11.5), we have

$$\sum_{i=1}^N \tilde{\alpha}_i b_{ij} = \sum_{i \in I_N \setminus J} \alpha_i p_i a_{ij} = \lambda_0 \alpha_j = \lambda_0 \tilde{\alpha}_j. \tag{11.10}$$

For $j \in J$ with $A_j p - b_j \geq \lambda_0$, (11.4) and (11.5) result in

$$\sum_{i=1}^N \tilde{\alpha}_i b_{ij} = \sum_{i \in I_N \setminus J} \alpha_i p_i a_{ij} + \hat{\alpha}_j (A_j p - b_j) = \lambda_0 \tilde{\alpha}_j + \varepsilon_j, \tag{11.11}$$

where $\varepsilon_j = \hat{\alpha}_j(A_j p - b_j) \geq \lambda_0 \hat{\alpha}_j > 0$. For $j \in J$ with $A_j p - b_j < \lambda_0$, as a consequence of (11.3), (11.4) and (11.6), we have

$$\begin{aligned}\sum_{i=1}^{N} \tilde{\alpha}_i b_{ij} &= \sum_{i \in I_N \setminus J} \alpha_i p_i a_{ij} + \tilde{\alpha}_j(A_j p - b_j)\\ &= [\lambda_0 - (A_j p - b_j)]\alpha_j + \tilde{\alpha}_j[(A_j p - b_j) - \lambda_0] + \lambda_0 \tilde{\alpha}_j\\ &= \lambda_0 \tilde{\alpha}_j + \varepsilon_j\,, \end{aligned} \tag{11.12}$$

where $\varepsilon_j = [\lambda_0 - (A_j p - b_j)](\alpha_j - \tilde{\alpha}_j) \geq 0$ ($= 0$ or ≤ 0) if p is above (on or below) $\gamma_j = \{y \in E\colon A_j y = b_j\}$.

Substituting (11.10)–(11.12) into $z_1'|_{(11.1)}$ and letting

$$Q(z) = z^T W^T D(\tilde{\alpha}) A W z = z_1 \Theta(z) + Q_1(z_2, \dots, z_N)\,, \tag{11.13}$$

where Θ is a linear function of z with $\Theta(0) = 0$, we obtain, setting $\tilde{z} = (z_2, \dots, z_N)^T$,

$$\begin{aligned} z_1'|_{(11.1)} &= -\lambda_0 z_1 - \sum_{j \in J} \varepsilon_j x_j(z) - Q(z)\\ &= -(\lambda_0 + \Theta(z)) z_1 - \sum_{j \in J} \varepsilon_j x_j(z) - Q_1(\tilde{z})\,. \end{aligned} \tag{11.14}$$

As $x = p$ and T_p are transformed to $z = 0$ and $\pi_1' = \{z \in \mathbb{R}^N\colon z_1 = 0\}$, respectively, the assumption (11.4) becomes

$$\forall z \in \pi_1' \setminus \{0\}\,, \quad Q_1(\tilde{z}) > 0\,. \tag{11.15}$$

Thus, Q_1 is a positive definite quadratic form of $\tilde{z} = (z_2, \dots, z_N)^T$.

Let $E' = \{W^{-1}(x - p)\colon x \in E\}$ and $\tilde{\pi}_1 = \{W^{-1}(x - p)\colon x \in T_p \cap E\}$. Thus, $\tilde{\pi}_1 = \pi_1' \cap E'$. Note that $x_j(z) \geq 0$ for all $j \in J$ and $z \in E'$. Then, from the assumption and the positive definiteness of Q_1, we obtain

$$\forall z \in E' \setminus \{0\}\,, \quad \sum_{j \in J} \varepsilon_j x_j(z) + Q_1(\tilde{z}) > 0\,. \tag{11.16}$$

It then follows from this and (11.14) that

$$\forall x \in (T_p \setminus \{p\}) \cap E\,, \quad (\tilde{\alpha}^T x)'|_{(11.1)} < 0\,.$$

By Theorem 11.1, $\Sigma \setminus \{p\}$ is below T_p.

Next, we show that p is a global attractor in E. Consider the function

$$V(x) = \prod_{i \in I_N \setminus J} x_i^{\alpha_i p_i}\,, \quad x \in E\,, \tag{11.17}$$

where the $\alpha_i > 0$ satisfy (11.2). Then,

$$\begin{aligned}\dot{V}(x)|_{(11.1)} &= V(x) \sum_{i\in I_N\setminus J} \alpha_i p_i (b_i - A_i x) \\ &= V(x) \sum_{i\in I_N\setminus J} \alpha_i p_i A_i (p - x) \\ &= V(x) \left(\sum_{i\in I_N\setminus J} \alpha_i p_i a_{i1} \cdots \sum_{i\in I_N\setminus J} \alpha_i p_i a_{iN} \right) (p - x) .\end{aligned}$$

From (11.2) and (11.5), we obtain

$$\forall j \in I_N \setminus J , \quad \sum_{i\in I_N\setminus J} \alpha_i p_i a_{ij} = \lambda_0 \alpha_j = \lambda_0 \tilde{\alpha}_j . \tag{11.18}$$

For $j \in J$ with $A_j p - b_j \geq \lambda_0$, it follows from (11.4) and (11.5) that

$$\sum_{i\in I_N\setminus J} \alpha_i p_i a_{ij} = \lambda_0 \hat{\alpha}_j = \lambda_0 \tilde{\alpha}_j . \tag{11.19}$$

For $j \in J$ with $A_j p - b_j < \lambda_0$, we have $A_j p - b_j \geq 0$ since p is saturated. Thus, from (11.3), (11.4) and (11.6), we have $\tilde{\alpha}_j \in [\hat{\alpha}_j, \alpha_j]$ so

$$\sum_{i\in I_N\setminus J} \alpha_i p_i a_{ij} = \lambda_0 \hat{\alpha}_j \leq \lambda_0 \tilde{\alpha}_j . \tag{11.20}$$

Then, substitution of (11.18)–(11.20) into $\dot{V}(x)|_{(11.1)}$ results in

$$\begin{aligned}\dot{V}(x)|_{(11.1)} &= V(x) \Big(\sum_{j\in I_N\setminus J} \lambda_0 \alpha_j (p_j - x_j) + \sum_{j\in J} \lambda_0 \hat{\alpha}_j (p_j - x_j) \Big) \\ &= V(x) \Big(\sum_{j\in I_N\setminus J} \lambda_0 \alpha_j (p_j - x_j) + \sum_{j\in J} \lambda_0 \hat{\alpha}_j (-x_j) \Big) \\ &\geq V(x) \lambda_0 \tilde{\alpha}^T (p - x) .\end{aligned}$$

Hence, for $x \in E$ below T_p, $V(x) > 0$ implies $\dot{V}(x)|_{(11.1)} > 0$. For any $x^0 \in \Sigma \setminus \{p\}$ with $x_i^0 > 0$ for $i \in I_N \setminus J$ and $x_j^0 \geq 0$ for $j \in J$, we have $x(t, x^0) \in \Sigma \setminus \{p\}$ and $V(x(t, x^0)) > 0$ for all $t \in \mathbb{R}$. Since $\Sigma \setminus \{p\}$ is below T_p, $V(x(t, x^0))$ is strictly increasing as t increases. This shows that $x(t, x^0)$ is not a periodic solution. By the boundedness of Σ, there is a $c > 0$ such that $\lim_{t\to+\infty} V(x(t, x^0)) = c$. Thus, $V(q) = c$ for all $q \in \omega(x^0)$. If $\omega(x^0) \neq \{p\}$, then there is a $q^0 \in \omega(x^0) \setminus \{p\} \subset \Sigma \setminus \{p\}$ so $\dot{V}(x(t, q^0))|_{(11.1)} > 0$ and $V(x(t, q^0)) > V(q^0) = c$ for $t > 0$, a contradiction to $V(q) = c$ for all $q \in \omega(x^0)$. Therefore, we must have $\lim_{t\to+\infty} x(t, x^0) = p$ for all $x^0 \in \Sigma \setminus \{p\}$, provided $x_i^0 > 0$ for all $i \in I_N \setminus J$. This shows that p is a global attractor relative to Σ.

We need to show that p is also a global attractor relative to E. For any $x^0 \in E \setminus \Sigma$ with $x_i^0 > 0$ for $i \in I_N \setminus J$ and $x_j^0 \geq 0$ for $j \in J$, Theorem 11.1 ensures that $\omega(x^0) \subset \Sigma$.

If $x(t,x^0)$ stays on or above T_p for all $t \geq 0$, then we must have $\omega(x^0) = \{p\}$. If there is a $T \geq 0$ such that $x(t,x^0)$ is below T_p for all $t \geq T$, then the same reasoning used in the previous paragraph leads to $\lim_{t\to+\infty} x(t,x^0) = p$. Otherwise, let $I^1 = \{t \geq 0\colon \tilde{\alpha}^T(x(t,x^0) - p) \geq 0\}$ and $I^2 = [0,+\infty) \setminus I^1$. Both I^1 and I^2 include divergent increasing sequences $\{t_n\}$. Since p is the only point of Σ on or above T_p, restricting t to I^1, we have $\lim_{t\to+\infty} x(t,x^0) = p$. For each sequence $\{t_n\} \subset I^2$ and every $n \geq 1$, there are $r_n, s_n \in I^1$ such that $r_n < t_n < s_n$, $x(r_n,x^0) \in T_p$, $x(s_n,x^0) \in T_p$, and $x(t,x^0)$ is below T_p for all $t \in (r_n,s_n)$. Thus, $\lim_{n\to\infty} V(x(r_n,x^0)) = V(p) = \lim_{n\to\infty} V(x(s_n,x^0))$. Note that $V(x(t,x^0))$ is increasing for $t \in (r_n,s_n)$ since $x(t,x^0)$ is below T_p. Hence, restricting t to $\cup_{n=1}^{\infty}[r_n,s_n]$, we have $\lim_{t\to+\infty} V(x(t,x^0)) = V(p)$ so $\lim_{t\to+\infty} x(t,x^0) = p$. This is also true for $t \in I^2$. Therefore, p is also a global attractor in E.

Finally, the proof for stability of the fixed point given for Theorem 10.1 is still valid here. □

A fixed point $p \in \Sigma$ is said to be *reverse-saturated* if $p_i = 0$ implies $b_i - A_i p \geq 0$ for any $i \in I_N$.

Theorem 11.3. *Assume that* (11.1) *is strongly competitive and* $p \in \Sigma_J^0$ *for some* $J \subsetneq I_N$ *is an isolated reverse-saturated fixed point of* (11.1). *If* $\tilde{\alpha}$ *is the positive vector defined by* (11.5) *and* (11.6), *and*

$$\forall x \in T_p \setminus \{p\}, \quad (x-p)^T D(\tilde{\alpha}) A(x-p) < 0\,, \tag{11.21}$$

then p *is a global repellor relative to* Σ.

Proof. We refer to the proof of Theorem 11.2 line by line. When p is an interior fixed point, $\tilde{\alpha} = \alpha \in E^0$ and by (11.21), $z_1'|_{(11.1)} > 0$ on $z_1 = 0$. When p belongs to the boundary of Σ, since p is reverse-saturated, $\varepsilon_j \leq 0$ in (11.12) and (11.14) and (11.16). By (11.21), "$>$" in (11.15), (11.16) and the inequality below (11.16) is replaced by "$<$". By Theorem 11.1, $\Sigma \setminus \{p\}$ is above T_p.

Since $A_j p - b_j \leq 0$ for all $j \in J$, instead of (11.20), from (11.3), (11.4) and (11.6) we have

$$\forall j \in J\,, \quad \tilde{\alpha}_j \in \left[\alpha_j, \hat{\alpha}_j\right]\,, \quad \sum_{i \in I_N \setminus J} \alpha_i p_i a_{ij} = \lambda_0 \hat{\alpha}_j \geq \lambda_0 \tilde{\alpha}_j\,. \tag{11.22}$$

Then,

$$\dot{V}(x)|_{(11.1)} \leq V(x)\lambda_0 \tilde{\alpha}^T(p - x)\,.$$

Hence, for $x \in E$ above T_p, $V(x) > 0$ implies $\dot{V}(x)|_{(11.1)} < 0$. By an argument similar to that used in the proof of Theorem 11.2, we obtain $\lim_{t\to-\infty} x(t,x^0) = p$ for all $x^0 \in \Sigma \setminus \{p\}$ as long as $x_i^0 > 0$ for all $i \in I_N \setminus J$. Therefore, p is a global repellor. □

Remark 11.1. In the applications of Theorems 11.2 and 11.3, the condition (11.8) (or (11.21)) can be converted to the positive (or negative) definiteness of a square matrix U, which can be computed by (10.15) or (10.18).

From Theorems 11.2 and 11.3, we know that for a strongly competitive system (11.1), we can always construct a strongly balanced plane T_p at a saturated (reverse saturated) isolated fixed point p. Then, Theorem 11.1 guarantees that every $\omega(x^0) \setminus \{p\}$, if not empty, is below (above) T_p if the trajectories pass through $(T_p \setminus \{p\}) \cap E$ in the downward (upward) direction. Unfortunately, these theorems cannot be applied to systems that are not strongly competitive. However, it is very tempting to modify these theorems to suit a larger class of systems than that of merely strongly competitive systems.

Conjecture 11.1. *If system (11.1) is competitive with $b_i > 0$ and $a_{ii} > 0$ for all $i \in I_N$, then every trajectory in $E \setminus \{O\}$ is asymptotic to one in $\overline{\Sigma}$; $\overline{\Sigma}$ is a balanced Lipschitz submanifold homeomorphic to the closed unit simplex in E via radial projection.*

This conjecture is true for $N = 2$ as it can be shown by a phase portrait. However, it needs to be clarified when $N \geq 3$.

Open Problem 11.1. *Assume that the following conditions hold:*
(i) *System (11.1) is competitive, though not strongly competitive.*
(ii) *For some $J \subsetneq I_N$, $p \in E_J^0$ is an isolated saturated (reverse-saturated) fixed point.*
(iii) *There is a strongly balanced plane T_p containing p such that the trajectories pass through $(T_p \setminus \{p\}) \cap E$ in the downward (upward) direction.*

Investigate whether $\omega(x^0) \setminus \{p\}$, if not empty, is below (above) T_p for every $x^0 \in E$ with $x_i^0 > 0$ for all $i \in I_N \setminus J$. Further, investigate whether the conclusion of Theorem 11.2 or Theorem 11.3 holds.

Open Problem 11.2. *Make modified assumptions (weaker than those of Theorems 10.1 and 10.2, but stronger than those of Theorem 11.2) so that the conclusion of Theorem 11.2 holds for a competitive (but not necessarily strongly competitive) system (11.1).*

Example 11.1. Consider a system (11.1) with

$$A = \begin{pmatrix} 1 & 0.4 & 0.5 \\ 0.4 & 1 & 0.1 \\ 0.2 & 2 & 1 \end{pmatrix}, \quad b = \begin{pmatrix} 1 \\ 1 \\ 1 \end{pmatrix}. \tag{11.23}$$

It has a fixed point $p = (\frac{5}{7}, \frac{5}{7}, 0)^T$ and $A_3 p = \frac{11}{7} > 1 = b_3$. Thus, p is saturated. Clearly, $\lambda_0 = 1$ and $\alpha_1 = \alpha_2 = 1$ satisfy (11.2) with $J = \{3\}$ and $N = 3$. As $A_3 p - b_3 = \frac{4}{7} < \lambda_0$, by (11.3) and (11.4), we have

$$\begin{aligned} \alpha_3 &= \frac{1}{\lambda_0 - (A_3 p - b_3)}(\alpha_1 p_1 a_{13} + \alpha_2 p_2 a_{23}) = 1, \\ \hat{\alpha}_3 &= \frac{1}{\lambda_0}(\alpha_1 p_1 a_{13} + \alpha_2 p_2 a_{23}) = \frac{3}{7}. \end{aligned}$$

By (11.5) and (11.6), $\tilde{\alpha}_1 = \alpha_1 = 1$, $\tilde{\alpha}_2 = \alpha_2 = 1$ and $\tilde{\alpha}_3 = 1/\rho \in [\frac{3}{7}, 1]$. Then,

$$\left(AD(\alpha)^{-1}\right)^S = \begin{pmatrix} 2 & 0.8 & 0.2 + 0.5\rho \\ 0.8 & 2 & 2 + 0.1\rho \\ 0.2 + 0.5\rho & 2 + 0.1\rho & 2\rho \end{pmatrix}.$$

By (10.18), we have

$$U = \begin{pmatrix} 1.6 + \rho & 1.4(\rho - 1) \\ \frac{19}{24} & 1.8\rho - 2 \end{pmatrix}.$$

For $\rho = 2 \in [1, \frac{7}{3}]$,

$$U = \begin{pmatrix} 3.6 & 1.4 \\ 1.4 & 1.6 \end{pmatrix}$$

is positive definite. By Theorem 11.2, p is globally asymptotically stable in $\mathbb{R}^3_+$.

12 Global attraction of competitive Lotka–Volterra systems

In this section, we demonstrate a geometric method in dealing with the global attraction of a saturated fixed point in competitive Lotka–Volterra systems. For convenience, we rewrite the system (11.1) as

$$x_i' = b_i x_i(1 - A_i x), \quad i \in I_N, \tag{12.1}$$

where $b_i > 0$ and $A_i = (a_{i1} \cdots a_{iN})$ with $a_{ii} > 0$ and $a_{ij} \geq 0$. For any $u, v \in \mathbb{R}^N$, by the inequality $u \leq v$, we mean $u_i \leq v_i$ for all $i \in I_N$. Let

$$\gamma_i = \{x \in E\colon A_i x = 1\}, \quad i \in I_N, \tag{12.2}$$

$$\pi_i(u) = \{x \in E\colon x_i = u_i, \quad \forall j \in I_N \setminus \{i\}, x_j \geq u_j\}, \quad i \in I_N, \tag{12.3}$$

and

$$[u, v] = \{x \in E\colon \forall i \in I_N, u_i \leq x_i \leq v_i\}, \quad u \in E, v \in E, u \leq v. \tag{12.4}$$

The core of the method is as follows. We first utilize the nullcline planes γ_i to determine an upper bound $v \in E$ for all ω-limit points of solutions in E^0. Then, the relative position of the nullcline planes on $\pi_i(0) \cap [0, v]$ for each $i \in I_N$ determines a lower bound $u \in E$, $u \neq 0$, for all ω-limit points. Then, $\omega(x^0) \subset [u, v]$ for all $x^0 \in E^0$ implies the existence of a smaller upper bound $v^1 \in [u, v]$ with $v^1 \neq v$ such that $\omega(x^0) \subset [u, v^1]$ for all $x^0 \in E^0$. The relative position of the nullcline planes on $\pi_i(u) \cap [u, v^1]$ is similar to that on $\pi_i(0) \cap [0, v]$ and hence determines a larger lower bound $u^1 \in [u, v^1]$ with $u^1 \neq u$ such that $\omega(x^0) \subset [u^1, v^1]$ for all $x^0 \in E^0$. If these cells $[u^{k+1}, v^{k+1}] \subset [u^k, v^k]$ are contracting to a point x^*, i. e. $\lim_{k\to\infty} u^k = x^* = \lim_{k\to\infty} v^k$, then x^* is a globally attracting fixed point.

We observe that the ith component $x_i(t)$ is decreasing (increasing) if $x(t)$ is above (below) γ_i, i. e. if $A_i x(t) > 1$ (< 1).

Lemma 12.1. *Let $\hat{x} \in E$ satisfy $\hat{x}_i = 0$ if $A_i\hat{x} > 1$ for any $i \in I_N$ and let $\hat{y} \in E$ be given by*

$$\hat{y}_j = \max\left\{0, \hat{x}_j + a_{jj}^{-1}\left(1 - A_j\hat{x}\right)\right\}, \quad j \in I_N . \tag{12.5}$$

If a solution of (12.1) in E satisfies $\underline{\lim}_{t\to+\infty} x(t) \geq \hat{x}$, then it also satisfies $\overline{\lim}_{t\to+\infty} x(t) \leq \hat{y}$.

Proof. Suppose the conclusion is not true. Then,

$$\exists i \in I_N , \quad \exists p > \hat{y}_i \text{ such that } \overline{\lim}_{t\to+\infty} x_i(t) = p . \tag{12.6}$$

Take $\delta = \frac{1}{4}a_{ii}(p - \hat{y}_i)$ and $\varepsilon \in (0,1)$ such that $\varepsilon A_i\hat{x} < \delta$. Then, by the assumption, there is a $T > 0$ such that $x(t) \geq (1-\varepsilon)\hat{x} \geq 0$ for all $t \geq T$. Hence, for any $\bar{t} > T$ such that $x_i(\bar{t}) \geq \frac{1}{4}(3p + \hat{y}_i) = 3a_{ii}^{-1}\delta + \hat{y}_i$, from (12.5), we have

$$\begin{aligned} A_i x(\bar{t}) &\geq (1-\varepsilon)[A_i\hat{x} - a_{ii}\hat{x}_i] + 3\delta + a_{ii}\hat{y}_i \\ &\geq (1-\varepsilon)[A_i\hat{x} - a_{ii}\hat{x}_i] + 3\delta + a_{ii}\hat{x}_i + 1 - A_i\hat{x} \\ &\geq 1 - \varepsilon A_i\hat{x} + 3\delta \\ &> 1 + 2\delta \end{aligned}$$

so that $x_i'(\bar{t}) < -2\delta b_i x_i(\bar{t})$. This shows that $x_i(t)$ is strictly decreasing as long as $x_i(t) \geq \frac{1}{4}(3p + \hat{y}_i)$. By (12.6), there is a $T' \geq T$ such that $x_i(t) > \frac{1}{4}(3p + \hat{y}_i)$ for all $t \geq T'$ so $x_i(t) \leq x_i(T')e^{-2\delta b_i(t-T')} \to 0$ as $t \to +\infty$, a contradiction to (12.6). Therefore, $\overline{\lim}_{t\to+\infty} x(t) \leq \hat{y}$. □

Lemma 12.2. *Assume that*

(i) *$\hat{x}, \hat{y} \in E$ as in Lemma 12.1 with $J = \{k \in I_N : \hat{x}_k > 0\}$;*

(ii) *every solution of (12.1) in E with $x_k(0) > 0$ for $k \in J$ satisfies*

$$\underline{\lim}_{t\to+\infty} x(t) \geq \hat{x} ;$$

(iii) *for some $i \in I_N$, $A_i\hat{x} < 1$ and $\gamma_i \cap \pi_i(\hat{x}) \cap [\hat{x}, \hat{y}]$, if not empty, is above every γ_j ($j \neq i$).*

Then, there is a $\delta > 0$ such that every solution of (12.1) in E with $x_k(0) > 0$ for $k \in J \cup \{i\}$ satisfies

$$\underline{\lim}_{t\to+\infty} x_i(t) \geq \hat{x}_i + \delta . \tag{12.7}$$

Proof. Put

$$\varepsilon = \sup\left\{A_i x \colon x \in \left(\bigcup_{j\in I_N\setminus\{i\}} \gamma_j \cup \{\hat{x}\}\right) \cap \pi_i(\hat{x}) \cap [\hat{x},\hat{y}]\right\}, \tag{12.8}$$

$$\Gamma_i = \{x \in E \colon A_i x - a_{ii}(x_i - \hat{x}_i) \ge (1+\varepsilon)/2\}. \tag{12.9}$$

Then, by (iii) and the compactness of the set in (12.8), $\varepsilon \in (0,1)$. We show that $\delta = \frac{1}{2}(1-\varepsilon)a_{ii}^{-1}$ meets the requirement of (12.7). For any solution x of (12.1) with $x_k(0) > 0$ for $k \in J \cup \{i\}$, we shall see in the next paragraph the existence of $T \ge 0$ such that

$$A_i x(t) + a_{ii}(\hat{x}_i - x_i(t)) < \tfrac{1}{2}(1+\varepsilon) = 1 - a_{ii}\delta \tag{12.10}$$

for all $t \ge T$. Then, for any $t \ge T$, if $x_i(t) \le \hat{x}_i + \delta$, then $A_i x(t) < 1$ so $x(t)$ is below γ_i. If (12.7) does not hold, since $x_i'(t) > 0$ if and only if $x(t)$ is below γ_i, we must have $x_i(t) \le \delta_0$ for some $\delta_0 < \hat{x}_i + \delta$ and all $t \ge T$. By (12.10),

$$A_i x(t) < \tfrac{1}{2}(1+\varepsilon) + a_{ii}(x_i(t) - \hat{x}_i) \le \tfrac{1}{2}(1+\varepsilon) + a_{ii}(\delta_0 - \hat{x}_i) \equiv \varepsilon_0 < 1.$$

Integration of the ith component equation of (12.1) leads to $x_i(t) \to +\infty$ as $t \to +\infty$, which contradicts $x_i(t) \le \delta_0$. Therefore, we must have (12.7).

If $A_i\hat{y} - a_{ii}(\hat{y}_i - \hat{x}_i) < \frac{1}{2}(1+\varepsilon)$, then the existence of $T \ge 0$ such that (12.10) holds for all $t \ge T$ follows from (i), (ii) and Lemma 12.1. Suppose $A_i\hat{y} - a_{ii}(\hat{y}_i - \hat{x}_i) \ge \frac{1}{2}(1+\varepsilon)$. Then, for $\hat{z} \in [\hat{x},\hat{y}]$ with $\hat{z}_i = \hat{x}_i$ and $\hat{z}_j = \hat{y}_j$ for $j \in I_N \setminus \{i\}$, we have $A_i\hat{z} \ge \frac{1}{2}(1+\varepsilon)$ so $\hat{z} \in [\hat{x},\hat{z}] \cap \Gamma_i \subset [\hat{x},\hat{y}] \cap \Gamma_i$. The definitions (12.8) and (12.9) suggest that $[\hat{x},\hat{y}] \cap \Gamma_i$ is above every γ_j $(j \ne i)$, so

$$\eta \equiv \inf\{A_j y \colon y \in [\hat{x},\hat{y}] \cap \Gamma_i, j \in I_N \setminus \{i\}\} > 1.$$

Then, by (i), (ii) and Lemma 12.1, since $x(t)$ is close to $[\hat{x},\hat{y}]$ for sufficiently large t, there is a $T_0 > 0$ such that for any $t \ge T_0$ and $j \in I_N \setminus \{i\}$, if $x(t) \in \Gamma_i$ then $A_j x(t) \ge \frac{1}{2}(1+\eta) > 1$ so $x_j'(t) \le -\frac{1}{2}b_j(\eta - 1)x_j(t)$. This indicates that either there is a $T \ge T_0$ such that $x(t) \notin \Gamma_i$ for all $t \ge T$ or $x(t) \in \Gamma_i$ for all $t \ge T_0$. In the latter case, $\lim_{t\to+\infty} x_j(t) = 0$ for all $j \in I_N \setminus \{i\}$ so that $a_{ii}\hat{x}_i \ge \frac{1}{2}(1+\varepsilon)$ by (12.9). This is impossible as, by (12.8), $a_{ii}\hat{x}_i \le A_i\hat{x} \le \varepsilon < \frac{1}{2}(1+\varepsilon)$. Hence, by the equivalence of $x(t) \notin \Gamma_i$ to (12.10), $x(t)$ satisfies (12.10) for all $t \ge T$. □

If we take $\hat{x} = 0$ and $\hat{y} = Y = (a_{11}^{-1} \cdots a_{NN}^{-1})^T$, then, by Lemma 12.1, every solution of (12.1) in E satisfies $0 \le \underline{\lim}_{t\to+\infty} x(t) \le \overline{\lim}_{t\to+\infty} x(t) \le Y$. For a fixed $i \in I_N$, if

$$\forall j \in I_N \setminus \{i\}, \quad \gamma_j \cap \pi_i(0) \cap [0,Y] \text{ is below } \gamma_i, \tag{12.11}$$

then either $\pi_i(0) \cap [0,Y]$ is below γ_i or $\gamma_i \cap \pi_i(0) \cap [0,Y]$ is above every γ_j $(j \ne i)$. By Lemma 12.2, (12.1) is $\{i\}$-permanent. If (12.11) holds for all $i \in I_N$, then (12.1) is permanent, so (12.1) has a unique fixed point x^* in $E^0 \cap [0,Y]$. We now derive an algebraic condition for (12.11). For any vector $v \in \mathbb{R}^N$ and subset $J \subset I_N$, $v^J \in \mathbb{R}^N$ is defined by $v_i^J = v_i$ for $i \in J$ and $v_j^J = 0$ for $j \in I_N \setminus J$. Thus, $v^\varnothing = 0$ and $v^{I_N} = v$.

Lemma 12.3. (i) *Let $J \subset I_N$ with cardinality $|J| > 1$. If* (12.11) *holds for each $i \in J$, then*

$$\max\left\{0, \frac{a_{ij}}{a_{jj}}\left(1 - A_j Y^{I_N\setminus\{i,j\}}\right)\right\} < 1 - A_i Y^{I_N\setminus\{i,j\}} \tag{12.12}$$

for all $i, j \in J$ with $i \neq j$.

(ii) *Conversely, for a fixed $i \in I_N$,* (12.12) *for all $j \in I_N \setminus \{i\}$ implies* (12.11).

(iii) *Hence,* (12.11) *holds for all $i \in I_N$ if and only if* (12.12) *holds for all $i, j \in I_N$ with $i \neq j$.*

Proof. (i) Suppose (12.11) holds for all $i \in J$ and there are $i, k \in J$ $(i \neq k)$ such that $A_i Y^{I_N\setminus\{i,k\}} \geq 1$. Since $A_i 0 < 1$, we have $\emptyset \neq [0, Y^{I_N\setminus\{i,k\}}] \cap \gamma_i \subset [0, Y^{I_N\setminus\{i\}}] \cap \gamma_i$. By (12.11) for i, $[0, Y^{I_N\setminus\{i,k\}}] \cap \gamma_i$ is above γ_k. On the other hand, since (12.11) also holds for k, $[0, Y^{I_N\setminus\{i,k\}}] \cap \gamma_i$ should be below γ_k. This contradiction shows that (12.11) for all $i \in J$ implies $A_i Y^{I_N\setminus\{i,j\}} < 1$ for all $i, j \in J\,(i \neq j)$. Since $A_j Y^{I_N\setminus\{i\}} \geq a_{jj} Y_j = 1$, we obtain $[Y^{I_N\setminus\{i,j\}}, Y^{I_N\setminus\{i\}}] \cap \gamma_j = \{z\}$ where $z_k = Y_k^{I_N\setminus\{i,j\}}$ for $k \in I_N \setminus \{j\}$ and

$$z_j = \frac{1}{a_{jj}}\left(1 - A_j Y^{I_N\setminus\{i,j\}}\right).$$

Then, by (12.11) for i, z is below γ_i, i. e. $A_i z = A_i Y^{I_N\setminus\{i,j\}} + a_{ij} z_j < 1$ so $a_{ij} z_j < 1 - A_i Y^{I_N\setminus\{i,j\}}$. Then, substitution of z_j into this gives (12.12).

(ii) If $A_i Y^{I_N\setminus\{i\}} < 1$, then $[0, Y^{I_N\setminus\{i\}}] = \pi_i(0) \cap [0, Y]$ is below γ_i and (12.11) follows. Suppose $A_i Y^{I_N\setminus\{i\}} \geq 1$. Since $A_i Y^{I_N\setminus\{i,j\}} < 1$ for all $j \in I_N \setminus \{i\}$ by (12.12), there is a unique $z^j \in E$ such that $[Y^{I_N\setminus\{i,j\}}, Y^{I_N\setminus\{i\}}] \cap \gamma_i = \{z^j\}$: $z_k^j = Y_k$ for all $k \in I_N \setminus \{i, j\}$, $z_i^j = 0$ and

$$0 < z_j^j = \frac{1}{a_{ij}}\left(1 - A_i T^{I_N\setminus\{i,j\}}\right) \leq Y_j\,.$$

From (12.12), we have

$$\frac{a_{jj}}{a_{ij}}\left(1 - A_i Y^{I_N\setminus\{i,j\}}\right) + A_j Y^{I_N\setminus\{i,j\}} > 1\,,$$

i. e. $A_j z^j > 1$. Thus,

$$\forall j \in I_N \setminus \{i\}, A_j Y^{I_N\setminus\{i\}} \geq A_j z^j > 1\,.$$

Also,

$$\forall j, k \in I_N \setminus \{i\}\ (j \neq k)\,, \quad A_k Y^{I_N\setminus\{i,j\}} \geq a_{kk} Y_k = 1\,.$$

Therefore, the line segment $[Y^{I_N\setminus\{i,j\}}, Y^{I_N\setminus\{i\}}] \setminus \{Y^{I_N\setminus\{i,j\}}\}$ is above γ_k. In particular, z^j is above γ_k for all $k \in I_N \setminus \{i\}$. It can be checked that $[0, Y^{I_N\setminus\{i\}}] \cap \gamma_i$ is the convex set with $\{z^j : j \in I_N \setminus \{i\}\}$ as its vertex set. Since $\{z^j : j \in I_N \setminus \{i\}\}$ is above γ_k for all $k \in I_N \setminus \{i\}$, the set $[0, Y^{I_N\setminus\{i\}}] \cap \gamma_i$ is above γ_j for all $j \in I_N \setminus \{i\}$. Thus, (12.11) holds.

(iii) This is a combination of (i) and (ii). □

Theorem 12.1. *Assume that* (12.12) *holds for all* $i, j \in I_N$ *with* $i \neq j$. *Then,* (12.1) *has a globally attracting fixed point* $x^* \in E^0$ *with* $0 < x_i^* \leq a_{ii}^{-1}$ *for all* $i \in I_N$.

Proof. Since (12.12) holds for all $i, j \in I_N$ $(i \neq j)$, by Lemma 12.3, (12.11) holds for all $i \in I_N$. By Lemma 12.2, (12.1) is permanent. Thus, (12.1) has a unique fixed point x^* in $E^0 \cap [0, Y]$. Then, for $\delta > 0$ small enough, we have $\underline{\lim}_{t\to\infty} x(t, x^0) \geq \delta x^*$ for all $x^0 \in E^0$. If we set $\hat{x} = \delta x^*$, then by Lemma 12.1, we have $\hat{y} = \delta x^* + (1 - \delta)Y$ and $\overline{\lim}_{t\to\infty} x(t, x^0) \leq \hat{y}$ for all $x^0 \in E^0$. If $0 < \delta < 1$, then, since the affine map $f\colon [0, Y] \to [\hat{x}, \hat{y}]$ given by $f(x) = \delta x^* + (1 - \delta)x$ shrinks each line segment $\overline{x^* x}$ into $\overline{x^* f(x)} \subset \overline{x^* x}$, x is below (on or above) γ_j if and only if $f(x)$ is below (on or above) γ_j. Thus, the relative position of the nullcline planes γ_j restricted to $[0, Y]$ is the same as that to $[\hat{x}, \hat{y}]$. Therefore, since (12.11) holds for all $i \in I_N$, we also have

$$\forall i \in I_N\,, \quad \forall j \in I_N \setminus \{i\}\,, \quad \gamma_j \cap \pi_i(\hat{x}) \cap [\hat{x}, \hat{y}] \text{ is below } \gamma_i\,.$$

By Lemma 12.2, there is a $\delta_1 > \delta$ such that $\underline{\lim}_{t\to+\infty} x(t, x^0) \geq \delta_1 x^*$ for all $x^0 \in E^0$. Take δ_1 as the supremum of such numbers. Then, the same reasoning as above shows that $\delta_1 = 1$. Then, $\hat{x} = \delta_1 x^* = x^*$ and $\hat{y} = \delta_1 x^* + (1 - \delta_1)Y = x^*$, so x^* is a global attractor. □

Thus far, we have demonstrated the geometric method of using the relative position of the nullcline planes γ_i to prove the global attraction of an interior fixed point $x^* \in E^0$. The method can also be applied to proving the global attraction of a boundary fixed point $x^* \in \partial E$. To this end, the following theorem is helpful. It asserts that if the ith nullcline plane γ_i is below a mean plane of the γ_j $(j \in I_N \setminus \{i\})$, then the ith component will vanish.

Theorem 12.2 ([19, Theorem 2]). *Assume that* (12.1) *satisfies the following conditions.*
(i) *There exists a nonempty* $J_0 \subset I_N$ *such that if* $J_0 \neq I_N$,

$$\forall k \in I_N \setminus J_0\,, \quad \forall x^0 \in E^0\,, \quad x_k(x^0, t) = o(1) \quad (t \to +\infty)\,. \tag{12.13}$$

(ii) *There exists* $i \in J_0$, *a nonempty* $J_1 \subset J_0 \setminus \{i\}$ *and* $c \in E$ *with* $c_j > 0$ *if and only if* $j \in J_1$ *and* $|c| = c_1 + \cdots + c_N = 1$ *such that*

$$\forall k \in J_0\,, \quad \sum_{j\in J_1} c_j a_{jk} < a_{ik}\,. \tag{12.14}$$

Then, there is a $\delta_0 > 0$ *such that the solution of* (12.1) *satisfies*

$$x_i\left(x^0, t\right) = o\left(e^{-\delta_0 t}\right) \quad (t \to \infty) \tag{12.15}$$

for all $x^0 \in E^0$ *and every* $x^0 \in \cap_{j\in I_N\setminus J_0} \pi_j(0)$ *with at least* $x_\ell^0 > 0$ *for all* $\ell \in J_1 \cup \{i\}$.

When $J_0 = I_N$, condition (i) requires nothing and (ii) is for (12.1) on E. However, if $J_0 \neq I_N$, then (i) says that system (12.1) on E^0 degenerates asymptotically to the subsystem of (12.1) on $\cap_{\ell\in I_N\setminus J_0}\pi_\ell(0)$ and condition (ii) is for this subsystem only. The geometric interpretation of (12.14) is that the ith nullcline plane γ_i intersects the x_k-axis for every $k\in J_0$ and, on $\cap_{\ell\in I_n\setminus J_0}\pi_\ell(0)$, γ_i is strictly below a mean plane $\gamma_c=\{x\in E\colon c^TAx=1\}$. Indeed, for any $x\in\gamma_i\cap(\cap_{\ell\in I_n\setminus J_0}\pi_\ell(0))$, (12.14) implies

$$c^TAx=\sum_{k\in J_0}\sum_{j\in J_1}c_ja_{jk}x_k<\sum_{k\in J_0}a_{ik}x_k=A_ix=1 \tag{12.16}$$

so $\gamma_i\cap(\cap_{\ell\in I_n\setminus J_0}\pi_\ell(0))$ is strictly below γ_c. Conversely, if γ_i intersects the x_k-axis for every $k\in J_0$, then (12.16) for all $x\in\gamma_i\cap(\cap_{\ell\in I_n\setminus J_0}\pi_\ell(0))$ implies (12.16).

13 Some notes

Section 10 is based on [7] and Section 11 is from [25] and [43]. For global attraction, Section 12 is based on [17] and further results can be found in [18]. Other related works can be found in [1–3, 11, 33, 42] and the references therein.

Bibliography

[1] S. Ahmad and A. C. Lazer, *One species extinction in an autonomous competitive model*, Proc. First World Congress on Nonlinear Analysis, pp. 359–368, Walter DeGruyter, Berlin, 1995.

[2] S. Ahmad and A. C. Lazer, *Necessary and sufficient average growth in a Lotka–Volterra system*, Nonlinear Anal. **34** (1998), 191–228.

[3] S. Ahmad and A. C. Lazer, *Average growth and extinction in a competitive Lotka–Volterra system*, Nonlinear Anal. **62** (2005), 545–557.

[4] S. Ahmad and A. C. Lazer, *Average growth and total permanence in a competitive Lotka–Volterra system*, Ann. Mat. Pura Appl. **185** (2006), S47–S67.

[5] S. Ahmad and I. M. Stamova, *Partial persistence and extinction in N-dimensional competitive systems*, Nonlinear Anal. **60** (2005), 821–836.

[6] M. Araki and B. Kondo, *Stability and transient behavior of composite nonlinear systems*, IEEE Trans. Automatic Control **17** (1972) 537–541.

[7] S. Baigent and Z. Hou, *Global stability of interior and boundary fixed points for Lotka–Volterra systems*, Differ. Equ. Dyn. Syst. **20** (2012), 53–66.

[8] A. Battauz and F. Zanolin, *Coexistence states for periodic competitive Kolmogorov systems*, J. Math. Anal. Appl. **219** (1998), 179–199.

[9] A. Berman and R. J. Plemmons, *Nonnegative Matrices in the Applied Mathematical Sciences*, Academic Press, New York, 1979.

[10] L. Chen, Z. Lu and W. Wang, *The effect of delays on the permanence for Lotka–Volterra systems*, Appl. Math. Lett. **8** (1995), 71–73.

[11] P. van den Driessche and M. L. Zeeman, *Three-dimensional competitive Lotka–Volterra systems with no periodic orbits*, SIAM J. Appl. Math. **58** (1998), 227–234.

[12] J. K. Hale, *Theory of Functional Differential Equations*, Springer-Verlag, 1977.

[13] M. W. Hirsch, *Systems of differential equations that are competitive or cooperative III: competing species*, Nonlinearity **1** (1988), 51–71.

[14] J. Hofbauer and S. J. Schreiber, *Robust permanence for interacting structured populations*, J. Differential Equations **248** (2010), 1955–1971.

[15] J. Hofbauer and K. Sigmund, *Evolutionary Games and Dynamical Systems*, Cambridge University Press, UK, 1998.
[16] R. A. Horn and C. R. Johnson, *Matrix Analysis*, Cambridge University Press, Cambridge, 1985.
[17] Z. Hou, *Global attractor in autonomous competitive Lotka–Volterra systems*, Proc. Amer. Math. Soc. **127** (1999), 3633–3642.
[18] Z. Hou, *Global attractor in competitive Lotka–Volterra systems*, Math. Nachr. **282** (2009), 995–1008.
[19] Z. Hou, *Vanishing components in autonomous competitive Lotka–Volterra systems*, J. Math. Anal. Appl. **359** (2009), 302–310.
[20] Z. Hou, *On permanence of all subsystems of competitive Lotka–Volterra systems with delays*, Nonlinear Anal. Real World Appl. **11** (2010), 4285–4301.
[21] Z. Hou, *On permanence of Lotka–Volterra systems with delays and variable intrinsic growth rates*, Nonlinear Anal. Real World Appl. **14** (2013), 960–975.
[22] Z. Hou, *Permanence and extinction in competitive Lotka–Volterra systems with delays*, Nonlinear Anal. Real World Appl. **12** (2011), 2130–2141.
[23] Z. Hou, *Oscillation and limit cycles in Lotka–Volterra systems with delays*, Nonlinear Anal. **75** (2012), 358–370.
[24] Z. Hou, *Permanence criteria for Kolmogorov systems with delays*, preprint.
[25] Z. Hou and S. Baigent, *Fixed point global attractors and repellors in competitive Lotka–Volterra systems*, Dyn. Syst. **26(4)** (2011), 367–390.
[26] W. Jansen, *A permanence theorem for replicator and Lotka–Volterra systems*, J. Math. Biol. **25** (1987), 411–422.
[27] Y. Kuang, *Delay Differential Equations with Applications in Population Dynamics*, Academic Press, New York, 1993.
[28] Y. Kuang and B. Tang, *Uniform persistence in nonautonomous delay differential Kolmogorov-type population models*, Rocky Mountain J. Math. **24** (1994), 165–186.
[29] S. Lang, *Real Analysis*, 2nd edn. Addison-Wesley, London, 1983.
[30] Z. Li and Z. Teng, *Permanence for non-autonomous food chain systems with delay*, J. Math. Anal. Appl. **286** (2003), 724–740.
[31] S. Liu and L. Chen, *Necessary-sufficient conditions for permanence and extinction in Lotka–Volterra system with distributed delays*, Appl. Math. Lett. **16** (2003), 911–917.
[32] G. Lu, Z. Lu and Y. Enatsu, *Permanence for Lotka–Volterra systems with multiple delays*, Nonlinear Anal. Real World Appl. **12** (2011), 2552–2560.
[33] F. Montes de Oca and M. L. Zeeman, *Balancing survival and extinction in nonautonomous competitive Lotka–Volterra systems*, J. Math. Anal. Appl. **192** (1995), 360–370.
[34] D. Mukherjee, *Permanence and global attractivity for facultative mutualism system with delay*, Math. Methods Appl. Sci. **26** (2003), 1–9.
[35] J. Mierczyński and S. J. Schreiber, *Kolmogorov vector fields with robustly permanent subsystems*, J. Math. Anal. Appl. **267** (2002), 329–337.
[36] H. L. Smith and H. R. Thieme, *Dynamical Systems and Population Persistence*, American Mathematical Society, 2011.
[37] Y. Takeuchi, *Global Dynamical Properties of Lotka–Volterra Systems*, World Scientific, Singapore, 1996.
[38] B. Tang and Y. Kuang, *Permanence in Kolmogorov-type systems of nonautonomous functional differential equations*, J. Math. Anal. Appl. **197** (1996), 427–447.
[39] Z. Teng, *Nonautonomous Lotka–Volterra systems with delays*, J. Differential Equations **179** (2002), 538–561.
[40] X. Yang, *The persistence of a general nonautonomous single-species Kolmogorov system with delays*, Nonlinear Anal. **70** (2009), 1422–1429.

[41] M. L. Zeeman, *Hopf bifurcation in competitive three-dimensional Lotka–Volterra systems*, Dynam. Stability Systems **8** (1993), 189–217.

[42] M. L. Zeeman, *Extinction in competitive Lotka–Volterra systems*, Proc. Amer. Math. Soc. **123** (1995), 87–96.

[43] E. C. Zeeman and M. L. Zeeman, *From local to global behavior in competitive Lotka–Volterra systems*, Trans. Amer. Math. Soc. **355** (2003), 713–734.

[44] X. Zhao, *Dynamical Systems in Population Biology*, Springer-Verlag, New York, 2003.

[45] J. Zhao and J. Jiang, *Average conditions for permanence and extinction in nonautonomous Lotka–Volterra system*, J. Math. Anal. Appl. **299** (2004), 663–675.

Benedetta Lisena

Competitive Lotka–Volterra systems with periodic coefficients

Benedetta Lisena: Dipartimento di Matematica, Università degli Studi di Bari, Bari, Italy

1 Introduction

The aim of this contribution is to give an account of some asymptotic stability questions concerning Lotka–Volterra competitive systems. Such differential systems are well known in mathematical biology as models describing the growth process of different species living in an isolated environment and competing for the same resources. From a mathematical point of view, the considered systems are nonlinear. As a consequence, we do not have techniques for an explicit knowledge of their solutions. Therefore, the study of the dynamic properties can be tackled by means of a qualitative analysis. In this general setting, our investigation is devoted to the nonautonomous case, precisely we suppose periodic and continuous coefficients. Periodic models occur when the biological parameters vary in a cyclic way, owing to the season alternation, for instance. For a differential equation, with periodic coefficients, we expect properties more specific than those holding for any nonautonomous case. In recent years, a few researchers investigated the relationship between a competitive system with periodic coefficients and the differential system whose coefficients are the integral mean values of the assigned coefficients. This last autonomous system is called the averaged system. Roughly speaking, an attempt was made to find a link between the periodic system and the corresponding averaged system ([10]). Such connection is not fruitful, but the discussion of the existence of a periodic solution and its asymptotic stability may lead to suitable conditions on coefficients in the form of integral averages. It is shown by the results of this paper, in which fixed point method, comparison results and the Lyapunov technique are employed.

Section 2 is devoted to the periodic logistic equation. In Sections 3–4, we consider the bidimensional Lotka–Volterra system, investigating either the existence and the global attractivity of the periodic solution or the competitive exclusion principle. For a model involving three different species, in Section 5, we deal with the case in which one species is biologically weaker than the remaining ones. Suitable assumptions on coefficients lead to the extinction of the weakest species and survival of the other two. Sections 6–8 treat competitive systems with impulses. In nature, the intrinsic growth law of an ecological model is often affected by short-term perturbations, owing to many exterior factors. Impulsive differential equations are effective mathematical tools to deal with such phenomena ([8],[22]). The presence of pulses is frequently a source of instability, and thus it is important to study the effects of impulses on the dynamical behavior of such ecosystems.

Our results mainly concentrate on two-dimensional and three-dimensional models starting from the logistic equation, following the scheme of previous sections. Many examples are given in order to show the feasibility of the method. The validity of the stated assumptions is checked by the help of numerical tools as well. We should point out that the reading of this material requires only elementary facts from ordinary differential equations and the Lyapunov method ([5],[1],[9]). Finally, we mention that the list of the proposed references is essential as its objective is to make the layout self-contained, as far as that is possible.

2 The autonomous model. The logistic equation

Suppose that in an isolated environment, two species live and there is a competitive-type interaction between them. A well-known mathematical model describing the evolution process is due to Lotka–Volterra. In the simplest case, it has the form

$$\begin{cases} u_1' = u_1(a_1 - b_{11}u_1 - b_{12}u_2) \\ u_2' = u_2(a_2 - b_{21}u_1 - b_{22}u_2) \end{cases} \tag{2.1}$$

where all coefficients are positive numbers.

If interaction coefficients b_{ij}, $i \neq j$ are zero, each species satisfies a growth model of logistic type.

In the dynamics of model (2.1), we are interested to the following situations:

(a) Coexistence of both populations for all values of time;

(b) Asymptotic extinction of the biologically weaker population. This phenomenon is known as competitive exclusion.

If we assume that

$$a_1 > b_{12}\left(\frac{a_2}{b_{22}}\right), \quad a_2 > b_{21}\left(\frac{a_1}{b_{11}}\right), \tag{H$_1$}$$

then system (2.1) admits a unique equilibrium point with positive coordinates. Indeed, (H$_1$) implies the inequality

$$b_{11}b_{22} > b_{12}b_{21}$$

so that the equilibrium point

$$\mathring{u}_1 = \frac{\begin{vmatrix} a_1 & b_{12} \\ a_2 & b_{22} \end{vmatrix}}{b_{11}b_{22} - b_{12}b_{21}}, \quad \mathring{u}_2 = \frac{\begin{vmatrix} b_{11} & a_1 \\ b_{12} & a_2 \end{vmatrix}}{b_{11}b_{22} - b_{12}b_{21}} \tag{2.2}$$

has positive coordinates. Under hypothesis (H_1), case (a) occurs because $(\mathring{u}_1, \mathring{u}_2)$ is globally attractive in the first quadrant ([5]), that is,

$$\lim_{t\to+\infty} u_1(t) = \mathring{u}_1 , \quad \lim_{t\to+\infty} u_2(t) = \mathring{u}_2$$

for any $(u_1(t), u_2(t))$ solution of (2.1) with a positive initial condition.

Let us consider case (b). Assume that

$$a_1 > b_{12}\left(\frac{a_2}{b_{22}}\right), \quad a_2 < b_{21}\left(\frac{a_1}{b_{11}}\right). \tag{H$_2$}$$

Then, population $u_1(t)$ aims to occupy the whole environment, driving species $u_2(t)$ to extinction. Precisely, for each positive solution $(u_1(t), u_2(t))$ of (2.1), we get

$$\lim_{t\to+\infty} u_2(t) = 0 , \quad \lim_{t\to+\infty} u_1(t) = \frac{a_1}{b_{11}} .$$

Observe that $\frac{a_1}{b_{11}}$ is the stationary solution of the logistic equation

$$u' = u(a_1 - b_{11}\, u) .$$

As mentioned in the introduction, we are interested in treating the periodic case, starting from the logistic equation

$$u' = u(a(t) - b(t)\, u) . \tag{2.3}$$

We suppose $a(t)$, $b(t)$ continuous and T-periodic functions, $b(t) > 0$ and $m[a] > 0$, where $m[a]$ denotes the integral average of $a(t)$, that is,

$$m[a] = \frac{1}{T}\int_0^T a(s)\, ds . \tag{2.4}$$

Theorem 2.1. *Under the above assumptions, logistic equation* (2.3) *admits a unique positive, T-periodic solution. Such a solution is globally asymptotically stable with respect to all other positive solutions to* (2.3).

Proof. Since $u(t) \equiv 0$ is a solution of (2.3), each solution $u(t)$ with initial condition $u(0) > 0$ is positive for all $t > 0$. There are different ways to prove this theorem. Our argument yields the explicit form of the periodic solution. Let $u(t)$ be a positive solution of (2.3). Introducing the new unknown function

$$y(t) = \frac{1}{u(t)} ,$$

equation (2.3) turns into the first order linear equation

$$y'(t) = -a(t)\, y(t) + b(t) . \tag{2.5}$$

If $y(0) = k$, then the variation of constants formula gives the following explicit form for $y(t)$, that is,

$$y(t) = e^{-\int_0^t a(s)\,ds}\left(k + \int_0^t e^{\int_0^s a(\tau)\,d\tau} b(s)\,ds\right),$$

and consequently

$$u(t) = \frac{e^{\int_0^t a(s)\,ds}}{k + \int_0^t e^{\int_0^s a(\tau)\,d\tau} b(s)\,ds}\,. \tag{2.6}$$

Requiring to (2.6) to verify the condition $u(T) = u(0) = \frac{1}{k}$, we obtain this unique value for k, namely,

$$k = \frac{\int_0^T e^{\int_0^s a(\tau)\,d\tau} b(s)\,ds}{e^{\int_0^T a(s)\,ds} - 1}\,. \tag{2.7}$$

It is positive as $m[a] > 0$. The unique T-periodic solution $\mathring{u}(t)$ has the explicit form (2.6) with k given by (2.7) ([1]).

Let us investigate the asymptotic behavior of (2.3). First, observe that all positive solutions $u(t)$ are bounded from above since

$$u(t) \le \max\left\{u(0), \max\frac{a(t)}{b(t)}\right\}.$$

If we consider two solutions $y_1(t)$, $y_2(t)$ of (2.5), the function $z(t) = y_1(t) - y_2(t)$ satisfies the linear equation

$$z'(t) = -a(t)\,z(t)$$

whose solutions have the form

$$z(t) = z(0)\,e^{-\int_0^t a(s)\,ds}\,.$$

Set

$$\bar{a}(t) = a(t) - m[a]\,, \quad p(t) = \int_0^t \bar{a}(s)\,ds\,.$$

Is easy to check that $p(t)$ is T-periodic because $\bar{a}(t)$ has zero average. In fact,

$$p(t+T) = \int_0^{t+T} \bar{a}(s)\,ds = \int_0^t \bar{a}(s)\,ds + \int_t^{t+T} \bar{a}(s)\,ds = p(t) + T\,m[\bar{a}] = p(t)\,.$$

Therefore,

$$\int_0^t a(s)\,ds = m[a]t + p(t) \tag{2.8}$$

and consequently

$$z(t) = z(0)P(t)\,e^{-m[a]t}$$

where $P(t) = e^{-p(t)}$. Since $z(t)$ vanishes at infinity, going back to logistic equation (2.3), we can state that

$$\lim_{t\to+\infty} |u(t) - \mathring{u}(t)| = 0 \tag{2.9}$$

for any positive solution $u(t)$ of equation (2.3).

The above simple technique showing the global attractivity of $\mathring{u}(t)$ does not work for Lotka–Volterra systems, and hence we illustrate another argument based on the Lyapunov method. In (2.3), make the substitution

$$x(t) = \frac{u(t)}{\mathring{u}(t)} - 1 .$$

Then,

$$x'(t) = \frac{u(t)}{\mathring{u}(t)}\left[\frac{u'(t)}{u(t)} - \frac{\mathring{u}'(t)}{\mathring{u}(t)}\right] = (1 + x(t))(-b(t)u(t) + b(t)\mathring{u}(t))$$

so that equation (2.3) becomes

$$x'(t) = -r(t)(1 + x(t))\,x(t) \tag{2.10}$$

where $x(t) > -1$ and $r(t) = b(t)\mathring{u}(t)$. The Lyapunov function

$$U(t) = x(t) - \ln(1 + x(t)) = \ln\left(\frac{e^{x(t)}}{1 + x(t)}\right)$$

has a derivative given by

$$U'(t) = \frac{x'(t)}{1 + x(t)}\,x(t) = -r(t)x^2(t) .$$

Using the inequality

$$\frac{e^x - (1 + x)}{e^x} \le x^2 , \quad x > -1 , \tag{2.11}$$

and the equality

$$1 - e^{-U(t)} = 1 - \frac{1 + x(t)}{e^{x(t)}},$$

we deduce that

$$1 - e^{-U(t)} \le x^2(t) .$$

As $r(t)$ is positive, we get that $U(t)$ satisfies the differential inequality

$$U'(t) \le -r(t)(1 - e^{-U(t)})$$

so that

$$\lim_{t\to+\infty} U(t) = 0 ,$$

which implies that $x(t)$ vanishes at infinity. Therefore, (2.9) holds. □

3 Two species periodic models

In the nonautonomous case, model (2.1) becomes

$$\begin{cases} u_1' = u_1(a_1(t) - b_{11}(t)u_1 - b_{12}(t)u_2) \\ u_2' = u_2(a_2(t) - b_{21}(t)u_1 - b_{22}(t)u_2) \,. \end{cases} \tag{3.1}$$

Throughout this section, we assume that coefficients $a_i(t)$ and $b_{ij}(t)$ are continuous and T-periodic functions, $b_{ij}(t) > 0$ and $m[a_i] > 0$, $i, j = 1, 2$. For each assigned initial condition in the first quadrant, there exists a unique solution defined in $[0, +\infty)$. Such solution lies in the first quadrant for all t. In fact, concerning $u_1(t)$, we have

$$\frac{u_1'(t)}{u_1(t)} = (a_1(t) - b_{11}(t)\, u_1(t) - b_{12}(t)\, u_2(t)),$$

and integrating from 0 to t,

$$\ln \frac{u_1(t)}{u_1(0)} = \int_0^t (a_1(s) - b_{11}(s)\, u_1(s) - b_{12}(s)\, u_2(s))\, ds \,.$$

Hence,

$$u_1(t) = u_1(0) e^{\int_0^t (a_1(s) - b_{11}(s)\, u_1(s) - b_{12}(s)\, u_2(s))\, ds} \,.$$

Analogously, $u_2(t) > 0$ for every $t > 0$. We will call such solutions positive.

Afterwards, we investigate the existence of a positive periodic solution to (3.1).

Let us denote, by $\overset{\circ}{U}_i(t)$, the positive periodic solution of the logistic equation

$$u' = u(a_i(t) - b_{ii}(t)\, u) \,, \quad i = 1, 2$$

and, by $r_i(t)$, the following periodic function

$$r_i(t) = a_i(t) - b_{ij}(t) \overset{\circ}{U}_j(t) \,, \quad i \neq j \,. \tag{3.2}$$

Theorem 3.1. *If the inequalities*

$$m[a_i] > m[b_{ij} \overset{\circ}{U}_j] \quad i \neq j \,, \tag{C$_1$}$$

hold, then the differential system (3.1) *has a T-periodic solution* $\overset{\circ}{u}(t)$ *such that*

$$\overset{\circ}{V}_i(t) \le \overset{\circ}{u}_i(t) \le \overset{\circ}{U}_i(t) \,, \quad t > 0 \,, \quad i = 1, 2 \,,$$

where $\overset{\circ}{V}_i(t)$ *is the positive periodic solution of the logistic equation*

$$v' = v(r_i(t) - b_{ii}(t)\, v) \,.$$

Proof. Fix $i \in \{1, 2\}$. Hypothesis (C_1) and (3.2) ensure that

$$m[r_i] > 0$$

so that, by Theorem 2.1, the periodic solution $\overset{\circ}{V}_i(t)$ there exists. The inequality

$$m[r_i] < m[a_i]$$

implies that

$$\overset{\circ}{V}_i(t) < \overset{\circ}{U}_i(t)\,, \quad t \geq 0\,.$$

If $u(t)$ is a solution of (3.1) with initial condition

$$\overset{\circ}{V}_i(0) \leq u_i(0) \leq \overset{\circ}{U}_i(0)\,,$$

then, by comparison results,

$$\overset{\circ}{V}_i(t) \leq u_i(t) \leq \overset{\circ}{U}_i(t)\,, \quad t > 0\,.$$

Set $D = [\overset{\circ}{V}_1(0), \overset{\circ}{U}_1(0)] \times [\overset{\circ}{V}_2(0), \overset{\circ}{U}_2(0)]$ and consider the map

$$F\colon D \longrightarrow D\,, \quad F(x_1, x_2) = (u_1(T), u_2(T))$$

where $u(t) = (u_1(t), u_2(t))$ is the solution of (3.1), satisfying the initial condition $u_1(0) = x_1$, $u_2(0) = x_2$.

By the previous argument,

$$\overset{\circ}{V}_i(0) = \overset{\circ}{V}_i(T) \leq u_i(T) \leq \overset{\circ}{U}_i(T) = \overset{\circ}{U}_i(0)\,, \quad i = 1, 2$$

so that F is well-defined. Since F is continuous, applying the Brouwer fixed-point theorem, there exists $(\bar{x}_1, \bar{x}_2) \in D$ such that $F(\bar{x}_1, \bar{x}_2) = (\bar{x}_1, \bar{x}_2)$. By construction, the solution $\overset{\circ}{u}(t)$ with initial condition $(\bar{x}_1, \bar{x}_2)$ verifies

$$\overset{\circ}{u}(0) = \overset{\circ}{u}(T)\,. \tag{3.3}$$

The differential system (3.1) has periodic coefficients, and hence equality (3.3) guarantees that $\overset{\circ}{u}(t)$ is the searched periodic solution ([9]). □

Remark 3.1. In the autonomous case $\overset{\circ}{U}_i(t) \equiv a_i/b_{ii}$, then condition ($C_1$) agrees with the hypothesis (H_1). On the contrary, in the nonautonomous case, assumption (C_1) is not alone sufficient to ensure the uniqueness and global attractivity of the periodic solution.

Many authors have obtained the attractivity property of $\overset{\circ}{u}(t)$ adding to (C_1) some inequalities, ensuring a sort of diagonal dominance property for matrix $(b_{ij}(t))$. For instance, in [25], the following hypothesis has been introduced:

$$\begin{aligned} &\exists \beta_1, \beta_2 > 0 \ \text{ such that, for all } \ t \in [0, T]\,, \\ &\beta_1\, b_{11}(t) > \beta_2\, b_{21}(t)\,, \quad \beta_2\, b_{22} > \beta_1\, b_{12}(t)\,. \end{aligned} \tag{3.4}$$

Note that (3.4) implies that matrix $(b_{ij}(t))$ has a positive determinant for each $t \in [0, T]$. Instead of pointwise inequalities (3.4), we are going to introduce a suitable average condition.

The following technical lemma yields an inequality in the plane which generalizes (2.11). It will play a central role in the proof of the next Lemma 3.2 and, in general, in the employed method.

Lemma 3.1. *Let*

$$\begin{aligned} A &= \{(x, y) \in \mathbb{R}^2 \mid x > -1\,, y > -1\}\,, \\ p(x, y) &= -a_{11}x^2 - (a_{12} + a_{21})xy - a_{22}y^2\,, \quad a_{ij} > 0 \end{aligned}$$

and

$$g(x, y) = \frac{e^{x+y} - (1 + x)(1 + y)}{e^{x+y}}\,.$$

Then, for all $(x, y) \in A$,

$$p(x, y) \le \alpha g(x, y) \tag{3.5}$$

where

$$\alpha = \max\left\{ \frac{(a_{12} + a_{21})^2}{4a_{11}} - a_{22}\,, \frac{(a_{12} + a_{21})^2}{4a_{22}} - a_{11} \right\}.$$

Proof. Here, we will only give the main steps of the proof. For a more detailed proof, see [14].

Firstly, observe that, using (2.11),

$$-a_{11}x^2 \le -a_{11}\frac{e^x - (1 + x)}{e^x} \le \alpha \frac{e^x - (1 + x)}{e^x}$$

which gives

$$p(x, 0) \le \alpha g(x, 0)\,, \quad x > -1\,.$$

In the same way, we get

$$p(0, y) \le \alpha g(0, y)\,, \quad y > -1\,.$$

Another step of the proof consists of showing the validity of (3.5) at the points of $\bar{A}$ (the closure of A) on the lines $x = -1$ and $y = -1$. Indeed, $p(-1, y)$ attains its maximum value at $I_1(-1, \frac{a_{12}+a_{21}}{2a_{22}})$ and

$$p(I_1) = -a_{11} + \frac{(a_{12} + a_{21})^2}{4a_{22}} \le \alpha\,.$$

As a consequence,

$$p(-1, y) \le p(I_1) \le \alpha g(-1, y)\,, \quad y > -1\,.$$

Analogously, $p(x,-1)$, $x > -1$ attains its maximum value at $I_2(\frac{a_{12}+a_{21}}{2a_{11}}, -1)$, and thus

$$p(x,-1) \le p(I_2) = -a_{22} + \frac{(a_{12}+a_{21})^2}{4a_{11}} \le \alpha = \alpha\, g(x,-1)\,, \quad x > -1\,.$$

Inequality (3.5) is also valid at the points of A belonging to the interior of the first and the third quadrant. In fact, assuming $x\,y > 0$, the inequality below

$$g(x,y) \le x^2 + y^2$$

implies that

$$p(x,y) < -a_{11}x^2 - a_{22}y^2 \le \max\{-a_{11}, -a_{22}\}\, g(x,y) < \alpha\, g(x,y)\,.$$

It remains to prove (3.5) at the points in A such that $x\,y < 0$.

Set $\Delta = (a_{12}+a_{21})^2 - 4a_{11}a_{22} \le 0$ and consider the following cases:

(i) $\Delta = (a_{12}+a_{21})^2 - 4a_{11}a_{22} = 0$.
In this situation, $\alpha = \max\left\{\frac{\Delta}{4a_{11}}, \frac{\Delta}{4a_{22}}\right\} = 0$ so that

$$p(x,y) \le 0 = \alpha\, g(x,y)\,.$$

(ii) $\Delta > 0$.
Note that

$$\frac{(a_{12}+a_{21})}{2a_{11}} > 1 \quad \text{or} \quad \frac{(a_{12}+a_{21})}{2a_{22}} > 1,$$

otherwise $\Delta \le 0$. Assume $a_{11} > a_{22}$ so that $\alpha = \frac{\Delta}{4a_{22}} = p(I_1)$ and $\frac{(a_{12}+a_{21})}{2a_{22}} > 1$. The maximum of $\frac{p(x,y)}{g(x,y)}$ in the set $\{(x,y) \in \bar{A} \mid x\,y < 0\}$ is attained at some point belonging to

$$B = \{(x,y) \in \bar{A} \mid y > -x\,, -1 \le x < 0\}\,.$$

We precisely show below that such a maximum point is $I_1 \in B$. Since

$$p(x,y) = -a_{22}\left(\frac{a_{12}+a_{21}}{2\,a_{22}}x + y\right)^2 + \alpha x^2 \le \alpha x^2$$

and the points in B can be written in the form $(-k, m\,k)$, $0 < k < 1$, $m > 1$, it is enough to prove that, for any fixed $m > 1$,

$$\alpha\, k^2 \le \alpha\, g(-k, m\,k)\,, \quad 0 < k < 1\,.$$

Taking into account the definition of $g(x,y)$, this last inequality is equivalent to

$$(1-k)(1+m\,k) \le (1-k^2)\, e^{(-1+m)k}\,, \quad 0 < k < 1\,, \quad m > 1\,,$$

whose validity can be verified by easy calculations.
When $a_{22} > a_{11}$, case ii) follows by analogous arguments.

(iii) $\Delta < 0$.

If $a_{11} > a_{22}$, we have $\alpha = p(I_2) < 0$. Writing $p(x, y)$ by

$$p(x, y) = -a_{11}\left(x + \frac{a_{12} + a_{21}}{2\,a_{11}}y\right)^2 + \alpha\, y^2\,,$$

we see that

$$p(x, y) \le \alpha\, y^2\,.$$

The points on the line

$$x = -\frac{a_{12} + a_{21}}{2\,a_{11}}y$$

have coordinates $(m\,k, -k)$, $m = \frac{a_{12}+a_{21}}{2\,a_{11}} < 1$. They lie in the fourth quadrant if $0 < k < 1$. To obtain (3.5), it is enough to prove that for any fixed $0 < m < 1$,

$$\alpha\, k^2 \le \alpha\, g(m\,k, -k)\,, \quad 0 < k < 1\,.$$

Taking into account the sign of α, the previous inequality becomes

$$(1 + k)\, e^{(m-1)k} \le (1 + m\,k)\,.$$

It is not difficult to check its validity for any $0 < k < 1$.

The other cases can be handled analogously.

We conclude that

$$p(x, y) \le \max\{p(I_1), p(I_2)\}\, g(x, y)\,, \quad (x, y) \in A\,.$$

This last estimate completes the proof. □

In order to employ (3.5) for $a_{ij}(t)$ T-periodic and positive functions, we need to take α depending on variable t, introducing the notation

$$\alpha(t) = \max\left\{\frac{(a_{12}(t) + a_{21}(t))^2}{4a_{11}(t)} - a_{22}(t)\,,\ \frac{(a_{12}(t) + a_{21}(t))^2}{4a_{22}(t)} - a_{11}(t)\right\}\,. \tag{3.6}$$

Lemma 3.2. *Let $(y_1(t), y_2(t))$ be a solution of the following differential system, namely,*

$$\begin{cases} y_1' = (1 + y_1)(-a_{11}(t)y_1 - a_{12}(t)y_2) \\ y_2' = (1 + y_2)(-a_{21}(t)y_1 - a_{22}(t)y_2) \end{cases} \tag{3.7}$$

with the initial condition $y_i(0) > -1$ and suppose $m[\alpha] < 0$, where $\alpha(t)$ is defined by (3.6). Then,

$$\lim_{t\to+\infty} y_1(t) = 0 = \lim_{t\to+\infty} y_2(t)\,.$$

Proof. It easy to verify that

$$y_i(t) > -1\,, \quad t > 0\,, \quad i = 1, 2\,.$$

For each fixed $t \geq 0$, set

$$V(t) = y_1(t) - \ln(1 + y_1(t)) + y_2(t) - \ln(1 + y_2(t))\,. \tag{3.8}$$

Hence,

$$\begin{aligned} V'(t) &= \frac{y_1'(t)\,y_1(t)}{1 + y_1(t)} + \frac{y_2'(t)\,y_2(t)}{1 + y_2(t)} \\ &= -a_{11}(t)\,y_1^2(t) - (a_{12}(t) + a_{21}(t))y_1(t)y_2(t) - a_{22}(t)\,y_2^2(t)\,. \end{aligned}$$

Using Lemma 3.1, (3.6) and the equality

$$g(y_1(t), y_2(t)) = 1 - e^{-V(t)}\,,$$

we deduce

$$V'(t) \leq \alpha(t)\,(1 - e^{-V(t)})\,. \tag{3.9}$$

Dividing both sides of (3.9) by $1 - e^{-V(t)}$ and integrating from 0 to t, we get

$$\ln \frac{e^{V(t)} - 1}{e^{V(0)} - 1} \leq \int_0^t \alpha(s)\,ds\,.$$

Using (2.8) for $\alpha(t)$, one deduces

$$\frac{e^{V(t)} - 1}{e^{V(0)} - 1} \leq P(t)\,e^{m[\alpha]t}$$

where $P(t)$ is a suitable positive, T-periodic function. The inequality $m[\alpha] < 0$ yields

$$\lim_{t\to+\infty} (e^{V(t)} - 1) = 0,$$

that is,

$$\lim_{t\to+\infty} V(t) = 0\,.$$

Taking into account (3.8), the proof is complete. □

We are ready to formulate the main result of this section.

Theorem 3.2. *Suppose that condition* (C_1) *holds and let* $\mathring{u}(t)$ *a periodic positive solution of* (3.1). *Define*

$$a_{ij}(t) = b_{ij}(t)\,\mathring{u}_j(t)\,, \quad i, j = 1, 2 \tag{3.10}$$

and assume

$$m[\alpha] < 0\,, \tag{C_2}$$

where $\alpha(t)$ *is the periodic function defined by* (3.6), *and* $a_{ij}(t)$ *are given by* (3.10). *Then, for any positive solution* $u(t)$ *of* (3.1), *we have*

$$\lim_{t\to+\infty} |u_1(t) - \mathring{u}_1(t)| = 0 = \lim_{t\to+\infty} |u_2(t) - \mathring{u}_2(t)|\,. \tag{3.11}$$

Proof. Take $(u_1(t), u_2(t))$ positive solution of (3.1) and make the substitution

$$y_i(t) = \frac{u_i(t)}{\overset{\circ}{u}_i(t)} - 1, \quad i = 1, 2.$$

The first equation of the differential system (3.1) turns into

$$\begin{aligned} y_1'(t) &= \frac{u_1{}'(t)}{\overset{\circ}{u}_1(t)} - u_1(t)\frac{\overset{\circ}{u}_1{}'(t)}{(\overset{\circ}{u}_1(t))^2} \\ &= \frac{u_1(t)}{\overset{\circ}{u}_1(t)}\left(-b_{11}(t)(u_1(t) - \overset{\circ}{u}_1(t)) - b_{12}(t)(u_2(t) - \overset{\circ}{u}_2(t))\right) \\ &= (1 + y_1(t))(-a_{11}(t)y_1(t) - a_{12}(t)y_2(t)). \end{aligned}$$

Analogously,

$$y_2'(t) = (1 + y_2(t))(-a_{21}(t)y_1(t) - a_{22}(t)y_2(t)),$$

that is, $(y_1(t), y_2(t))$ satisfies the system (3.7). By (C_2) and Lemma 3.2, we get

$$\lim_{t\to+\infty} y_i(t) = 0, \quad i = 1, 2,$$

so that (3.11) is proved. □

Remark 3.2. Hypothesis (C_2) can be replaced by the stronger pointwise inequality

$$(a_{12}(t) + a_{21}(t))^2 - 4a_{11}(t)a_{22}(t) < 0, \quad t \in [0, T]$$

which ensures that $\alpha(t)$ is negative for every $t \in [0, T]$.

Remark 3.3. Going back to the autonomous case, easy calculations yield

$$-\left(a_1 - b_{12}\frac{a_2}{b_{22}}\right) \le \frac{(a_{12} + a_{21})^2}{4a_{11}} - a_{22}, \quad -\left(a_2 - b_{21}\frac{a_1}{b_{11}}\right) \le \frac{(a_{12} + a_{21})^2}{4a_{22}} - a_{11},$$

where $a_{ij} = b_{ij}\overset{\circ}{u}_j$ and $\overset{\circ}{u}_1$, $\overset{\circ}{u}_2$ are given by (2.2). We deduce that the assumption $\alpha < 0$ implies the hypothesis (H_1).

Example 3.1. Consider the following differential system with 2π-periodic coefficients

$$\begin{cases} u_1' = u_1[(6.1 + 3\cos t) - 3u_1 - (3.1 + 3\cos t)u_2] \\ u_2' = u_2[(5.1 + \cos t) - (1.1 + \cos t)u_1 - 4u_2]. \end{cases} \tag{3.12}$$

We have

$$m[a_1] = 6.1, \quad m[a_2] = 5.1$$

and

$$b_{12}(t) = 3.1 + 3\cos t, \quad b_{21}(t) = 1.1 + \cos t.$$

Following (2.6) and (2.7), the positive 2π-periodic solution of the logistic equation

$$U' = U[(5.1 + \cos t) - 4U]$$

has the form

$$\overset{\circ}{U}_2(t) = \frac{e^{5.1t+\sin t}}{k + G(t)}$$

where

$$G(t) = 4\int_0^t e^{5.1x+\sin x}\, dx$$

and the constant k is given by

$$k = \frac{G(2\pi)}{e^{10.2\pi} - 1}\,.$$

By numerical calculations, we obtain

$$m[b_{12}\overset{\circ}{U}_2] < 4.4 < m[a_1]\,.$$

Moreover, denote by $\overset{\circ}{U}_1$ the positive 2π-periodic solution of the logistic equation

$$U' = U(a_1(t) - b_{11}(t)\,U),$$

that is,

$$U' = U[(6.1 + 3\cos t) - 3U]\,.$$

Observe that

$$\int_0^{2\pi} \frac{\overset{\circ}{U}_1{}'(s)}{\overset{\circ}{U}_1(s)}\, ds = \ln \overset{\circ}{U}_1(2\pi) - \ln \overset{\circ}{U}_1(0) = 0$$

so that

$$m[a_1] = m[b_{11}\overset{\circ}{U}_1]\,.$$

We get

$$m[b_{21}\overset{\circ}{U}_1] \le \sup\left(\frac{b_{21}}{b_{11}}\right) m[b_{11}\overset{\circ}{U}_1] = \sup\left(\frac{b_{21}}{b_{11}}\right) m[a_1] = \frac{2.1}{3}6.1 < m[a_2]$$

so that condition (C_1) holds.

Since $\overset{\circ}{u}(t) = (1, 1)$ is a *periodic* solution of (3.12), we can take this solution to verify condition (C_2). We have

$$\frac{(a_{12}(t) + a_{21}(t))^2}{4a_{11}(t)} - a_{22}(t) = \frac{(4.2 + 4\cos t)^2}{12} - 4$$

and

$$\frac{(a_{12}(t) + a_{21}(t))^2}{4a_{22}(t)} - a_{11}(t) = \frac{(4.2 + 4\cos t)^2}{16} - 3\,.$$

A numerical estimate of the average of the function

$$\alpha(t) = \max\left\{\frac{(4.2+4\cos t)^2}{12} - 4\,,\ \frac{(4.2+4\cos t)^2}{16} - 3\right\}$$

yields

$$m[\,\alpha\,] < -1.3\,.$$

Using Theorem 3.2 we can conclude that $\overset{\circ}{u}(t)$ is globally asymptotically stable with respect to the other positive solutions of (3.12).

Few nice results are known for the N-dimensional case

$$u_i'(t) = u_i(t)\Big[a_i(t) - \sum_{j=1}^{N} b_{ij}(t)\,u_j(t)\Big]\,, \quad i = 1,\ldots,N\,. \tag{3.13}$$

We state the one due to Zhang, Jiang and Lazer [27].

Theorem 3.3. *Suppose that periodic system* (3.13) *satisfies the following conditions:*

(i)

$$\int_0^T \Big[a_i(t) - \sum_{j=1, j\neq i}^{N} b_{ij}(t)\,\overset{\circ}{U}_j(t)\Big]dt > 0\,, \quad i = 1,\ldots,N$$

and

(ii) *there exist positive constants* γ_j, $j = 1,\ldots,N$ *and* δ *such that*

$$\gamma_j\,b_{jj}(t) > \sum_{i=1, i\neq j}^{N} \gamma_i\,b_{ij}(t) + \delta\,, \quad j = 1,\ldots,N\,.$$

Then, system (3.13) *has a unique T-periodic and positive solution which is globally attractive.*

4 Competitive exclusion

Our purpose is to study the phenomenon of *competitive exclusion* for Lotka–Volterra systems (3.1). In Section 2, we saw that the autonomous system (2.1) obeys this principle under hypothesis (H_2). It is legitimate to wonder if the following average inequalities

$$m[a_1] > m[b_{12}\overset{\circ}{U}_2]\,, \quad m[a_2] < m[b_{21}\overset{\circ}{U}_1] \tag{4.1}$$

are sufficient to guarantee the validity of such a principle in the periodic case. In this direction, a first local result can be obtained.

Theorem 4.1. *Suppose that inequalities* (4.1) *hold. Then, for the competitive system* (3.1), *solution* $(\overset{\circ}{U}_1(t), 0)$ *is locally stable, whereas* $(0, \overset{\circ}{U}_2(t))$ *is unstable* ([11],[15]).

On the other hand, inequalities (4.1) are not sufficient for the validity of the competitive exclusion as it is shown by de Mottoni e Schiaffino by the following example ([10]).

Example 4.1. Take $0 < \epsilon < 1$, and define

$$a_1(t) = b_{11}(t) = 1 + \epsilon + \epsilon \sin t\,, \qquad b_{12}(t) = \frac{a_1(t)}{1 + \epsilon \sin t}\,,$$

$$a_2(t) = b_{22}(t) = a_1(t) + \frac{\epsilon \cos t}{1 + \epsilon \sin t}\,, \qquad b_{21}(t) = 2a_1(t) - a_2(t)(1 + \epsilon \sin t)\,.$$

It is easily seen that $\mathring{U}_1(t) \equiv 1, \mathring{U}_2(t) \equiv 1$,

$$m[a_1 - b_{12}\mathring{U}_2] = m[a_1 - b_{12}] = m\left[a_1 - \frac{a_1}{1 + \epsilon \sin t}\right] = m\left[a_1 \frac{\epsilon \sin t}{1 + \epsilon \sin t}\right]$$
$$= m\left[\epsilon \sin t + \frac{\epsilon^2 \sin t}{1 + \epsilon \sin t}\right] = \epsilon^2 m\left[\frac{\sin t}{1 + \epsilon \sin t}\right] < 0\,.$$

Moreover, we see that

$$m[a_2] = m[a_1] + \frac{1}{2\pi}[\ln(1 + \epsilon \sin t)]_0^{2\pi} = m[a_1]\,,$$

$$m[b_{21}\mathring{U}_1] = m[b_{21}] = m[2a_1 - a_1(1 + \epsilon \sin t) - \epsilon \cos t] = m[a_1 - a_1\epsilon \sin t]$$
$$= m[a_1] - m[(1 + \epsilon)\epsilon \sin t + \epsilon^2 \sin^2 t] < m[a_1] = m[a_2]\,.$$

Therefore, $m[a_1] < m[b_{12}\mathring{U}_2]$, $m[a_2] > m[b_{21}\mathring{U}_1]$ and yet there are positive solutions $(u_1(t), u_2(t))$ with no component vanishing at infinity. Indeed, with the above indicated choice of the coefficients, system (3.1) possesses the solution

$$\left(\frac{1}{2}, \frac{1 + \epsilon \sin t}{2}\right)$$

which is positive and 2π-periodic.

Next, the lemma shows that inequalities (4.1) ensure that the rectangle

$$Q_t = [0, \mathring{U}_1(t)] \times [0, \mathring{U}_2(t)]$$

attracts all solutions of (3.1) with positive initial conditions.

Lemma 4.1. *Let $(u_1(t), u_2(t))$ be a positive solution of* (3.1) *whose coefficients verify* (4.1). *Then, one of the following two cases occurs:*

(i) $\lim_{t\to+\infty}(u_1(t) - \mathring{U}_1(t)) = 0 = \lim_{t\to+\infty} u_2(t)\,;$

(ii) *there exists $\bar{t} > 0$ such that, for each $t > \bar{t}$,*

$$(u_1(t), u_2(t)) \in \,]0, \mathring{U}_1(t)[\, \times \,]0, \mathring{U}_2(t)[\, = \mathring{Q}_t\,.$$

Proof. See [15]. □

The theorem below provides the main result of this section. For its proof, we need some new notations.

Fix $\alpha, \beta > 0$ and $t \geq 0$. Let

$$f_t(u,v) = (-\alpha\, a_1(t) + \beta\, a_2(t)) + (\alpha\, b_{11}(t) - \beta\, b_{21}(t))\, u + (\alpha\, b_{12}(t) - \beta\, b_{22}(t))\, v \tag{4.2}$$

and

$$A(t) = \max_{(u,v)\in Q_t} f_t(u,v)\,. \tag{4.3}$$

Since $f_t(u,v)$ is a linear function, its maximum value in Q_t will be achieved in one of its vertices, that is,

$$A(t) = \max\left\{f_t(0,0),\ f_t(\mathring{U}_1(t),0),\ f_t(0,\mathring{U}_2(t)),\ f_t(\mathring{U}_1(t),\mathring{U}_2(t))\right\}.$$

Theorem 4.2. *Assume that inequalities* (4.1) *hold and take* α, β *real numbers, verifying*

$$\frac{m[a_2]}{m[a_1]} < \frac{\alpha}{\beta} < \min\left\{\frac{m[a_2]}{m[b_{12}\mathring{U}_2]},\ \frac{m[b_{21}\mathring{U}_1]}{m[a_1]}\right\}. \tag{4.4}$$

Suppose that

$$m[A] < 0 \tag{4.5}$$

for this choice of α, β. *Then,*

$$\lim_{t\to\infty} u_2(t) = 0\,, \quad \lim_{t\to\infty} (u_1(t) - \mathring{U}_1(t)) = 0$$

for any $(u_1(t), u_2(t))$ *positive solution of* (3.1).

Proof. First, observe that it is possible to choose α, β as in (4.4) since (4.1) holds.

Let $(u_1(t), u_2(t))$ be a positive solution of (3.1). By Lemma 4.1, it is not restrictive to suppose that there exists $\bar{t} > 0$ such that

$$(u_1(t), u_2(t)) \in \mathring{Q}_t\,, \quad t > \bar{t}\,.$$

For $t > \bar{t}$, consider the function

$$Z(t) = (u_1(t))^{-\alpha}(u_2(t))^{\beta}\,.$$

Then,

$$\begin{aligned} Z'(t) &= -\alpha\,(u_1(t))^{-\alpha-1}u_1'(t)\,(u_2(t))^{\beta} + \beta\,(u_1(t))^{-\alpha}(u_2(t))^{\beta-1}u_2'(t) \\ &= Z(t)\left[-\alpha\frac{u_1'(t)}{u_1(t)} + \beta\frac{u_2'(t)}{u_2(t)}\right]. \end{aligned}$$

By using (4.2) and (4.3), we have

$$Z'(t) = Z(t)\, f_t(u_1(t), u_2(t)) \le Z(t)\, A(t)\,.$$

Our choice of α, β permits to obtain

$$\begin{aligned} m[f_t(0,0)] &= m[-\alpha a_1 + \beta a_2] < 0\,, \\ m[f_t(\mathring{U}_1, 0)] &= m[-\alpha a_1 + \beta a_2] + \alpha m[a_1] - \beta m[b_{21}\mathring{U}_1] < 0\,, \\ m[f_t(0, \mathring{U}_2)] &= m[-\alpha a_1 + \beta a_2] + \alpha m[b_{12}\mathring{U}_2] - \beta m[a_2] < 0\,, \\ m[f_t(\mathring{U}_1, \mathring{U}_2)] &= m[-\alpha a_1 + \beta a_2] + (\alpha m[a_1] - \beta m[b_{21}\mathring{U}_1]) \\ &\qquad + (\alpha m[b_{12}\mathring{U} - 2] - \beta m[a_2]) < 0, \end{aligned}$$

that is, in each vertex of rectangle Q_t, $f_t(u, v)$ has negative mean value.

When the coefficients are positive numbers, the above property ensures $A(t)$ is a negative constant. Therefore, in this simpler case, hypothesis (H_2) is sufficient to guarantee the extinction of species $u_2(t)$, as stated in Section 2.

For periodic coefficients, adding to (4.4) assumption (4.5), we get

$$\lim_{t\to\infty} Z(t) = 0\,.$$

The equality $(u_2(t))^\beta = Z(t)\,(u_1(t))^\alpha$ yields that

$$\lim_{t\to+\infty} u_2(t) = 0\,.$$

Then, $u_1(t)$ is a positive solution of the logistic equation

$$u' = u\,(c(t) - b_{11}(t)\,u)$$

where $c(t) = (a_1(t) - b_{12}(t)\,u_2(t))$ and

$$\lim_{t\to+\infty} |c(t) - a_1(t)| = 0\,.$$

As a consequence ([20]), $u_1(t)$ has the same asymptotic behavior of the solutions to the logistic equation

$$u' = u\,(a_1(t) - b_{11}(t)\,u)$$

so that, by Theorem 2.1,

$$\lim_{t\to+\infty} |u_1(t) - \mathring{U}_1(t)| = 0\,.$$

The proof is complete. □

Remark 4.1. If system (3.1) satisfies (4.1) and admits a positive T-periodic solution $(v_1(t), v_2(t))$, by Lemma 4.1, such a solution has to lie in the interior of Q_t for each t.

Moreover,

$$m[a_1 - b_{11} v_1 - b_{12} v_2] = m\left[\frac{v_1'}{v_1}\right] = 0\,,$$

$$m[a_2 - b_{21} v_1 - b_{22} v_2] = m\left[\frac{v_2'}{v_2}\right] = 0\,,$$

so that

$$m[f_t(v_1(t), v_2(t))] = 0.$$

Therefore, condition (4.3) cannot be verified. This situation happens for the differential system considered in Example 4.1.

Theorem 4.2 can be applied to the next example.

Example 4.2. Consider the model

$$\begin{cases} u_1' = u_1((4 + 2\sin t) - (1 + 0.5\sin t)\, u_1 - 2\, u_2) \\ u_2' = u_2((1 + 0.7\cos t) - u_1 - (1 + 0.7\cos t)\, u_2)\,. \end{cases}$$

We have $\mathring{U}_1 = 4, \mathring{U}_2 = 1$,

$$m[a_1] = 4 > 2 = m[b_{12}\mathring{U}_2]$$

$$m[a_2] = 1 < 4 = m[b_{21}\mathring{U}_1]$$

so condition (4.1) holds. Since

$$\frac{m[a_2]}{m[a_1]} = \frac{1}{4}\,, \quad \frac{m[a_2]}{m[b_{12}\mathring{U}_2]} = \frac{1}{2}\,, \quad \frac{m[b_{21}\mathring{U}_1]}{m[a_1]} = 1$$

in the definition of $f_t(u, v)$, we can take $\alpha = 4$, $\beta = 10$. This choice gives

$$f_t(0,0) = -6 - 8\sin t + 7\cos t \qquad f_t(4,0) = -30 + 7\cos t$$

$$f_t(0,1) = -8 - 8\sin t\,, \qquad f_t(4,1) = -32\,,$$

and consequently

$$A(t) = \max\{-6 - 8\sin t + 7\cos t,\ -8 - 8\sin t\}\,.$$

A numerical computation leads to

$$m[A(t)] < -4\,,$$

and therefore inequality (4.5) is verified as well. We can state that

$$\lim_{t\to\infty} u_1(t) = 4\,, \quad \lim_{t\to\infty} u_2(t) = 0\,.$$

The argument used in Theorem 4.2 also works for proving the following theorem. It requires assumptions which are stronger than (4.1) and (4.5), though more feasible.

Theorem 4.3. *Assume that the inequalities*

$$-b_{22}(t)m[a_1] + b_{12}(t)m[a_2] < 0 \tag{4.6}$$

$$-b_{21}(t)m[a_1] + b_{11}(t)m[a_2] < 0 \tag{4.7}$$

hold for all $t > 0$. If $(u_1(t), u_2(t))$ is any positive solution of (3.1), then

$$\lim_{t\to\infty}(u_1(t) - \overset{\circ}{U}_1(t)) = 0\,, \quad \textit{and} \quad \lim_{t\to\infty} u_2(t) = 0\,.$$

Proof. Let $(u_1(t), u_2(t))$ be a positive solution of (3.1).

First, note that inequalities (4.6) and (4.7) imply (4.1). Indeed, (4.6) implies that

$$m[a_1] > \max\left(\frac{b_{12}(t)}{b_{22}(t)}\right) m[a_2]$$

and

$$m[b_{12}\overset{\circ}{U}_2] = m\left[\frac{b_{12}}{b_{22}} b_{22}\overset{\circ}{U}_2\right] \le \max\left(\frac{b_{12}(t)}{b_{22}(t)}\right) m[b_{22}\overset{\circ}{U}_2] = \max\left(\frac{b_{12}(t)}{b_{22}(t)}\right) m[a_2],$$

so that

$$m[a_1] > m[b_{12}\overset{\circ}{U}_2]\,.$$

Analogously, inequality (4.7) yields

$$m[a_2] < m[b_{21}\overset{\circ}{U}_1]\,.$$

Furthermore, we can choose two positive constants α, β, satisfying the condition

$$\frac{m[a_2]}{m[a_1]} < \frac{\alpha}{\beta} < \min\left\{\frac{b_{21}}{b_{11}}(t), \frac{b_{22}}{b_{12}}(t)\right\}, \quad t > 0\,. \tag{4.8}$$

The Lyapunov function

$$Z(t) = (u_1(t))^{-\alpha}(u_2(t))^{\beta}\,, \quad t > 0$$

satisfies the differential inequality

$$Z'(t) < Z(t)(-\alpha a_1(t) + \beta a_2(t))$$

where $m[-\alpha a_1 + \beta a_2] < 0$, owing to (4.8). Arguing as in Theorem 4.2, we conclude the proof. □

We point out that the model proposed in Example 4.2 does not satisfy condition (4.6). In fact, the function

$$-b_{22}(t)m[a_1] + b_{12}(t)m[a_2] = -2 - 2.8\cos t$$

has positive value for some value of t.

In [6], the authors investigate the phenomenon of extinction of only one species in a Lotka–Volterra system (3.13) involving N-species, where all coefficients are continuous and T-periodic, $b_{ij}(t) > 0$ and $m[a_i] > 0$, $i, j = 1, \dots, N$.

They show that if the coefficients satisfy certain inequalities, generalizing (4.6) and (4.7), then any solution with positive initial conditions has the property that all but one of its components vanish at infinity while the remaining component approaches the positive periodic solution of a logistic equation. They precisely prove the following result.

Theorem 4.4. *Assume that, for each $k > 1$, there exists $i_k < k$ such that for any $j \le k$, the inequality*

$$m[a_k]\, b_{i_k j}(t) - m[a_{i_k}]\, b_{kj}(t) < 0$$

holds. If $u(t) = (u_1(t), \dots, u_N(t))$ is any positive solution of (3.13)*, then, for $i = 2, \dots, N$ $u_i(t) \to 0$ as $t \to +\infty$ and $u_1(t) - \overset{\circ}{U}_1(t) \to 0$ as $t \to +\infty$.*

5 One species extinction in three-dimensional models

The results showed in Sections 2–3 constitute an essential tool for a further contribution to a periodic Lotka–Volterra system involving three competing species occupying a common environment. Precisely, we consider the differential system

$$\begin{cases} u_1' = u_1(a_1(t) - b_{11}(t)u_1 - b_{12}(t)u_2 - b_{13}(t)u_3) \\ u_2' = u_2(a_2(t) - b_{21}(t)u_1 - b_{22}(t)u_2 - b_{23}(t)u_3) \\ u_3' = u_3(a_3(t) - b_{31}(t)u_1 - b_{32}(t)u_2 - b_{33}(t)u_3)\,, \end{cases} \tag{5.1}$$

where $m[a_i] > 0$, $b_{ij}(t) > 0$, $i, j = 1, 2, 3$.

We introduce average conditions which force species $u_3(t)$ to succumb from a biological point of view. Such conditions lead to survival of species $u_1(t)$, $u_2(t)$ and extinction of the third species.

Obviously, in a three-dimensional competitive system, many other situations may occur, even for autonomous systems ([7, 12]).

Let us start recalling some properties of (5.1)([2],[3]).

(P_1) If $u(t)$ is any solution of (5.1) with initial value $u_i(0) > 0$, $i = 1, 2, 3$, then it is positive for all $t > 0$;

(P_2) If $u(t)$ is a positive solution of (5.1), then $u(t)$ is defined in $[0, +\infty)$ and

$$u_i(t) \le \max\left\{u_i(0), \frac{\max a_i(t)}{\min b_{ii}(t)}\right\} \quad \text{for } t > 0,\ i = 1, 2, 3\,.$$

As in the previous sections, we denote, by $\overset{\circ}{U}_i(t)$, the positive periodic solution of the logistic equation

$$u' = u\,(a_i(t) - b_{ii}(t)\,u)\,, \quad i = 1,2,3\,.$$

The following theorem gives suitable average conditions under which the three-dimensional rectangle

$$R_t = [0, \overset{\circ}{U}_1(t)] \times [0, \overset{\circ}{U}_2(t)] \times [0, \overset{\circ}{U}_3(t)] \tag{5.2}$$

attracts all positive solutions inside.

Theorem 5.1. *Assume that the coefficients of* (5.1) *satisfy*

$$m[a_1] > m[b_{12}\overset{\circ}{U}_2]\,, \quad m[a_2] > m[b_{21}\overset{\circ}{U}_1] \tag{5.3}$$

and either $m[a_1] > m[b_{13}\overset{\circ}{U}_3]$ *or* $m[a_2] > m[b_{23}\overset{\circ}{U}_3]$ *holds. If* $u(t)$ *is any positive solution of* (5.1), *then, for sufficiently large* t,

$$u(t) \in \overset{\circ}{R}_t\,.$$

Proof. Let $u(t) = (u_1(t), u_2(t), u_3(t))$ be a positive solution of (5.1). We want to prove that there exists $\bar{t} > 0$ such that, for $t > \bar{t}$,

$$0 < u_i(t) < \overset{\circ}{U}_i(t)\,, \quad i = 1,2,3\,. \tag{5.4}$$

Fix $i \in \{1,2,3\}$, for instance, $i = 2$. Taking into account the asymptotic behavior of the positive solutions to the logistic equation

$$u' = u\,(a_2(t) - b_{22}(t)\,u)\,,$$

and applying comparison properties for differential equations, we deduce that the following two cases may occur, namely,

(i) $\lim_{t\to+\infty} |u_2(t) - \overset{\circ}{U}_2(t)| = 0$;

(ii) there exists $\bar{t} > 0$ such that

$$u_2(t) < \overset{\circ}{U}_2(t)\,, \quad t > \bar{t}\,.$$

Let us show that under our hypotheses, case (i) cannot occur. By contradiction, suppose that (i) holds. Since

$$-b_{21}(t)u_1(t) - b_{23}(t)u_3(t) = \left(\frac{u_2'(t)}{u_2(t)} - \frac{\overset{\circ}{U}_2{}'(t)}{\overset{\circ}{U}_2(t)}\right) + b_{22}(t)(u_2(t) - \overset{\circ}{U}_2(t))$$

and the right hand-side vanishes at infinity, it follows

$$\lim_{t\to\infty} (u_1(t), (u_2 - \overset{\circ}{U}_2)(t), u_3(t)) = (0,0,0)\,.$$

By (5.3), $m[a_1] > m[b_{12}\overset{\circ}{U}_2]$, so that it is possible to find an $\epsilon > 0$, sufficiently small, such that

$$m[a_1] > m[b_{12}\overset{\circ}{U}_2] + \epsilon\, m[b_{11} + b_{12} + b_{13}]\,. \tag{5.5}$$

and there exists $t_0 > 0$ such that

$$(u_1(t), u_2(t), u_3(t)) \in [0,\epsilon] \times [\overset{\circ}{U}_2(t) - \epsilon, \overset{\circ}{U}_2(t) + \epsilon] \times [0,\epsilon] \quad t > t_0\,.$$

Then, for $t > t_0$,

$$u_1'(t) > u_1(t)(a_1 - b_{11}\,\epsilon - b_{12}(\overset{\circ}{U}_2 + \epsilon) - b_{13}\,\epsilon)\,.$$

By using (5.5),

$$\lim_{t\to\infty} u_1(t) = +\infty,$$

which contradicts property (P_2) above.

Similar arguments yield estimate (5.4) for the other components of $u(t)$. □

In the next step, we will find average conditions ensuring the extinction of species $u_3(t)$.

For fixed $t \geq 0$ and $\alpha, \beta, \gamma > 0$, set

$$\begin{aligned} f_t(u,v,w) = {} & (-\alpha\, a_1(t) - \beta\, a_2(t) + 2\gamma\, a_3(t)) \\ & + (\alpha\, b_{11}(t) + \beta\, b_{21}(t) - 2\gamma\, b_{31}(t))\, u \\ & + (\alpha\, b_{12}(t) + \beta\, b_{22}(t) - 2\gamma\, b_{32}(t))\, v \\ & + (\alpha\, b_{13}(t) + \beta\, b_{23}(t) - 2\gamma\, b_{33}(t))\, w \end{aligned}$$

and

$$F(t) = \max_{(u,v,w)\in R_t} f_t(u,v,w) \tag{5.6}$$

where R_t is defined by (5.2). Since $f_t(u,v,w)$ is linear with respect to u, v, w, this maximum value will be achieved in one of the eight vertices of R_t. Therefore,

$$\begin{aligned} F(t) = \max\{ & f_t(0,0,0), f_t(\overset{\circ}{U}_1(t),0,0), f_t(0,\overset{\circ}{U}_2(t),0), f_t(0,0,\overset{\circ}{U}_3(t)), \\ & f_t(\overset{\circ}{U}_1(t),\overset{\circ}{U}_2(t),0), f_t(\overset{\circ}{U}_1(t),0,\overset{\circ}{U}_3(t)), f_t(0,\overset{\circ}{U}_2(t),\overset{\circ}{U}_3(t)), \\ & f_t(\overset{\circ}{U}_1(t),\overset{\circ}{U}_2(t),\overset{\circ}{U}_3(t))\}\,. \end{aligned}$$

Theorem 5.2. *Assume that*

$$\begin{cases} m[a_1] > m[b_{12}\overset{\circ}{U}_2]\,, & m[a_1] > m[b_{13}\overset{\circ}{U}_3] \\ m[a_2] > m[b_{21}\overset{\circ}{U}_1]\,, & m[a_2] > m[b_{23}\overset{\circ}{U}_3] \\ m[a_3] < m[b_{31}\overset{\circ}{U}_1]\,, & m[a_3] < m[b_{32}\overset{\circ}{U}_2] \end{cases} \tag{5.7}$$

and take

$$\frac{m[a_3]}{m[a_1]} < \frac{\alpha}{\gamma} < \min\left\{\frac{m[b_{31}\mathring{U}_1]}{m[a_1]}, \frac{m[a_3]}{m[b_{12}\mathring{U}_2]}, \frac{m[a_3]}{m[b_{13}\mathring{U}_3]}\right\}, \tag{5.8}$$

$$\frac{m[a_3]}{m[a_2]} < \frac{\beta}{\gamma} < \min\left\{\frac{m[b_{32}\mathring{U}_2]}{m[a_2]}, \frac{m[a_3]}{m[b_{21}\mathring{U}_1]}, \frac{m[a_3]}{m[b_{23}\mathring{U}_3]}\right\}. \tag{5.9}$$

If, for such value of α, β, γ, one gets

$$m[F] < 0\,, \tag{5.10}$$

then, for any positive solution $u(t)$ of (5.1), *we have*

$$\lim_{t\to+\infty} u_3(t) = 0\,.$$

Proof. Let $(u_1(t), u_2(t), u_3(t))$ be a positive solution of (5.1). By Theorem 5.1, there exists $\bar{t} > 0$ such that

$$(u_1(t), u_2(t), u_3(t)) \in \mathring{R}_t\,, \quad t > \bar{t}\,.$$

Inequalities (5.7) permit to find three real numbers α, β, γ as in (5.8) and (5.9). For $t > \bar{t}$, define

$$V(t) = (u_1(t))^{-\alpha}\,(u_2(t))^{-\beta}\,(u_3(t))^{2\gamma}\,.$$

Easy calculations yield

$$V'(t) = f_t(u_1(t), u_2(t), u_3(t))\,V(t)\,.$$

Using (5.6), we get

$$V'(t) \le F(t)\,V(t)\,.$$

It is not difficult to verify that our choice of constants α, β, γ makes the average of $f_t(u, v, w)$ negative at each vertex of R_t. The stronger hypothesis (5.10) implies that

$$\lim_{t\to+\infty} V(t) = 0\,.$$

Hence, the extinction of population $u_3(t)$ easily follows. □

Our method can be applied in a more general way. If we are able to find three appropriate constants α, β, γ in such a way that the corresponding periodic function $F(t)$, defined by (5.6), has negative mean value, then the conclusion of Theorem 5.2 is still valid.

An application of the above comment is provided by the following system

$$\begin{cases} u_1' = u_1\,[(4+\sin t) - 3u_1 - u_2 - u_3\,] \\ u_2' = u_2\,[(4-\cos t) - u_1 - 3u_2 - u_3\,] \\ u_3' = u_3\,[(2+2\sin^2 t) - 2u_1 - 2u_2 - 2u_3\,]\,. \end{cases}$$

Note that

$$m[a_1] = 4 > 1 \cdot \frac{4}{3} = b_{12}\frac{m[a_2]}{b_{22}} = m[b_{12}\overset{\circ}{U}_2]\,,$$
$$m[a_1] = 4 > 1 \cdot \frac{3}{2} = b_{13}\frac{m[a_3]}{b_{33}} = m[b_{13}\overset{\circ}{U}_3]$$

and analogously $m[a_2] > m[b_{21}\overset{\circ}{U}_1]$ and $m[a_2] > m[b_{23}\overset{\circ}{U}_3]$, but

$$m[a_3] = 3 > 2 \cdot \frac{4}{3} = b_{31}\frac{m[a_1]}{b_{11}}.$$

Therefore, (5.7) is not completely satisfied. However, choosing $\alpha = \beta = \gamma = 1$ and proceeding as in the proof of Theorem 5.2, we get

$$f_t(u, v, w) = -4 - \sin t + \cos t + 4\sin^2 t - 2w$$

so that

$$V'(t) < (-4 - \sin t + \cos t + 4\sin^2 t)\,V(t)\,.$$

Since

$$m[-4 - \sin t + \cos t + 4\sin^2 t] = -2 < 0\,,$$

$V(t)$ vanishes as t goes to infinity. We can conclude that for any positive solution $u(t)$ of the above differential system,

$$\lim_{t\to+\infty} u_3(t) = 0\,.$$

At this point, it is natural to investigate the asymptotic behavior of the remaining species $u_1(t)$ and $u_2(t)$ when extinction of species $u_3(t)$ occurs. This will be our next objective. Let us begin by assuming b_{ij} to be real numbers.

Theorem 5.3. *Suppose that in system (5.1), b_{ij} are positive constants and the following inequalities*

$$\begin{cases} m[a_1] > \frac{b_{12}}{b_{22}} m[a_2]\,, & m[a_2] > \frac{b_{21}}{b_{11}} m[a_1] \\ m[a_1] > \frac{b_{13}}{b_{33}} m[a_3]\,, & m[a_2] > \frac{b_{23}}{b_{33}} m[a_3] \\ m[a_3] < \frac{b_{31}}{b_{11}} m[a_1]\,, & m[a_3] < \frac{b_{32}}{b_{22}} m[a_2] \end{cases} \tag{5.11}$$

hold. Then, for any positive solution $(u_1(t), u_2(t), u_3(t))$ of (5.1), one gets

$$\lim_{t\to+\infty} u_3(t) = 0\,, \quad \lim_{t\to+\infty} |u_1(t) - \overset{\circ}{u}_1(t)| = 0 = \lim_{t\to+\infty} |u_2(t) - \overset{\circ}{u}_2(t)|$$

where $(\overset{\circ}{u}_1(t), \overset{\circ}{u}_2(t))$ is the unique positive periodic solution of the two-dimensional system

$$\begin{cases} u_1' = u_1(a_1(t) - b_{11}\,u_1 - b_{12}\,u_2) \\ u_2' = u_2(a_2(t) - b_{21}\,u_1 - b_{22}\,u_2)\,. \end{cases} \tag{5.12}$$

Proof. Note that in our case

$$m[\mathring{U}_i] = \frac{m[a_i]}{b_{ii}} \quad i = 1, 2, 3\,.$$

Inequalities (5.11) imply

$$\frac{m[a_3]}{m[b_{12}\mathring{U}_2]} < \frac{b_{32}/b_{22}\, m[a_2]}{b_{12}/b_{22}\, m[a_2]} = \frac{b_{32}}{b_{12}}$$

$$\frac{m[a_3]}{m[b_{21}\mathring{U}_1]} < \frac{b_{31}/b_{11}\, m[a_1]}{b_{21}/b_{11}\, m[a_1]} = \frac{b_{31}}{b_{21}}$$

so that it is allowed to take three constants α, β, γ as follows:

$$\frac{m[a_3]}{m[a_1]} < \frac{\alpha}{\gamma} < \min\left\{\frac{b_{31}}{b_{11}}, \frac{b_{32}}{b_{12}}, \frac{b_{33}}{b_{13}}\right\}$$

$$\frac{m[a_3]}{m[a_2]} < \frac{\beta}{\gamma} < \min\left\{\frac{b_{31}}{b_{21}}, \frac{b_{32}}{b_{22}}, \frac{b_{33}}{b_{23}}\right\}\,.$$

This choice of α, β, γ yields the inequalities

$$\begin{aligned} \alpha\, b_{11} + \beta\, b_{21} - 2\gamma\, b_{31} &< 0\,, \\ \alpha\, b_{12} + \beta\, b_{22} - 2\gamma\, b_{32} &< 0\,, \\ \alpha\, b_{13} + \beta\, b_{23} - 2\gamma\, b_{33} &< 0\,, \end{aligned}$$

and therefore, in rectangle R_t, the function $f_t(u, v, w)$ is less than $(-\alpha\, a_1(t) - \beta\, a_2(t) + 2\gamma\, a_3(t))$. Arguing as in the proof of Theorem 5.2, we get

$$V'(t) < V(t)\,(-\alpha\, a_1(t) - \beta\, a_2(t) + 2\gamma\, a_3(t))$$

where

$$m[-\alpha\, a_1 - \beta\, a_2 + 2\gamma\, a_3] < 0\,,$$

owing to our choice of α, β, γ. As a consequence,

$$\lim_{t\to\infty} V(t) = 0$$

from which we deduce that $u_3(t) \to 0$ as $t \to +\infty$.

By Theorem 3.1, system (5.12) admits a positive periodic solution $(\mathring{u}_1(t), \mathring{u}_2(t))$. In fact, by (5.11),

$$m[a_1] > b_{12}\frac{m[a_2]}{b_{22}} = m[b_{12}\,\mathring{U}_2]$$

and analogously

$$m[a_2] > m[b_{21}\,\mathring{U}_1]\,.$$

Now, take

$$U(t) = m[a_2]\left|\log\frac{u_1(t)}{\mathring{u}_1(t)}\right| + m[a_1]\left|\log\frac{u_2(t)}{\mathring{u}_2(t)}\right| .$$

From the first and the second differential equation in (5.1), it follows

$$\begin{aligned} U'(t) \le\, & m[a_2](-b_{11}\,|u_1(t) - \mathring{u}_1(t)| + b_{12}\,|u_2(t) - \mathring{u}_2(t)|) \\ & + m[a_1](b_{21}\,|u_1(t) - \mathring{u}_1(t)| - b_{22}\,|u_2(t) - \mathring{u}_2(t)|) \\ & + (m[a_2]b_{13} + m[a_1]b_{23})|u_3(t)| . \end{aligned}$$

The previous differential inequality can be rewritten as

$$U'(t) \le -\nu_1\,|u_1(t) - \mathring{u}_1(t)| - \nu_2\,|u_2(t) - \mathring{u}_2(t)| + h(t)$$

where

$$\nu_1 = m[a_2]b_{11} - m[a_1]b_{21}\,, \quad \nu_2 = m[a_1]b_{22} - m[a_2]b_{12}\,,$$

and

$$h(t) = (m[a_2]b_{13} + m[a_1]b_{23})|u_3(t)| .$$

Note that, by (5.11), ν_1 and ν_2 are positive constants. A well-established procedure ([2]) yields the existence of an appropriate $\delta > 0$ such that

$$U'(t) \le -\delta\, U(t) + h(t) .$$

Since $h(t)$ vanishes at infinity, we conclude that

$$\lim_{t\to+\infty} U(t) = 0$$

which implies

$$u_i(t) - \mathring{u}_i(t) \to 0 \quad \text{as} \quad t \to +\infty\,, \quad i = 1, 2 .$$

The previous property ends the proof. □

Example 5.1. Consider the Lotka–Volterra system

$$\begin{cases} u_1' = u_1\,[(3 + 2\sin t) - 2u_1 - u_2 - 2u_3\,] \\ u_2' = u_2\,[(2 + 2\cos^2 t) - u_1 - 2u_2 - 2u_3\,] \\ u_3' = u_3\,[(2 - 5\cos t) - 2u_1 - 2u_2 - 2u_3\,] . \end{cases}$$

Here,

$$m[a_1] = 3 > 1\cdot\tfrac{3}{2} = b_{12}\frac{m[a_2]}{b_{22}}\,,$$

$$m[a_1] = 3 > 2\cdot\tfrac{2}{2} = b_{13}\frac{m[a_3]}{b_{33}}\,.$$

Analogously, $m[a_2] > b_{21}\frac{m[a_1]}{b_{11}}$, $m[a_2] > b_{23}\frac{m[a_3]}{b_{33}}$.

Moreover,

$$m[a_3] = 2 < 2 \cdot \tfrac{3}{2} = b_{31}\frac{m[a_1]}{b_{11}},$$

$$m[a_3] = 2 < 2 \cdot \tfrac{3}{2} = b_{32}\frac{m[a_2]}{b_{22}},$$

and hence inequalities (5.11) hold. From Theorem 5.3, the extinction of species $u_3(t)$ and the survival of the other two species follow.

It is possible to extend the validity of Theorem 5.3 without assuming $b_{ij} \in \mathbb{R}$.

Lemma 5.1 ([16]). *Let $p(t), g(t)$ be continuous functions for $t > t_0$, satisfying*
(i) $\int_s^t p(\tau)\,d\tau \le -k(t-s)$ *for some fixed $k > 0$ and sufficiently large t, $t \ge s \ge t_0$,*
(ii) $\lim_{t\to\infty} g(t) = 0$.
Then, the solution $z(t)$ of the following initial value problem

$$\begin{cases} z' = p(t)\,z + g(t) \\ z(t_0) = z_0 > 0 \end{cases}$$

vanishes as $t \to \infty$.

Theorem 5.4. *Suppose that all assumptions of Theorem 5.2 hold. If $(\mathring{u}_1(t), \mathring{u}_2(t))$ denotes the positive periodic solution of system (3.1) whose corresponding $\alpha(t)$, defined by (3.6), satisfies*

$$m[\alpha] < 0,$$

then

$$\lim_{t\to+\infty} |u_1(t) - \mathring{u}_1(t)| = 0 = \lim_{t\to+\infty} |u_2(t) - \mathring{u}_2(t)|, \quad \lim_{t\to+\infty} u_3(t) = 0$$

for any positive solution $u(t)$ of (5.1).

Proof. Take $(u_1(t), u_2(t), u_3(t))$ positive solution of (5.1). The extinction of species $u_3(t)$ follows from Theorem 5.2.

Under our hypotheses, condition (C_1) holds so that by Theorem 3.1, we can consider $(\mathring{u}_1(t), \mathring{u}_2(t))$ a positive periodic solution of system (3.1). Following the argument of Theorem 3.2, the functions

$$x(t) = \frac{u_1(t)}{\mathring{u}_1(t)} - 1, \quad y(t) = \frac{u_2(t)}{\mathring{u}_2(t)} - 1 \tag{5.13}$$

yield a solution of the two-dimensional system

$$\begin{cases} x' = (1+x)(-a_{11}(t)x - a_{12}(t)y - b_{13}(t)u_3) \\ y' = (1+y)(-a_{21}(t)x - a_{22}(t)y - b_{23}(t)u_3), \end{cases} \qquad a_{ij}(t) = b_{ij}(t)\mathring{u}_j(t).$$

The Lyapunov function

$$V(t) = (x(t) - \ln(1 + x(t))) + (y(t) - \ln(1 + y(t)))$$

satisfies

$$V'(t) \leq \alpha(t)(1 - e^{-V(t)}) + g(t)$$

where

$$g(t) = -(b_{13}x + b_{23}y)(t)u_3(t)$$

vanishes as $t \to +\infty$. It follows that the further function

$$Z(t) = e^{V(t)} - 1$$

verifies the differential inequality

$$Z' \leq (\alpha(t) + g(t))Z + g(t)\,.$$

We claim that

$$\lim_{t\to\infty} Z(t) = 0\,.$$

Define

$$p(t) = \alpha(t) + g(t)$$

and observe that, since $\lim_{t\to\infty} g(t) = 0$, one obtains

$$\lim_{t\to\infty} \frac{1}{t-s}\int_s^t p(\tau)\,d\tau = \lim_{\omega\to\infty} \frac{1}{\omega}\int_s^{s+\omega} (\alpha(\tau) + g(\tau))\,d\tau = m[\alpha] < 0$$

uniformly with respect to $s > 0$.

Therefore, there exists $k > 0$ such that, for sufficiently large t, we have

$$\int_s^t p(\tau)\,d\tau \leq -k(t-s)\,.$$

By Lemma 5.1, the solution $z(t)$ of the initial value problem

$$\begin{cases} z' = p(t)z + g(t) \\ z(0) = Z(0) \end{cases}$$

satisfies

$$\lim_{t\to\infty} z(t) = 0\,.$$

Since $Z(t)$ is positive, the claim is proved.

The relationship between $Z(t)$ and $V(t)$ leads to the property

$$\lim_{t\to+\infty} V(t) = 0$$

and, consequently, $x(t)$ and $y(t)$ approach zero as $t \to +\infty$. Taking into account (5.13), the proof is complete. □

6 The impulsive logistic equation

Let us begin with the autonomous logistic equation

$$\begin{cases} u'(t) = u(t)(a - b\,u(t))\,, & t \neq t_k, \\ u(t_k^+) - u(t_k^-) = \gamma_k\, u(t_k)\,, & k \in \mathbb{N}\,. \end{cases} \tag{6.1}$$

As usual, $a, b > 0$, $\{t_k\}_{k\in\mathbb{N}}$ is a strictly increasing sequence of positive numbers, $\lim_{k\to+\infty} t_k = +\infty$ and $(1 + \gamma_k) > 0$.

A solution $u(t)$ of (6.1) is a piecewise continuous function, with discontinuity points of the first type at t_k and $u(t_k) = u(t_k^-)$, that is, $u(t)$ is continuous on the left at each t_k.

For any assigned initial condition $u(\tau^+) = u_0$, (6.1) has a unique solution in $(\tau, +\infty)$.

Assuming a certain periodicity in the impulses, we can seek a periodic solution for (6.1). Let $T > 0$ and assume that there exists $q \in \mathbb{N}$ such that

$$t_{k+q} = t_k + T\,, \quad \gamma_{k+q} = \gamma_k\,.$$

Then, (6.1) admits a unique T-periodic solution as it is shown below in a particular case.

Example 6.1. Consider

$$\begin{cases} u'(t) = u(t)\left(3 - \frac{3}{2}u(t)\right)\,, & t \neq t_k, \\ u(t_k^+) = e^{-\frac{1}{2}}\, u(t_k) \end{cases} \tag{6.2}$$

where $T = 1$, $t_k = k$. Here, $\gamma_k = e^{-\frac{1}{2}} - 1\,,\quad k \in \mathbb{N}$. Following (2.6), in $(0.1]$, the solution of (6.2) with the initial condition $u(0^+) = u_0$ is given by

$$u(t) = \frac{2u_0}{u_0 + (2 - u_0)e^{-3t}}\,, \quad 0 < t \leq 1\,, \tag{6.3}$$

and

$$u(1^+) = e^{-\frac{1}{2}}\, u(1) = \frac{2u_0 e^{-\frac{1}{2}}}{u_0 + (2 - u_0)e^{-3}}\,.$$

This solution is 1-periodic if

$$u(1^+) = u(0^+) = u_0\,.$$

The equation

$$\frac{2u_0 e^{-\frac{1}{2}}}{u_0 + (2 - u_0)e^{-3}} = u_0$$

has the unique solution

$$u_0 = \frac{2(e^{-\frac{1}{2}} - e^{-3})}{(1 - e^{-3})}\,.$$

Substituting this value of u_0 in (6.3) for the periodic solution $\overset{\circ}{U}(t)$, we get

$$\overset{\circ}{U}(t) = \begin{cases} \dfrac{2(e^{-1/2} - e^{-3})}{(e^{-1/2} - e^{-3}) - (e^{-1/2} - 1)e^{-3t}} & 0 < t \le 1\,, \\ \overset{\circ}{U}(t-k) & t \in (k, k+1]\,, \quad k \ge 1\,. \end{cases}$$

Note that although our differential equation is autonomous, the presence of impulses makes this periodic solution not constant, but piecewise continuous.

We are more interested in logistic equations with periodic coefficients.

Let $J = [0, +\infty)$ and $PC(J, \mathbb{R})$ be the set of piecewise continuous functions on J with discontinuity of the first type in t_k. If $h(t) \in PC(J, \mathbb{R})$, we assume

$$h(t_k) = \lim_{t \to t_k^-} h(t) = h(t_k^-)\,.$$

For a fixed $T > 0$, $PC_T = \{h \in PC(J, \mathbb{R}) \mid h(t+T) = h(t)\,, t > 0\}$. Given $a(t) \in PC_T$, $b(t)$ continuous and T-periodic function, let us consider the following logistic equation with impulsive perturbations at t_k, that is,

$$\begin{cases} u'(t) = u(t)(a(t) - b(t)u(t))\,, & t \neq t_k, \\ u(t_k^+) = (1 + b_k)\,u(t_k)\,, & k \in \mathbb{N}\,. \end{cases} \tag{6.4}$$

The following restrictions on equation (6.4) are usual

$$b(t) > 0\,, \quad (1 + b_k) > 0,\, k \in \mathbb{N}\,,$$

together with the following periodicity assumptions on the jumps:

there exists $q \in \mathbb{N}$ such that

$$\{t_1, t_2, \ldots, t_q\} \in (0, T)\,, \quad \text{and} \quad t_{k+q} = t_k + T\,, \quad b_{k+q} = b_k\,, \quad k \ge 1\,.$$

A solution $u(t)$ of (6.4) is said T-periodic if $u(t) \in PC_T$.

Theorem 6.1. *Under previous conditions, assume that*

$$m[a] + \frac{1}{T}\sum_{k=1}^{q} \ln(1 + b_k) > 0\,. \tag{6.5}$$

Then, equation (6.4) has a unique T-periodic solution $\overset{\circ}{u}(t)$ for which $\overset{\circ}{u}(t) > 0$ for $t > 0$.

Proof. The proof will employ some known results for linear impulsive equations ([8]). In (6.4), we carry out the change of variable $x(t) = \frac{1}{u(t)}$ and obtain the linear nonhomogeneous impulsive equation

$$\begin{cases} x'(t) = -a(t)x(t) + b(t)\,, & t \neq t_k, \\ x(t_k^+) = \frac{1}{(1+b_k)}\,x(t_k)\,. \end{cases} \tag{6.6}$$

The solution of (6.6) with initial condition x_0 is given by

$$x(t) = x_0 \prod_{0<t_k<t} \frac{1}{1+b_k} \exp\left(-\int_0^t a(s)\,ds\right) + \int_0^t \prod_{s<t_k<t} \frac{1}{1+b_k} \left(\exp\int_s^t a(z)\,dz\right) b(s)\,ds\,. \tag{6.7}$$

This solution is T-periodic if $x(T) = x(0)$, or if

$$\left(1 - \prod_{0<t_k<T} \frac{1}{1+b_k} \exp\left(-\int_0^T a(s)\,ds\right)\right) x_0 = \int_0^T \prod_{s<t_k<T} \frac{1}{1+b_k} \left(\exp\int_s^T a(z)\,dz\right) b(s)\,ds\,,$$

that is,

$$\left(1 - e^{-(Tm[a]+\sum_{k=1}^{q}\ln(1+b_k))}\right) x_0 = \int_0^T \prod_{s<t_k<T} \frac{1}{1+b_k} \left(\exp\int_s^T a(z)\,dz\right) b(s)\,ds\,. \tag{6.8}$$

From (6.5), equation (6.8) admits a unique solution $x_0 > 0$. To such initial value there corresponds the unique T-periodic solution $p(t)$ of (6.6) which is positive in view of (6.7). Then, the function $\mathring{u}(t) = \frac{1}{p(t)}$ is the searched periodic solution. □

Theorem 6.2. *If all assumptions of Theorem 6.1 hold, then periodic solution $\mathring{u}(t)$ of (6.4) is globally attractive in the first quadrant.*

Proof. If $u(t)$ is a positive solution of (6.4), we have to demonstrate that

$$\lim_{t\to+\infty} |u(t) - \mathring{u}(t)| = 0\,. \tag{6.9}$$

The change of variable

$$x(t) = \frac{u(t)}{\mathring{u}(t)} - 1\,, \quad x(t) > -1$$

turns (6.4) into the differential equation

$$x'(t) = -r(t)\,x(t)(1 + x(t))\,, \quad t \neq t_k\,,$$

where $r(t) = b(t)\mathring{u}(t) \in PC_T$. For any $k \in \mathbb{N}$,

$$x(t_k^+) = \frac{(1+b_k)u(t_k)}{(1+b_k)\mathring{u}(t_k)} - 1 = x(t_k),$$

and therefore $x(t)$ is continuous for all $t \in J$.

Choose the Lyapunov function

$$V(t) = x(t) - \ln(1 + x(t)) .$$

For $t \neq t_k$,

$$V'(t) = \frac{x'(t)}{1 + x(t)} x(t) = -r(t)\, x^2(t) .$$

Using (2.11), one deduces that $V(t)$ satisfies the differential inequality

$$V'(t) \leq -r(t)(1 - e^{-V(t)}) , \quad t \neq t_k .$$

Observe that $V(t)$ is continuous and positive for $t > 0$, but $V'(t)$ is only piecewise continuous and bounded on J. Although $r(t)$ has discontinuity at t_k, the asymptotic behavior of the differential equation

$$v'(t) = -r(t)(1 - e^{-v(t)})$$

does not change with respect to the case of the $r(t)$ continuous function. Hence, $V(t) \to 0$ as it goes to infinity. As a consequence, (6.9) easily follows. □

Remark 6.1. If $b_k = 0$, $k \in \mathbb{N}$, then $u(t_k^+) = u(t_k^-)$, that is, the solutions of (6.4) have no jumps. In this case, assumption (6.5) becomes

$$m[a] > 0,$$

accordingly to Theorem 2.1.

Remark 6.2. When Theorem 6.1 holds, for periodic solution $\mathring{u}(t)$, the following equality is valid, that is,

$$m[b \cdot \mathring{u}] = m[a] + \frac{1}{T} \sum_{k=1}^{q} \ln(1 + b_k) . \tag{6.10}$$

In Section 7, we will need a comparison result for solutions of impulsive logistic equations. It is the subject of the theorem below.

Theorem 6.3 ([8]). *Let $\omega_0 > 0$ and denote by $u(t)$ the solution of (6.4) with $u(0^+) = \omega_0$. Suppose that $v(t) \in PC(J, \mathbb{R})$ verifies*

$$\begin{cases} v'(t) \leq v(t)\,(a(t) - b(t)v(t)) , & t \neq t_k, \\ v(t_k^+) \leq (1 + b_k)v(t_k), \\ v(0^+) \leq \omega_0 . \end{cases}$$

Then, $v(t) \leq u(t)$ for all $t > 0$. A similar result holds when all the inequalities are reversed.

In Section 8, among the properties of impulsive logistic equations, Theorem 6.4 will also be recalled.

Theorem 6.4. *Under the assumptions of Theorem 6.1, let $\beta(t) \in PC(J, \mathbb{R})$ such that*

$$\lim_{t \to +\infty} |\beta(t) - a(t)| = 0 .$$

Then, for any solution $v(t)$ of the impulsive equation

$$\begin{cases} v'(t) = v(t)(\beta(t) - b(t)v(t)) , & t \neq t_k, \\ v(t_k^+) = (1 + b_k)v(t_k) \end{cases}$$

with a positive initial condition, we have

$$\lim_{t \to +\infty} |v(t) - \mathring{u}(t)| = 0 .$$

Proof. See [21, Lemma 2.5]. □

7 Two species systems with impulsive effects. A look at the N-dimensional case

As before, let $\{t_k\}_{k \in \mathbb{N}}$ be a sequence of positive numbers such that

$$t_k < t_{k+1} , \quad \lim_{k \to +\infty} t_k = +\infty .$$

We consider the following impulsive Lotka–Volterra system:

$$\begin{cases} u_1'(t) = u_1(t)[a_1(t) - b_{11}(t)\, u_1(t) - b_{12}(t)\, u_2(t)] , & t \neq t_k, \\ u_2'(t) = u_2(t)[a_2(t) - b_{21}(t)\, u_1(t) - b_{22}(t)\, u_2(t)] , & t \neq t_k, \\ u_i(t_k^+) = (1 + c_{ik})u_i(t_k) , \quad i = 1, 2 , \quad k \in \mathbb{N} , \end{cases} \tag{7.1}$$

where for $i, j = 1, 2$, $a_i(t)$, $b_{ij}(t)$ are continuous and T-periodic, $b_{ij}(t) > 0$. Furthermore, there exists $q \in \mathbb{N}$ such that

$$\{t_1, \ldots, t_q\} \in (0, T) , \quad t_{k+q} = t_k + T , \quad c_{i(k+q)} = c_{ik} .$$

Taking into account the biological meaning of $u_i(t)$, henceforth we take

$$(1 + c_{ik}) > 0 , \quad i = 1, 2 , \quad k \in \mathbb{N} .$$

In fact, such restriction on the jumps ensures the validity of the following lemma.

Lemma 7.1. *Let $u(t)$ be the solution of system (7.1) with initial condition $(\bar{u}_1, \bar{u}_2)$, $\bar{u}_i > 0$. Then, $u(t)$ is positive for all $t > 0$.*

Proof. It is known that each component $u_i(t)$ belongs to the space $PC(J, \mathbb{R})$. The first component of $u(t)$ satisfies

$$u_1'(t) = u_1(t)\, \gamma(t) , \quad t \neq t_k \quad \text{and} \quad u_1(t_k^+) = (1 + c_{1k})u_1(t_k) , \quad k \in \mathbb{N} ,$$

where $y(t) = a_1(t) - b_{11}(t)\,u_1(t) - b_{12}(t)\,u_2(t)$. By (6.7),

$$u_1(t) = \bar{u}_1 \prod_{0<t_k<t} (1 + c_{1k}) \exp \int_0^t y(s)\, ds\,.$$

Since $\bar{u}_1 > 0$ and $(1 + c_{1k}) > 0$ for any $k \in \mathbb{N}$, one gets $u_1(t) > 0$ for all $t > 0$. The same property holds for $u_2(t)$. □

The first aim of this section is to show the existence of at least one positive periodic solution of system (7.1). Under the hypothesis

$$m[a_i] + \frac{1}{T}\sum_{k=1}^{q} \ln(1 + c_{ik}) > 0\,, \quad i = 1, 2\,, \tag{7.2}$$

Theorem 6.1 ensures the existence of the $\mathring{U}_i(t)$ positive periodic solution to the impulsive logistic equation

$$\begin{cases} u'(t) = u(t)(a_i(t) - b_{ii}(t)u(t))\,, & t \neq t_k, \\ u(t_k^+) = (1 + c_{ik})u(t_k)\,, & k \in \mathbb{N}\,. \end{cases}$$

Theorem 7.1. *Suppose that the following inequalities*

$$m[a_i] + \frac{1}{T}\sum_{k=1}^{q} \ln(1 + c_{ik}) > m[b_{ij} \cdot \mathring{U}_j] \quad i \neq j \tag{7.3}$$

are satisfied. Then, impulsive differential system (7.1) *has at least a positive periodic solution.*

Proof. It is possible to adapt the technique employed in Theorem 3.1 to our impulsive case. Let us show the main steps of the proof. Denote by

$$r_i(t) = a_i(t) - b_{ij}(t)\mathring{U}_j(t)\,, \quad i, j = 1, 2\,, \quad i \neq j$$

and consider the logistic equation

$$v'(t) = v(t)(r_i(t) - b_{ii}(t)v(t))\,, \quad t \neq t_k \tag{7.4}$$

with impulses at t_k, that is,

$$v(t_k^+) = (1 + c_{ik})v(t_k)\,.$$

Hypothesis (7.3) and Theorem 6.1 yield the $\mathring{V}_i(t)$ positive periodic solution of (7.4). Take $(u_1(t), u_2(t))$ positive solution of (7.1) with an initial condition verifying

$$\mathring{V}_i(0^+) \leq u_i(0^+) \leq \mathring{U}_i(0^+)\,, \quad i = 1, 2\,.$$

Thanks to Theorem 6.3, the argument of Theorem 3.1 also works in this situation, giving the estimate

$$\overset{\circ}{V}_i(t) \le u_i(t) \le \overset{\circ}{U}_i(t)\,, \quad t > 0\,, \quad i = 1,2\,.$$

The application of the Brouwer fixed point theorem yields the expected result. □

Using the asymptotic stability of $\overset{\circ}{V}_i(t)$ and $\overset{\circ}{U}_i(t)$, and the comparison results for impulsive differential equations ([8]), we obtain the permanence of system (7.1). Indeed, if we take

$$0 < \delta < \min_{t\in[0,T]} \overset{\circ}{V}_i(t)\,, \quad \Delta > \max_{t\in[0,T]} \overset{\circ}{U}_i(t)\,, \quad i = 1,2$$

and the $u(t)$ positive solution of (7.1), we have

$$\delta \le u_i(t) \le \Delta\,, \quad i = 1,2$$

for sufficiently large t.

It is not difficult to extend Theorem 7.1 to the N-dimensional case because our arguments also work for the impulsive system

$$\begin{cases} u_i'(t) = u_i(t)\left[\, a_i(t) - \sum_{j=1}^{N} b_{ij}(t)\, u_j(t)\,\right], & i = 1,\dots,N\,, \quad t \neq t_k, \\ u_i(t_k^+) = (1 + c_{ik})\, u_i(t_k)\,, & k \in \mathbb{N}, \end{cases} \tag{7.5}$$

assuming on the coefficients and the jumps are the same assumptions formulated above for system (7.1).

Theorem 7.2. *Suppose that*

$$m[a_i] + \frac{1}{T}\sum_{k=1}^{q} \ln(1 + c_{ik}) > \sum_{j=1, j\neq i}^{N} m[b_{ij} \cdot \overset{\circ}{U}_j]\,, \quad i = 1,\dots,N\,.$$

Then, impulsive system (7.5) is permanent and admits a positive T-periodic solution.

Proof. See [17]. □

To improve the knowledge of the dynamical properties of system (7.1), we need to investigate the stability of the periodic solution. In the next theorem, we discuss the global attractivity of a two-dimensional system (7.1) by means of certain inequalities recalling (3.4).

For a given $f \in PC_T$, let $f_M = \sup_{t\in[0,T]} f(t)$ and introduce the constants

$$\nu_1 = m[a_1] + \frac{1}{T}\sum_{k=1}^{q} \ln(1 + c_{1k})\,, \quad \nu_2 = m[a_2] + \frac{1}{T}\sum_{k=1}^{q} \ln(1 + c_{2k})\,.$$

Theorem 7.3. *Suppose* $\nu_1, \nu_2 > 0$ *and*

$$\nu_1 > \nu_2 \left(\frac{b_{12}}{b_{22}}\right)_M , \quad \nu_2 > \nu_1 \left(\frac{b_{21}}{b_{11}}\right)_M . \tag{7.6}$$

Then, system (7.1) admits a unique positive solution $(\overset{\circ}{u}_1(t), \overset{\circ}{u}_2(t))$, *satisfying*

$$\lim_{t\to+\infty} |u_1(t) - \overset{\circ}{u}_1(t)| = 0 = |u_2(t) - \overset{\circ}{u}_2(t)|$$

for any positive solution $u(t)$ *of* (7.1).

Proof. Using (6.10) we can rewrite ν_2 in the form

$$\nu_2 = m[b_{22} \cdot \overset{\circ}{U}_2] .$$

From (7.6), we deduce

$$\nu_1 > \left(\frac{b_{12}}{b_{22}}\right)_M \cdot m[b_{22} \cdot \overset{\circ}{U}_2] \geq m[b_{12} \cdot \overset{\circ}{U}_2],$$

and analogously

$$\nu_2 > m[b_{21} \cdot \overset{\circ}{U}_1]$$

so that (7.3) holds. By Theorem 7.1, it makes sense to consider the $(\overset{\circ}{u}_1(t), \overset{\circ}{u}_2(t))$ positive periodic solution of (7.1).

Now, take $(u_1(t), u_2(t))$ a positive solution of (7.1). The change of unknown functions

$$x(t) = \frac{u_1(t)}{\overset{\circ}{u}_1(t)} - 1 , \quad y(t) = \frac{u_2(t)}{\overset{\circ}{u}_2(t)} - 1 \tag{7.7}$$

turns system (7.1) into

$$\begin{cases} x'(t) = (1 + x(t))(-a_{11}(t)\,x(t) - a_{12}(t)\,y(t)) \\ y'(t) = (1 + y(t))(-a_{21}(t)\,x(t) - a_{22}(t)\,y(t)) \end{cases} \tag{7.8}$$

where

$$a_{ij}(t) = b_{ij}(t)\overset{\circ}{u}_j(t) , \quad i, j = 1, 2 . \tag{7.9}$$

It is relevant to note that system (7.8) is not impulsive because

$$x(t_k^+) = x(t_k^-) , \quad y(t_k^+) = y(t_k^-) .$$

New coefficients $a_{ij}(t)$ belong to the space PC_T, and hence $x(t)$, $y(t)$ are continuous, though their derivatives have jumps at t_k.

Choosing the Lyapunov function

$$V(t) = \nu_2 |\ln(1 + x(t)| + \nu_1 |\ln(1 + y(t)| ,$$

for $t \neq t_k$, we get

$$V'(t) \leq -\nu_2 a_{11}(t)|x(t)| + \nu_2 a_{12}(t)|y(t)| + \nu_1 a_{21}(t)|x(t)| - \nu_1 a_{22}(t)|y(t)| .$$

Since

$$\nu_2 a_{11}(t) - \nu_1 a_{21}(t) = \left(\nu_2 - \frac{b_{21}(t)}{b_{11}(t)}\nu_1\right) a_{11}(t) ,$$

using (7.6), there exists $\gamma_1 > 0$ such that for all $t > 0$,

$$\nu_2 a_{11}(t) - \nu_1 a_{21}(t) \geq \gamma_1 .$$

In a similar way, there exists $\gamma_2 > 0$ such that

$$\nu_1 a_{22}(t) - \nu_2 a_{12}(t) \geq \gamma_2 , \quad t > 0 ,$$

so that we can state that

$$V'(t) \leq -\gamma_1|x(t)| - \gamma_2|y(t)| , \quad t \neq t_k . \tag{7.10}$$

Integrating both sides of (7.10) between 0 and t, we obtain

$$\gamma_1 \int_0^t |x(s)|\, ds + \gamma_2 \int_0^t |y(s)|\, ds \leq V(0)$$

and, taking the limit as $t \to +\infty$, one gets

$$\gamma_1 \int_0^{+\infty} |x(s)|\, ds + \gamma_2 \int_0^{+\infty} |y(s)|\, ds < +\infty .$$

The functions $x(t)$, $y(t)$ are uniformly continuous since they are continuous, piecewise differentiable and $x'(t)$, $y'(t)$ are bounded. It follows that

$$\lim_{t\to+\infty} |x(t)| = 0 = \lim_{t\to+\infty} |y(t)|$$

([23]). Using (7.7), the proof is complete. □

Example 7.1. We consider the following system with b_{ij} real numbers

$$\begin{cases} u_1'(t) = u_1(t)[(7 + \cos(2\pi t)) - 5\,u_1(t) - 3\,u_2(t)] , & t \neq t_k, \\ u_2'(t) = u_2(t)[(9/2 + 5\sin(2\pi t)) - 2\,u_1(t) - 3\,v(t)] , & t \neq t_k, \\ u_1(t_k^+) = e^{-1}u_1(t_k) , \quad u_2(t_k^+) = e^{-1/2}u_2(t_k) , \end{cases} \tag{7.11}$$

where

$$T = 1 , \quad t_k = \tfrac{1}{4} + (k-1) , \quad c_{1k} = e^{-1} - 1 , \quad c_{2k} = e^{-1/2} - 1 \quad \text{for all } k \geq 1 .$$

It easy to see that

$$\nu_1 = m[a_1] + \ln(e^{-1}) = 7 - 1 = 6 > 0, \quad \nu_2 = m[a_2] + \ln(e^{-1/2}) = \tfrac{9}{2} - \tfrac{1}{2} = 4 > 0.$$

In addition,

$$\nu_1 = 6 > \tfrac{3}{3}4 = \frac{b_{12}}{b_{22}}\nu_2, \quad \nu_2 = 4 > \tfrac{2}{5}6 = \frac{b_{21}}{b_{11}}\nu_1,$$

and hence inequalities (7.6) are verified. According to Theorem 7.3, system (7.11) has a positive solution in the space PC_T. It is globally asymptotically stable and, consequently, it is unique.

Previous assumption (7.6) can be rewritten in the form of the pointwise inequalities

$$\frac{b_{12}(t)}{b_{22}(t)} < \frac{\nu_1}{\nu_2}, \quad \frac{b_{21}(t)}{b_{11}(t)} < \frac{\nu_2}{\nu_1} \quad \text{for all } t \in [0, T].$$

On the other hand, for the nonimpulsive case, in Theorem 3.2, the global attractivity of the periodic solution has only been obtained by means of average estimates. Fortunately, the technique employed in Theorem 3.2 can be adapted to the impulsive system (7.1) as in the following result.

Theorem 7.4. *Suppose that inequalities* (7.3) *hold and let* $\mathring{u}(t)$ *be a periodic solution of* (7.1). *Furthermore, assume that*

$$m[\alpha] < 0$$

where $\alpha(t)$ *is defined by* (3.6), *and* $a_{ij}(t)$ *are given by* (7.9). *Then,* $\mathring{u}(t)$ *is globally asymptotically stable with respect to all positive solutions of system* (7.1).

Proof. By Theorem 7.1, it is possible to consider the periodic solution $(\mathring{u}_1(t), \mathring{u}_2(t))$.

Arguing, as in the previous theorem, substitution (7.7) transforms system (7.1) into the nonlinear system (7.8) with piecewise continuous coefficients. Our aim becomes proving that

$$\lim_{t\to+\infty} |x(t)| = 0 = \lim_{t\to+\infty} |y(t)| \tag{7.12}$$

for any solution $(x(t), y(t))$ of (7.8) with $x(t) > -1, y(t) > -1$.

Since the differential system (7.8) is not impulsive and has coefficients in the space PC_T, it is possible to obtain (7.12) by means of the same tools employed in Lemma 3.2. Introduce the Lyapunov function

$$V(t) = x(t) - \ln(1 + x(t)) + y(t) - \ln(1 + y(t)).$$

$V(t)$ is continuous and, for $t \neq t_k$,

$$V'(t) = -a_{11}(t)\,x^2(t) - a_{12}(t)\,y(t)\,x(t) - a_{21}(t)\,x(t)\,y(t) - a_{22}(t)\,y^2(t).$$

It is proved in (3.9) that

$$V'(t) \le \alpha(t)\,(1 - e^{-V(t)}) \tag{7.13}$$

where $\alpha(t)$ is given by (3.6). The circumstance that coefficients $a_{ij}(t)$ are only piecewise continuous (and not continuous) restricts the validity of (7.13) at the values $t \neq t_k$.

Integrating both sides of (7.13) between 0 and t, and using (2.8), one get

$$\ln\left(\frac{e^{V(t)}-1}{e^{V(0)}-1}\right) = \int_0^t \frac{e^{V(s)}}{e^{V(s)}-1} V'(s)\,ds \le m[\alpha]\,t + p(t)$$

where $p(t)$ is a suitable continuous T-periodic function.

Since we have supposed that $\alpha(t)$ has a negative average, the previous inequality yields

$$\lim_{t\to+\infty} \ln(e^{V(t)}-1) = -\infty\,,$$

from which we deduce

$$\lim_{t\to+\infty} V(t) = 0\,.$$

Taking into account the definition of $V(t)$, we conclude that property (7.12) holds, as desired. □

In order to apply the previous theorem, one needs to evaluate the integral average of $\alpha(t)$, depending not only on coefficients $b_{ij}(t)$, $j = 1, 2$, but also on the periodic solution $\overset{\circ}{u}(t)$, which is generally not explicitly known. Numerical estimates may help to control the value of $m[\alpha]$, as shown by the next example.

Example 7.2. In this example, the application of Theorem 7.3 will not be possible, whereas all hypotheses of Theorem 7.4 are verified.

Considering the system with 1-periodic coefficients

$$\begin{cases} u_1'(t) = u_1(t)[3 - 1.5\,u_1(t) - (2 - 0.5\cos(2\pi t))\,u_2(t)]\,, & t \neq t_k, \\ u_2'(t) = u_2(t)[1 - (1.5 + \sin(2\pi t))\,u_1(t) - 3\,u_2(t)]\,, & t \neq t_k, \\ u_1(t_k^+) = e^{-1/2}u_1(t_k)\,, \quad u_2(t_k^+) = e^2 u_2(t_k)\,, \end{cases} \tag{7.14}$$

we fix $t_k = k$, $k \ge 1$. We have $b_{12}(t) > 0$, $b_{21}(t) > 0$ and

$$\nu_1 = 3 + \ln\left(e^{-1/2}\right) = \tfrac{5}{2} > 0\,, \quad \nu_2 = 1 + \ln\left(e^2\right) = 3 > 0\,.$$

However,

$$b_{11} = \tfrac{3}{2}\,, \quad (b_{12})_M = \tfrac{5}{2}\,, \quad b_{22} = 3\,, \quad (b_{21})_M = \tfrac{5}{2}\,,$$

and hence

$$\nu_1 \left(\frac{b_{21}}{b_{11}}\right)_M = \tfrac{5}{2}\cdot\tfrac{5}{3} = \tfrac{25}{6} > 3 = \nu_2\,,$$

so that (7.6) fails.

In order to employ Theorem 7.4, let us introduce $\overset{\circ}{U}_1(t)$ and $\overset{\circ}{U}_2(t)$.

By definition, $\mathring{U}_1(t)$ is the positive periodic solution to the following impulsive logistic equation

$$\begin{cases} U'(t) = U(t)[3 - 1.5\,U(t)]\,, & t \neq t_k, \\ U(t_k^+) = e^{-1/2}U(t_k)\,. \end{cases}$$

Following Example 6.1, in interval $(0, 1]$, $\mathring{U}_1(t)$ is given by

$$\mathring{U}_1(t) = \frac{2(e^{-1/2} - e^{-3})}{(e^{-1/2} - e^{-3}) - (e^{-1/2} - 1)e^{-3t}}\,, \qquad 0 < t \leq 1\,.$$

A numerical calculus of the integral

$$\int_0^1 (1.5 + \sin(2\pi t))\,\mathring{U}_1(t)\,dt$$

yields the estimate

$$m[b_{21} \cdot \mathring{U}_1] < 2.39\,.$$

$\mathring{U}_2(t)$ is the positive periodic solution to

$$\begin{cases} V'(t) = V(t)[3 - 1.5\,V(t)]\,, & t \neq t_k, \\ V(t_k^+) = e^{2}V(t_k)\,. \end{cases}$$

It has the following explicit form, that is, for $0 < t \leq 1$,

$$\mathring{U}_2(t) = \frac{1/3(e^2 - e^{-1})}{(e^2 - e^{-1}) - (e^2 - e^{-1})e^{-t}}\,.$$

Since a numerical estimate shows that

$$\int_0^1 b_{12}(t)\mathring{U}_2(t)\,dt < 1.91\,,$$

inequalities (7.3) are satisfied. According to Theorem 7.1, a positive periodic solution $(\mathring{u}_1(t), \mathring{u}_2(t))$ exists. Taking the periodic condition on the jumps into account, it has to happen that

$$\mathring{u}_1(1^-) = \frac{\mathring{u}_1(1^+)}{e^{-1/2}} = \frac{\mathring{u}_1(0^+)}{e^{-1/2}}\,, \quad \mathring{u}_2(1^-) = \frac{\mathring{u}_2(1^+)}{e^2} = \frac{\mathring{u}_2(0^+)}{e^2}\,. \tag{7.15}$$

Numerical techniques lead to the solution of (7.14) with the initial value

$$\mathring{u}(0^+) \approxeq 0.82\,, \quad \mathring{v}(0^+) \approxeq 1.72,$$

satisfying (7.15) with a *good* approximation. Therefore, taking as a numerical approximation of $(\mathring{u}_1(t), \mathring{u}_2(t))$ the solution with initial condition $(0.82, 1.72)$, we may handle coefficients $a_{ij}(t)$, $i, j = 1, 2$, defined by (7.9), as numerical objects. Further numerical calculations show that

$$m[\alpha] = \int_0^1 \alpha(t)\, dt < -0.22\,.$$

By Theorem 7.4, system (7.14) admits a unique positive periodic solution which attracts all other positive solutions.

Next, the corollary shows that under the assumptions of Theorem 7.4, taken as $u(t)$, a positive solution of (7.1), the Euclidean norm $\|u(t) - \mathring{u}(t)\|$ tends exponentially to zero. In addition, the rate of exponential decay is exactly $\frac{m[\alpha]}{2}$.

Corollary 7.1. *Assume all hypotheses of Theorem 7.4 hold. If $u(t)$ is any positive solution of (7.1), then there exist $M > 0$ and $t_0 > 0$ such that*

$$\|u(t) - \mathring{u}(t)\| \leq M\, e^{\frac{m[\alpha]}{2} t}, \quad t > t_0.$$

Proof. Let us fix $u(t) = (u_1(t), u_2(t))$, the positive solution of (7.1). As before, introducing

$$x(t) = \frac{u_1(t)}{\mathring{u}_1(t)} - 1\,, \quad y(t) = \frac{u_2(t)}{\mathring{u}_2(t)} - 1\,,$$

and arguing as in Theorem 7.4, we get

$$V'(t) \leq \alpha(t)\,(1 - e^{-V(t)})\,.$$

Setting $Z(t) = e^{V(t)} - 1$, the above differential inequality becomes

$$Z'(t) \leq \alpha(t)\, Z(t)\,.$$

The solution of the corresponding linear equation with initial condition $z(0) = Z(0)$ is given by

$$z(t) = Z(0) e^{\int_0^t \alpha(s)\, ds}.$$

Then, using the comparison result and (2.8), we deduce, for an appropriate $c > 0$,

$$Z(t) \leq c\, Z(0) e^{m[\alpha] t}\,. \tag{7.16}$$

Since $V(t)$ vanishes as t goes to infinity, we have

$$Z(t) = O(V(t)) = O(x^2(t) + y^2(t))\,, \quad \text{as } t \to +\infty\,. \tag{7.17}$$

Putting together (7.16) and (7.17), one yields

$$(u_1(t) - \mathring{u}_1(t))^2 + (u_2(t) - \mathring{u}_2(t))^2 \leq M^2\, e^{m[\alpha] t}\,, \quad t > t_0\,.$$

for suitable $M > 0$ and $t_0 > 0$. Since, by the definition of the Euclidean norm

$$\|u(t) - \mathring{u}(t)\| = ((u_1(t) - \mathring{u}_1(t))^2 + (u_2(t) - \mathring{u}_2(t))^2)^{1/2}\,,$$

our statement easily follows. □

In the further considerations, we shall look again at the N-species impulsive system (7.5)

$$\begin{cases} u_i'(t) = u_i(t)[a_i(t) - \sum_{j=1}^N b_{ij}(t)\, u_j(t)]\,, & i = 1,\dots,N\,,\ t \neq t_k,\\ u_i(t_k^+) = (1 + c_{ik})\, u_i(t_k)\,, & k \in \mathbb{N}\,. \end{cases}$$

Its global asymptotic stability has been studied, introducing some pointwise estimates which provide a sort of diagonal dominance for matrix $(b_{ij}(t))_{1\le i,j\le N}$. The theorems below show two recent results in this direction.

Theorem 7.5. *Assume that inequalities*

$$m[a_i] + \frac{1}{T}\sum_{k=1}^{q} \ln(1 + c_{ik}) > \sum_{j=1, j\neq i}^{N} m[b_{ij} \cdot \mathring{U}_j]\,, \quad i = 1,\dots,N$$

hold. Let $\mathring{u}(t)$ be a positive periodic solution of (7.5) *and set*

$$a_{ij}(t) = b_{ij}(t)\mathring{u}_j(t)\,, \quad i,j = 1,\dots,N\,. \tag{7.18}$$

If the quadratic form

$$\sum_{i=1}^{N} a_{ii}(t)\, x_i^2 + \sum_{i,j=1, i\neq j}^{N} \frac{a_{ij}(t) + a_{ji}(t)}{2}\, x_i x_j \tag{7.19}$$

is positive definite for all $t \in [0,T]$, then $\mathring{u}(t)$ is attractive with respect to any other positive solution of impulsive system (7.5).

Proof. The existence of $\mathring{u}(t)$ is ensured by Theorem 7.2. Take $(u_1(t),\dots,u_N(t))$, that is, the positive solution of (7.5). Making the substitution

$$x_i(t) = \frac{u_i(t)}{\mathring{u}_i(t)} - 1\,, \quad i = 1,\dots,N, \tag{7.20}$$

system (7.5) turns into

$$x_i'(t) = -(1 + x_i(t)) \sum_{j=1}^{N} a_{ij}(t)\, x_j(t) \quad i = 1,\dots,N\,, \tag{7.21}$$

where new coefficients $a_{ij}(t)$ are defined by (7.18). They have jumps at points t_k, and thus they are piecewise continuous. On the other hand, substitution (7.20) has changed impulsive system (7.5) into nonimpulsive system (7.21). Indeed, unknown functions $x_i(t)$ do not have jumps because

$$x_i(t_k^+) = \frac{u_i(t_k^+)}{\mathring{u}_i(t_k^+)} - 1 = \frac{(1 + c_{ik})\, u_i(t_k)}{(1 + c_{ik})\mathring{u}_i(t_k)} - 1 = \frac{u_i(t_k^-)}{\mathring{u}_i(t_k^-)} - 1 = x_i(t_k^-)\,.$$

In conclusion, each $x_i(t)$ is continuous, but its derivative belongs to the space $PC(J, \mathbb{R})$.

Let us introduce the Lyapunov function

$$V(t) = \sum_{i=1}^{N} (x_i(t) - \ln(1 + x_i(t))) .$$

For $t \neq t_k$, its derivative has the form

$$\begin{aligned} V'(t) &= -\sum_{i=1}^{N} \frac{x_i(t)}{1 + x_i(t)} (1 + x_i(t)) \sum_{j=1}^{N} a_{ij}(t)\, x_j(t) \\ &= -\sum_{i=1}^{N} a_{ii}(t)\, x_i^2(t) - \sum_{i,j=1, i\neq j}^{N} \frac{(a_{ij}(t) + a_{ji}(t))}{2} x_i(t) x_j(t) . \end{aligned} \tag{7.22}$$

Let $A(t)$ be the matrix $(a_{ij}(t))_{1\leq i,j\leq N}$, and $A^\star(t)$ be its adjoint, and thus the matrix $[A(t) + A^\star(t)]/2$ is symmetric and

$$\sum_{i=1}^{N} a_{ii}(t)\, x_i^2 + \sum_{i,j=1, i\neq j}^{N} \frac{(a_{ij}(t) + a_{ji}(t))}{2} x_i x_j$$

is the quadratic form associated to it. By our hypothesis on quadratic form (7.19),

$$\sum_{i=1}^{N} a_{ii}(t)\, x_i^2 + \sum_{i,j=1, i\neq j}^{N} \frac{(a_{ij}(t) + a_{ji}(t))}{2} x_i x_j \geq \lambda(t) \sum_{i=1}^{N} x_i^2 , \quad \lambda(t) > 0$$

where $\lambda(t)$ is the smallest eigenvalue of matrix $[A(t) + A^\star(t)]/2$. The $\lambda(t)$ is a continuous, T-periodic function. It is positive since quadratic form (7.19) is positive definite for all $t \in [0, T]$ by our hypothesis. From (7.22), it follows that

$$V'(t) \leq -\lambda(t) \sum_{i=1}^{N} x_i(t)^2 , \quad t \neq t_k .$$

Integrating the previous inequality from 0 to t, we get

$$V(t) - V(0) \leq -\int_0^t \lambda(s) \sum_{i=1}^{N} x_i^2(s)\, ds$$

which gives, for $m = \min_{t\in[0,T]} \lambda(t)$,

$$m \int_0^t \sum_{i=1}^{N} x_i^2(s)\, ds \leq \int_0^t \lambda(s) \sum_{i=1}^{N} x_i^2(s)\, ds \leq V(0) .$$

Taking the limit, as t goes to infinity, one yields

$$\int_0^{+\infty} \sum_{i=1}^{N} x_i^2(s)\, ds < +\infty .$$

Since $\sum_{i=1}^{N} x_i^2(s)$ is uniformly continuous,

$$\lim_{t\to+\infty} \sum_{i=1}^{N} x_i^2(t) = 0\,.$$

Then, each $x_i(t)$ vanishes at infinity.

Using (7.20), we conclude that

$$\lim_{t\to+\infty} |u_i(t) - \mathring{u}_i(t)| = 0\,, \quad i = 1,\dots,N$$

in accordance with our statement. □

In a more general setting, J. Hou et al. [13] studied the permanence and global attractivity of system (7.5), assuming bounded and continuous coefficients such that for each $i = 1,\dots,N$,

$$\liminf_{t\to+\infty} \left(\int_t^{t+\omega} b_{ii}(s)\,ds \right) > 0\,,$$

$$\liminf_{t\to+\infty} \left(\int_t^{t+\omega} (a_i(s) - \sum_{j\neq i}^{N} b_{ij}(s)u_i^\star(s))\,ds + \sum_{t\le t_k\le t+\omega} \ln(1 + c_{ik}) \right) > 0\,.$$

In the above assumption, ω is a fixed positive constant and $u^\star(t)$ is some positive solution of system (7.5).

Theorem 7.6. *Under previous assumptions, system (7.5) is permanent, that is, there are constants $m > 0$ and $M > 0$ such that*

$$m \le \liminf_{t\to+\infty} u_i(t) \le \limsup_{t\to+\infty} u_i(t) \le M\,, \quad i = 1,\dots,N \tag{7.23}$$

for any positive solution $u(t)$. If, in addition, we suppose that there are constants $k_i > 0$ and a nonnegative continuous function $g(t)$, satisfying

$$k_i b_{ii}(t) - \sum_{j=1,\ \ j\neq i}^{N} k_j b_{ji}(t) \ge g(t)\,, \quad t \ge 0\,, \quad i = 1,\dots,N \tag{7.24}$$

and

$$\int_0^{+\infty} g(t)\,dt = +\infty\,,$$

then system (7.5) is globally attractive.

Proof. For the permanence result, see [13].

Let $u(t), v(t)$ be two positive solutions of system (7.5). To prove its global attractivity, we have to demonstrate that $u(t)$ and $v(t)$ satisfy

$$\lim_{t\to+\infty} |u_i(t) - v_i(t)| = 0\,, \quad i = 1,2,\dots,N\,. \tag{7.25}$$

From (7.23), it follows that there exist positive numbers r, R such that

$$r \leq u_i(t)\,, \quad v_i(t) \leq R\,, \quad t \geq 0\,, \quad i = 1, \ldots, N\,. \tag{7.26}$$

We consider the Lyapunov function

$$V(t) = \sum_{i=1}^{N} k_i |\ln u_i(t) - \ln v_i(t)|\,. \tag{7.27}$$

Since for any impulsive time t_k, we have

$$V(t_k^+) = \sum_{i=1}^{N} k_i \left| \ln \frac{(1 + c_{ik})u_i(t_k)}{(1 + c_{ik})v_i(t_k)} \right| = V(t_k)\,,$$

$V(t)$ is continuous for all $t \geq 0$. Calculating its derivative, for $t \neq t_k$, we obtain

$$\begin{aligned}
V'(t) &= \sum_{i=1}^{N} k_i \left(\frac{u_i'(t)}{u_i(t)} - \frac{v_i'(t)}{v_i(t)} \right) \operatorname{sgn}(u_i(t) - v_i(t)) \\
&= \sum_{i=1}^{N} k_i \operatorname{sgn}(u_i(t) - v_i(t)) \Bigg[- b_{ii}(t)(u_i(t) - v_i(t)) \\
&\qquad\qquad - \sum_{j=1, j\neq i}^{N} b_{ij}(t)(u_j(t) - v_j(t)) \Bigg] \\
&\leq \sum_{i=1}^{N} k_i \left[-b_{ii}(t)|u_i(t) - v_i(t)| + \sum_{j=1, j\neq i}^{N} b_{ij}(t)|u_j(t) - v_j(t)| \right] \\
&= \sum_{i=1}^{N} \left(-k_i b_{ii}(t) + \sum_{j=1, j\neq i}^{N} k_j b_{ji}(t) \right) |u_i(t) - v_i(t)| \\
&\leq -g(t) \sum_{i=1}^{N} |u_i(t) - v_i(t)|\,.
\end{aligned}$$

Note that by the mean value theorem, for $x, y > 0$, the following inequality holds

$$\theta |\ln x - \ln y| = |x - y|$$

where θ is between x and y. Therefore, using (7.26), there exist constants r, $R > 0$ such that for each $i = 1, \ldots, N$ and $t \geq 0$, $t \neq t_k$,

$$\frac{1}{R} |u_i(t) - v_i(t)| \leq |\ln u_i(t) - \ln v_i(t)| \leq \frac{1}{r} |u_i(t) - v_i(t)|\,. \tag{7.28}$$

Coming back to $V'(t)$, we see that

$$V'(t) \leq -\frac{g(t) r}{\alpha} \sum_{i=1}^{N} k_i |\ln u_i(t) - \ln v_i(t)| \leq -g(t) \frac{r}{\alpha} V(t)\,, \quad t \neq t_k\,,$$

where $\alpha = \max\{k_i \mid 1 \le i \le N\}$. Integrating from 0 to t, we have

$$V(t) \le V(0) \exp\left(-\frac{r}{\alpha}\int_0^t g(s)\,ds\right).$$

Since

$$\lim_{t\to+\infty}\int_0^t g(s)\,ds = \int_0^{+\infty} g(s)\,ds = +\infty\,,$$

we deduce

$$\lim_{t\to+\infty} V(t) = 0\,.$$

Finally, from (7.27), we get (7.25). □

Remark 7.1. In the previous argument, the positivity of the functions

$$k_i\, b_{ii}(t) - \sum_{\substack{j=1\\ j\ne i}}^{N} k_j\, b_{ji}(t)\,, \quad i = 1,\dots,N$$

is an essential property. Otherwise, estimate (7.28) cannot be employed.

The use of Lyapunov function (7.27) is not new. For instance, Ahmad and Lazer introduce (7.27) for nonimpulsive systems in [2, 3]. Afterwards, Ahmad and Stamova [4] take a Lyapunov function in the same form for the impulsive case.

The permanence property for system (7.5) can be obtained by means of inequalities which do not involve any positive solution $u^\star(t)$, adding a restrictive condition on the jumps. Such a result is showed in [4]. It is stated below in the periodic case.

Theorem 7.7. *Let the following conditions hold:*
(1) $m[a_i] > \sum_{j=1,\,j\ne i}^{N} \frac{\sup b_{ij}(t)}{\inf b_{jj}(t)} m[a_j]\,, \quad 1 \le i \le N,$
(2) $(1 + c_{ik}) \le 1\,, \quad 1 \le i \le N\,, \quad k \in \mathbb{N}.$
Then, system (7.5) is permanent.

Example 7.3. The following three-dimensional system is proposed in [4]. We study its permanence using Theorem 7.7 and its global stability by means of Theorem 7.6. Consider

$$\begin{cases} u_1' = u_1[\,(8 + 7\cos t) - (5 + \sin t)u_1 - u_2 - u_3\,] \\ u_2' = u_2[\,(6 + 5\sin t) - u_1 - (3 + \cos t)u_2 - u_3\,] \\ u_3' = u_3[\,(3 + 2\cos 2t) - u_1 - 0.1u_2 - (4 + \sin t)u_3\,]\,. \end{cases} \tag{7.29}$$

Here, we take $\{t_k\}_{k\in\mathbb{N}}$ such that

$$0 < t_1 < 2\pi\,, \quad t_{k+1} = t_k + 2\pi\,, \quad k \ge 1$$

and the following impulsive jumps

$$u_i(t_k^+) = (1-0.2)u_1(t_k), \quad u_2(t_k^+) = (1-0.5)u_2(t_k), \quad u_3(t_k^+) = (1-0.3)u_3(t_k).$$

It is easy to verify that

$$\begin{aligned} m[a_1] &= 8 > \tfrac{1}{2}\cdot 6 + \tfrac{1}{3}\cdot 3 = \frac{b_{12}}{\inf b_{22}(t)} m[a_2] + \frac{b_{13}}{\inf b_{33}(t)} m[a_3] \\ m[a_2] &= 6 > \tfrac{1}{4}\cdot 8 + \tfrac{1}{3}\cdot 3 = \frac{b_{21}}{\inf b_{11}(t)} m[a_1] + \frac{b_{23}}{\inf b_{33}(t)} m[a_3] \\ m[a_3] &= 3 > \tfrac{1}{4}\cdot 8 + \tfrac{0.1}{2}\cdot 6 = \frac{b_{31}}{\inf b_{11}(t)} m[a_1] + \frac{b_{32}}{\inf b_{22}(t)} m[a_2], \end{aligned}$$

and thus, by Theorem 7.7, system (7.29) is permanent.

In order to use (7.24), we take

$$k_1 = k_2 = k_3 = 1$$

and we see that

$$\begin{aligned} k_1 b_{11}(t) - k_2 b_{21} - k_3 b_{31} &= (5 + \sin t) - 1 - 1 = 3 + \sin t \\ k_2 b_{22}(t) - k_1 b_{12} - k_3 b_{32} &= (3 + \cos t) - 1 - 0.1 = 1.9 + \cos t \\ k_3 b_{33}(t) - k_1 b_{13} - k_2 b_{23} &= (4 + \sin t) - 1 - 1 = 2 + \sin t. \end{aligned}$$

Set

$$g(t) = \min\{2 + \sin t, 1.9 + \cos t\} > 0.9,$$

then, by Theorem 7.6, system (7.29) is globally asymptotically stable.

8 The influence of impulsive perturbations on extinction in three-species models

In this last section, we will discuss the phenomenon of extinction in three-dimensional systems, extending the results in Section 5 to the impulsive case. For two-species systems, Wang et al. [26] investigated the phenomenon of competitive exclusion for impulsive system (7.1), adapting the ideas introduced in Section 4.

Our investigation concerns the following Lotka–Volterra system with continuous, T-periodic coefficients and impulses at t_k

$$\begin{cases} u_1' = u_1(a_1(t) - b_{11}(t)u_1 - b_{12}(t)u_2 - b_{13}(t)u_3), & t \neq t_k, \\ u_2' = u_2(a_2(t) - b_{21}(t)u_1 - b_{22}(t)u_2 - b_{23}(t)u_3), & t \neq t_k, \\ u_3' = u_3(a_3(t) - b_{31}(t)u_1 - b_{32}(t)u_2 - b_{33}(t)u_3), & t \neq t_k, \\ u_i(t_k^+) = (1 + c_{ik})\, u_i(t_k) & i = 1,2,3, \quad k \in \mathbb{N}. \end{cases} \tag{8.1}$$

Henceforth, we assume that for $i, j = 1, 2, 3$ and $k \in \mathbb{N}$,

$$b_{ij}(t) > 0\,, \quad 1 + c_{ik} > 0\,,$$

and there exists an integer $q \geq 1$ such that

$$c_{i(k+q)} = c_{ik}\,, \quad t_{k+q} = t_k + T\,, \quad \{t_1, \ldots, t_q\} \in (0, T)\,.$$

Moreover, for each $i = 1, 2, 3$, we set

$$\nu_i = m[a_i] + \frac{1}{T} \sum_{k=1}^{q} \ln(1 + c_{ik})\,. \tag{8.2}$$

We suppose

$$\nu_i > 0$$

in order to guarantee the existence of the positive periodic function $\mathring{U}_i(t)$, as in Section 7, according to Theorem 6.1.

The arguments of Lemma 7.1 ensure the positivity of every solution to (8.1) with positive initial conditions. The first step of our analysis concerns the attractivity of rectangle R_t introduced by (5.2), that is,

$$R_t = [0, \mathring{U}_1(t)] \times [0, \mathring{U}_2(t)] \times [0, \mathring{U}_3(t)]\,,$$

with respect to all positive solutions of (8.1).

Theorem 8.1. *Suppose that*

$$\nu_1 > m[b_{12} \cdot \mathring{U}_2]\,, \quad \nu_2 > m[b_{21} \cdot \mathring{U}_1]\,. \tag{8.3}$$

If, either $\nu_1 > m[b_{13} \cdot \mathring{U}_3]$ or $\nu_2 > m[b_{23} \cdot \mathring{U}_3]$ holds as well, then any positive solution of (8.1) *lies in the interior of R_t for sufficiently large t.*

Proof. Let $u(t) = (u_1(t), u_2(t), u_3(t))$ be a positive solution of (8.1). Our objective is to show the existence of $\tau_0 > 0$ such that

$$u(t) \in \mathring{R}_t\,, \quad t > \tau_0\,.$$

If its initial condition lies in R_t, by comparison results for impulsive equations, we have

$$0 < u_i(t) < \mathring{U}_i(t)\,, \quad i = 1, 2, 3$$

for all $t > 0$, so that $u(t) \in \mathring{R}_t$, $t > 0$.

Now, suppose that $u_1(0) > \mathring{U}_1(0)$. We will prove that there exists $\tau_1 > 0$ such that

$$u_1(t) < \mathring{U}_1(t)\,, \quad t > \tau_1\,. \tag{8.4}$$

By contradiction, assume that for all $t > 0$,

$$u_1(t) \geq \overset{\circ}{U}_1(t)\,. \tag{8.5}$$

Let $\bar{u}_1(t)$ be the solution of the impulsive logistic equation

$$u' = u(a_1(t) - b_{11}(t)u)\,, \quad t \neq t_k\,, \quad u(t_k^+) = (1 + c_{1k})\,u(t_k)$$

with

$$\bar{u}_1(0) = u_1(0)\,.$$

By Theorems 6.2 and 6.3, we get

$$0 < u_1(t) < \bar{u}_1(t)\,, \quad t > 0 \quad \text{and} \quad \lim_{t\to+\infty} |\bar{u}_1(t) - \overset{\circ}{U}_1(t)| = 0\,,$$

and thus using our contradiction hypothesis (8.5), also for $u_1(t)$, we have

$$\lim_{t\to+\infty} |u_1(t) - \overset{\circ}{U}_1(t)| = 0\,. \tag{8.6}$$

We claim that another consequence of (8.5) is that

$$\lim_{t\to+\infty} u_2(t) = 0 = \lim_{t\to+\infty} u_3(t)\,. \tag{8.7}$$

Observe that

$$\frac{u_1'(s)}{u_1(s)} - \frac{\overset{\circ}{U}{}_1'(s)}{\overset{\circ}{U}_1(s)} = -[b_{11}(u_1 - \overset{\circ}{U}_1) + b_{12}u_2 + b_{13}u_3](s)\,,$$

therefore, integrating between 0 and t, one gets

$$\int_0^t [b_{11}(u_1 - \overset{\circ}{U}_1) + b_{12}u_2 + b_{13}u_3](s)\,ds = \ln\left(\frac{u_1(0)}{\overset{\circ}{U}_1(0)}\right) - \ln\left(\frac{u_1(t)}{\overset{\circ}{U}_1(t)}\right).$$

Taking the limit as t goes to infinity, one yields

$$\int_0^{+\infty} [b_{11}(u_1 - \overset{\circ}{U}_1) + b_{12}u_2 + b_{13}u_3](s)\,ds < +\infty\,.$$

On the other hand, using (8.5),

$$\begin{aligned}
&\int_0^{+\infty} [b_{11}(u_1 - \overset{\circ}{U}_1) + b_{12}u_2 + b_{13}u_3](s)\,ds \\
&= \int_0^{+\infty} \left[b_{11}\overset{\circ}{U}_1\left(\frac{u_1}{\overset{\circ}{U}_1} - 1\right) + b_{12}\overset{\circ}{U}_2\left(\frac{u_2}{\overset{\circ}{U}_2}\right) + b_{13}\overset{\circ}{U}_3\left(\frac{u_3}{\overset{\circ}{U}_3}\right)\right](s)\,ds \\
&\geq \int_0^{+\infty} \left[c_1\left(\frac{u_1}{\overset{\circ}{U}_1} - 1\right) + c_2\left(\frac{u_2}{\overset{\circ}{U}_2}\right) + c_3\left(\frac{u_3}{\overset{\circ}{U}_3}\right)\right](s)\,ds
\end{aligned}$$

where c_1, c_2, c_3 are suitable positive constants. It follows that

$$\int_0^{+\infty}\left[c_1\left(\frac{u_1}{\overset{\circ}{U}_1}-1\right)+c_2\left(\frac{u_2}{\overset{\circ}{U}_2}\right)+c_3\left(\frac{u_3}{\overset{\circ}{U}_3}\right)\right](s)\,ds<+\infty\,. \tag{8.8}$$

Since in (8.8) the function inside the integral is positive, continuous and has a bounded derivative on $[0,+\infty[$, we deduce

$$\lim_{t\to+\infty}\left[c_1\left(\frac{u_1}{\overset{\circ}{U}_1}-1\right)+c_2\left(\frac{u_2}{\overset{\circ}{U}_2}\right)+c_3\left(\frac{u_3}{\overset{\circ}{U}_3}\right)\right](t)=0$$

and, taking into account (8.5) and (8.6), our claim (8.7) is proved. On the contrary, inequalities (8.3) lead to the property that $u_2(t)$ is bounded from below.

In fact, $u_2(t)$ can be seen as a solution of the logistic equation

$$u'=u[(a_2-b_{21}u_1-b_{23}u_3)(t)-b_{22}(t)u]\,,\quad u(t_k^+)=(1+c_{2k})u(t_k)$$

where

$$\lim_{t\to+\infty}(b_{21}u_1+b_{23}u_3-b_{21}\overset{\circ}{U}_1)(t)=0\,.$$

By Theorem 6.4, $u_2(t)$ has the same asymptotic behavior of any solution to impulsive logistic equation

$$v'=v[(a_2(t)-b_{21}(t)\overset{\circ}{U}_1(t))-b_{22}(t)\,v]\,,\quad v(t_k^+)=(1+c_{2k})v(t_k)\,.$$

The assumption

$$\nu_2>m[b_{21}\,\overset{\circ}{U}_1] \tag{8.9}$$

and Theorem 6.2 ensure the existence of a positive constant h such that

$$u_2(t)>h\,,\quad t>0,$$

contradicting (8.7). We conclude that (8.4) holds.

Analogously, using the inequality $\nu_1>m[b_{12}\,\overset{\circ}{U}_2]$ instead of (8.9), one can prove the existence of $\tau_2>0$ such that

$$u_2(t)<\overset{\circ}{U}_2(t)\,,\quad t>\tau_2\,.$$

Finally, hypothesis $\nu_2>m[b_{23}\overset{\circ}{U}_3]$ (or alternatively $\nu_1>m[b_{13}\overset{\circ}{U}_3]$) permits us to verify that there exists $\tau_3>0$ such that

$$u_3(t)<\overset{\circ}{U}_3(t)\,,\quad t>\tau_3\,. \tag{8.10}$$

In fact, if (8.10) is not true, arguing as above, we have

$$\lim_{t\to+\infty}(u_3(t)-\overset{\circ}{U}_3(t))=0\,,$$

and consequently

$$\lim_{t\to+\infty} u_1(t) = 0 = \lim_{t\to+\infty} u_2(t)\,. \tag{8.11}$$

Considering that

$$u_2' = u_2[(a_2 - b_{21}u_1 - b_{23}u_3)(t) - b_{22}(t)u_2]$$

and

$$\lim_{t\to+\infty} (b_{21}u_1 + b_{23}u_3 - b_{23}\overset{\circ}{U}_3)(t) = 0\,,$$

Theorem 6.4 and inequality $\nu_2 > m[b_{23}\overset{\circ}{U}_3]$ yield the boundness of $u_2(t)$ from below. This last consequence contradicts (8.11), and thus (8.10) is proved.

Now, taking $\tau_0 = \max\{\tau_1, \tau_2, \tau_3\}$, we get

$$0 < u_i(t) < \overset{\circ}{U}_i(t)\,, \quad t > \tau_0\,, \quad i = 1, 2, 3$$

as desired. □

In the proof of the main results of this section, we shall need the following lemma.

Lemma 8.1. *Let $u(t) \in PC(J, \mathbb{R})$ satisfy the inequalities*

$$u'(t) \le p(t)\,u(t)\,, \quad t > 0,\, t \ne t_k\,,$$
$$u(t_k^+) \le (1 + \gamma_k)\,u(t_k)$$

where $p(t)$ is continuous and T-periodic, $(1 + \gamma_k) > 0$ and $\gamma_{k+q} = \gamma_k$. Then, there exists a positive, T-periodic, continuous function $q(t)$ such that, for $t > 0$,

$$u(t) \le u(0)\,q(t)\exp(\nu\, t)\,, \quad \nu = m[p] + \frac{1}{T}\sum_{k=1}^{q}\ln(1 + \gamma_k)\,.$$

Proof. Let $v(t)$ be the unique solution of the impulsive linear equation

$$\begin{cases} u't) = p(t)\,u(t)\,, & t \ne t_k, \\ u(t_k^+) = (1 + \gamma_k)\,u(t_k) \end{cases}$$

with initial condition $v(0) = u(0)$. Then, by (6.7),

$$v(t) = u(0) \prod_{0<t_k<t} (1 + \gamma_k)\exp\left(\int_0^t p(s)\,ds\right)\,.$$

Since $(1 + \gamma_k) > 0$, $v(t)$ can be rewritten in the form

$$v(t) = u(0)\exp\left(\sum_{0<t_k<t} \ln(1 + \gamma_k) + \int_0^t p(s)\,ds\right)\,.$$

Using (2.8) for $p(t)$, we have

$$\int_0^t p(s)\,ds = m[p]\,t + \tilde{p}(t)$$

where $\tilde{p}(t)$ is a suitable periodic function. Using the properties

$$\{t_1, \dots, t_q\} \subset (0, T)\,, \quad t_{k+q} = t_k + T\,, \quad \gamma_{k+q} = \gamma_k\,,$$

it easy to check that the function

$$Q(t) = \sum_{0<t_k<t} \ln(1+\gamma_k) - \frac{t}{T}\sum_{k=1}^{q} \ln(1+\gamma_k)$$

is T-periodic. Therefore,

$$\sum_{0<t_k<t} \ln(1+\gamma_k) = \frac{1}{T}\sum_{k=1}^{q} \ln(1+\gamma_k)\,t + Q(t)$$

and consequently

$$v(t) = u(0)\,q(t)\exp\left(\frac{1}{T}\sum_{k=1}^{q} \ln(1+\gamma_k) + m[p]\right)t$$

where $q(t) = \exp(\tilde{p}(t) + Q(t))$. The comparison results for impulsive equations give the inequality

$$u(t) \le v(t) \quad t \ge 0\,,$$

completing the proof. □

The last part of this section is devoted to establishing certain conditions leading to the extinction of species $u_3(t)$ and, in general, to analyze the asymptotic behavior of (8.1) when such extinction takes place. We start by assuming interaction coefficients b_{ij} are real numbers.

Theorem 8.2. *Suppose that in system* (8.1), b_{ij}, $i, j = 1, 2, 3$ *are positive constants and the below inequalities*

$$\nu_1 > \frac{b_{12}}{b_{22}}\nu_2\,, \quad \nu_2 > \frac{b_{21}}{b_{11}}\nu_1\,, \tag{8.12}$$

$$\nu_1 > \frac{b_{13}}{b_{33}}\nu_3\,, \quad \nu_2 > \frac{b_{23}}{b_{33}}\nu_3\,, \tag{8.13}$$

$$\nu_3 < \frac{b_{31}}{b_{11}}\nu_1\,, \quad \nu_3 < \frac{b_{32}}{b_{22}}\nu_2 \tag{8.14}$$

are verified. Then, for each positive solution $u(t)$ *of* (8.1), *one yields*

$$\lim_{t\to+\infty} u_3(t) = 0\,, \quad \lim_{t\to+\infty} |u_1(t) - \mathring{u}_1(t)| = 0 = \lim_{t\to+\infty} |u_2(t) - \mathring{u}_2(t)|$$

where $(\mathring{u}_1(t), \mathring{u}_2(t))$ *is the unique positive periodic solution of the two-dimensional system*

$$\begin{cases} u_1'(t) = u_1(t)[a_1(t) - b_{11}\, u_1(t) - b_{12}\, u_2(t)]\,, & t \neq t_k, \\ u_2'(t) = u_2(t)[a_2(t) - b_{21}\, u_1(t) - b_{22}\, u_2(t)]\,, & t \neq t_k, \\ u_i(t_k^+) = (1 + c_{ik})u(t_k)\,, & i = 1, 2\,, \quad k \in \mathbb{N}\,. \end{cases}$$

Proof. Using (8.12) and (8.14), we get the further inequalities

$$\frac{\nu_3}{\nu_1} < \frac{b_{32}}{b_{12}}\,, \quad \frac{\nu_3}{\nu_2} < \frac{b_{31}}{b_{21}}\,.$$

At this point, we are allowed to take two positive constants α, β as follows

$$\frac{\nu_3}{\nu_1} < 2\alpha < \min\left\{\frac{b_{31}}{b_{11}}, \frac{b_{32}}{b_{12}}, \frac{b_{33}}{b_{13}}\right\}\,,$$
$$\frac{\nu_3}{\nu_2} < 2\beta < \min\left\{\frac{b_{31}}{b_{21}}, \frac{b_{32}}{b_{22}}, \frac{b_{33}}{b_{23}}\right\}\,.$$

This choice of α, β yields the inequalities

$$\alpha\, b_{11} + \beta\, b_{21} - b_{31} < 0\,,$$
$$\alpha\, b_{12} + \beta\, b_{22} - b_{32} < 0\,,$$
$$\alpha\, b_{13} + \beta\, b_{23} - b_{33} < 0\,,$$

and

$$-\alpha\,\nu_1 - \beta\,\nu_2 + \nu_3 < 0\,.$$

Let $(u_1(t), u_2(t), u_3(t))$ be a positive solution of (8.1) and consider the Lyapunov function

$$Z(t) = (u_1(t))^{-\alpha}(u_2(t))^{-\beta}u_3(t)\,.$$

For $t \neq t_k$, a direct calculation gives

$$Z'(t) \leq f(t; u_1(t), u_2(t), u_3(t))\, Z(t)$$

where

$$f(t; u, v, w) = (-\alpha\, a_1(t) - \beta\, a_2(t) + a_3(t)) + (\alpha\, b_{11} + \beta\, b_{21} - b_{31})\, u$$
$$+ (\alpha\, b_{12} + \beta\, b_{22} - b_{32})\, v + (\alpha\, b_{13} + \beta\, b_{23} - b_{33})\, w\,.$$

Taking the previous inequalities into account, we deduce

$$Z'(t) \leq p(t)\, Z(t)\,, \quad t \neq t_k\,, \quad p(t) = -\alpha\, a_1(t) - \beta\, a_2(t) + a_3(t)\,.$$

At $t = t_k$, $Z(t)$ has jumps given by

$$Z(t_k^+) = (1 + c_{1k})^{-\alpha}(1 + c_{2k})^{-\beta}(1 + c_{3k})Z(t_k)\,.$$

By Lemma 8.1, all solutions of the impulsive differential inequality

$$\begin{cases} u'(t) \le p(t)\,u(t)\,, & t \ne t_k, \\ u(t_k^+) = (1+c_{1k})^{-\alpha}(1+c_{2k})^{-\beta}(1+c_{3k})\,u(t_k) \end{cases}$$

tend to zero as $t \to +\infty$ because

$$m[p] + \frac{1}{T}\sum_{k=1}^{q} \ln((1+c_{1k})^{-\alpha}(1+c_{2k})^{-\beta}(1+c_{3k})) = -\alpha\,\nu_1 - \beta\,\upsilon_2 + \upsilon_3 < 0\,.$$

Then,

$$\lim_{t\to+\infty} Z(t) = 0$$

and consequently

$$\lim_{t\to+\infty} u_3(t) = 0\,.$$

It remains to be proven that

$$\lim_{t\to+\infty} |u_1(t) - \mathring{u}_1(t)| = 0 = \lim_{t\to+\infty} |u_2(t) - \mathring{u}_2(t)|\,. \tag{8.15}$$

The existence and uniqueness of $(\mathring{u}_1(t), \mathring{u}_2(t))$ is ensured by Theorem 7.3. Define

$$\Phi(t) = \Phi(u_1(t), u_2(t)) = \nu_2|\ln u_1(t) - \ln\mathring{u}_1(t)| + \nu_1|\ln u_2(t) - \ln\mathring{u}_2(t)|\,.$$

For each t_k, we have

$$\Phi(t_k^+) = \Phi(t_k)$$

so that $\Phi(t)$ is continuous and, for $t \ne t_k$. Moreover $\Phi(t)$ is differentiable almost everywhere and

$$\Phi'(t) \le (-\nu_2\,b_{11} + \nu_1\,b_{21})|u_1(t) - \mathring{u}_1(t)| + (-\nu_1\,b_{22} + \nu_2\,b_{12})|u_2(t) - \mathring{u}_2(t)| + \phi(t)$$

where $\phi(t)$ is a positive function vanishing at infinity for the presence of factor $u_3(t)$. Since $\nu_2\,b_{11} - \nu_1\,b_{21} > 0$ and $\nu_1\,b_{22} - \nu_2\,b_{12} > 0$, usual arguments ([24]) employing the mean-value theorem lead to

$$\Phi'(t) \le -m\,\Phi(t) + \phi(t)\,, \quad a.e. \text{ in } J$$

for some positive constant m.

Let $\epsilon > 0$ and fix $\tau_0 > 0$ such that $\phi(t) < \epsilon$ for $t \ge \tau_0$. Then,

$$\begin{aligned} \Phi(t) &\le \Phi(\tau_0)\,e^{-m(t-\tau_0)} + \int_{\tau_0}^{t} e^{-m(t-s)}\phi(s)\,ds \\ &\le \Phi(\tau_0)\,e^{-m(t-\tau_0)} + \epsilon\,m^{-1}(1 - e^{-m(t-\tau_0)}) \end{aligned}$$

and (8.15) easily follows. □

At this point, a natural question arises. Are we able to establish a similar result for the $b_{ij}(t)$ T-periodic functions? Next, Theorems 8.3 and 8.4 give a possible positive answer.

Let us introduce some notations employed in the next theorems.

For fixed $t > 0$ and $\alpha, \beta > 0$, set

$$f(t; u, v, w) = (-\alpha a_1(t) - \beta a_2(t) + a_3(t)) + (\alpha b_{11}(t) + \beta b_{21}(t) - b_{31}(t))\, u \\ + (\alpha b_{12}(t) + \beta b_{22}(t) - b_{32}(t))\, v + (\alpha b_{13}(t) + \beta b_{23}(t) - b_{33}(t))\, w$$

and denote the maximum value on rectangle R_t by $F(t)$, that is,

$$F(t) = \max_{(u,v,w) \in R_t} f(t; u, v, w)\,. \tag{8.16}$$

We are ready to establish the following result concerning the extinction of species $u_3(t)$.

Theorem 8.3. *Assume that all hypotheses of Theorem 8.1 hold. In addition, suppose that there exist two suitable constants $\alpha, \beta > 0$ such that*

$$m[F] + \frac{1}{T} \sum_{k=1}^{q} (-\alpha \ln(1 + c_{1k}) - \beta \ln(1 + c_{2k}) + \ln(1 + c_{3k})) < 0 \tag{8.17}$$

where $F(t)$ is defined by (8.16).

If $(u_1(t), u_2(t), u_3(t))$ is any positive solution of (8.1), then

$$\lim_{t \to +\infty} u_3(t) = 0\,.$$

Proof. Take $((u_1(t), u_2(t), u_3(t))$ positive solution of (8.1). From Theorem 8.1, the existence of $\tau_0 > 0$ such that

$$((u_1(t), u_2(t), u_3(t)) \in \mathring{R}_t\,, \quad t > \tau_0$$

follows. For $t > \tau_0$, consider the function

$$Z(t) = (u_1(t))^{-\alpha} (u_2(t))^{-\beta} u_3(t)\,.$$

When $t \neq t_k$, easy calculations lead to

$$Z'(t) = Z(t)\, f(t; u_1(t), u_2(t), u_3(t)),$$

from which we obtain

$$Z'(t) \leq F(t)\, Z(t)\,.$$

$Z(t)$ has jumps in t_k, that is,

$$Z(t_k^+) = (1 + c_{1k})^{-\alpha} (1 + c_{2k})^{-\beta} (1 + c_{3k}) Z(t_k)\,,$$

and

$$\sum_{k=1}^{q} \ln((1+c_{1k})^{-\alpha}(1+c_{2k})^{-\beta}(1+c_{3k}))$$
$$= \sum_{k=1}^{q} (-\alpha \ln(1+c_{1k}) - \beta \ln(1+c_{2k}) + \ln(1+c_{3k})) \,.$$

Using Lemma 8.1 and (8.17), one yields

$$Z(t) \to 0 \quad \text{as} \quad t \to +\infty \,.$$

Since

$$u_3(t) = Z(t)(u_1(t))^{\alpha}(u_2(t))^{\beta},$$

our proof is complete. □

The following corollary indicates an easy way to verify condition (8.17).

Corollary 8.1. *Hypothesis* (8.17) *of Theorem* 8.3 *is satisfied when it is possible to find two real numbers* $\alpha, \beta > 0$ *such that, for all* $t > 0$,

$$\begin{aligned} &\alpha b_{11}(t) + \beta b_{21}(t) - b_{31}(t) < 0 \,, \\ &\alpha b_{12}(t) + \beta b_{22}(t) - b_{32}(t) < 0 \,, \\ &\alpha b_{13}(t) + \beta b_{23}(t) - b_{33}(t) < 0 \,, \end{aligned} \tag{8.18}$$

and

$$-\alpha v_1 - \beta v_2 + v_3 < 0$$

where constants v_i *are defined by* (8.2).

Proof. For every $t > 0$, $f(t; u, v, w)$ is a linear function, and therefore its maximum value on rectangle R_t will be achieved in one of its eight vertices. It is not difficult to verify that inequalities (8.18) make the coefficients of variables u, v, w in function $f(t; u, v, w)$ negative. As a consequence,

$$F(t) = f(t; 0, 0, 0) = -\alpha a_1(t) - \beta a_2(t) + a_3(t) \,.$$

Therefore,

$$\begin{aligned} m[F] &+ \frac{1}{T} \sum_{k=1}^{q} (-\alpha \ln(1+c_{1k}) - \beta \ln(1+c_{2k}) + \ln(1+c_{3k})) \\ &= -\alpha\Big(m[a_1] + \frac{1}{T}\sum_{k=1}^{q} \ln(1+c_{1k})\Big) - \beta\Big(m[a_2] + \frac{1}{T}\sum_{k=1}^{q} \ln(1+c_{2k})\Big) \\ &\quad + (m[a_3] + \frac{1}{T}\sum_{k=1}^{q} \ln(1+c_{3k})) = -\alpha v_1 - \beta v_2 + v_3 < 0 \,. \end{aligned}$$

By Theorem 8.3, the above equality provides our statement. □

Example 8.1. Let us consider the following competitive system with 2π-periodic coefficients

$$\begin{cases} u_1' = u_1[(4+3\sin t) - 2u_1 - u_2 - 2u_3] \\ u_2' = u_2[(4-\cos t) - u_1 - u_2 - (1+\frac{sin^2(2t)}{2})u_3] \\ u_3 = u_3[(2+2\sin^2 t) - 2u_1 - (\frac{3}{2} - \frac{\cos t}{4})u_2 - \frac{5}{2}u_3]\,. \end{cases}$$

About the jumps, we suppose $q = 2$ and the following values for constants $1 + c_{ik}$, $i = 1, 2, 3$ and $k = 1, 2$, namely,

$$1 + c_{11} = 0.8\,, \quad 1 + c_{21} = 2.5\,, \quad 1 + c_{31} = 0.5\,,$$
$$1 + c_{12} = 2.3\,, \quad 1 + c_{22} = 0.6\,, \quad 1 + c_{32} = 1.4\,.$$

Taking $\alpha = \frac{1}{2}$ and $\beta = \frac{1}{2}$, it easy to see that all inequalities (8.18) are satisfied. Moreover,

$$\begin{aligned} \nu_1 &= 4 + \frac{1}{2\pi}\ln(1.84) = m[2\mathring{U}_1]\,, \\ \nu_2 &= 4 + \frac{1}{2\pi}\ln(1.5) = m[\mathring{U}_2]\,, \\ \nu_3 &= 3 + \frac{1}{2\pi}\ln(0.7) = m[\frac{5}{2}\mathring{U}_3]\,, \end{aligned}$$

and

$$-\frac{1}{2}\nu_1 - \frac{1}{2}\nu_2 + \nu_3 < 0\,.$$

Let us check that the hypotheses of Theorem 8.1 holds as well. We get

$$\begin{aligned} \nu_1 &= 4 + \frac{1}{2\pi}\ln(1.84) > 4 + \frac{1}{2\pi}\ln(1.5) = m[\mathring{U}_2] = m[b_{12}\mathring{U}_2]\,, \\ \nu_2 &= 4 + \frac{1}{2\pi}\ln(1.5) > 2 + \frac{1}{4\pi}\ln(1.84) = m[\mathring{U}_1] = m[b_{21}\mathring{U}_1]\,, \\ \nu_3 &= 4 + \frac{1}{2\pi}\ln(1.84) > \frac{4}{5}(3 + \frac{1}{2\pi}\ln(0.7)) = m[2\mathring{U}_3] = m[b_{13}\mathring{U}_3]\,. \end{aligned}$$

By Theorem 8.3 and Corollary 8.1, we may conclude that in our model, the third species goes to extinction.

To complete the knowledge of the dynamics of the impulsive system (8.1), we have to investigate the asymptotic behavior of species $u_1(t)$ and $u_2(t)$ when the extinction of species $u_3(t)$ occurs.

Theorem 8.4. *Suppose that all hypotheses of Theorems 7.4 and 8.3 are satisfied. Then, for every positive solution $u(t)$ of (8.1), we have*

(i) $\lim\limits_{t\to+\infty} u_3(t) = 0.$

(ii) $\lim\limits_{t\to+\infty} |u_1(t) - \mathring{u}_1(t)| = 0 = \lim\limits_{t\to+\infty} |u_2(t) - \mathring{u}_2(t)|,$

where $(\mathring{u}_1(t), \mathring{u}_2(t))$ is the unique positive periodic solution of the two-dimensional system (7.1) whose existence is guaranteed by Theorem 7.4.

Proof. Property (i) has been proved in Theorem 8.3.

Now, take the $u(t)$ positive solution of (8.1). It is obvious that $(u_1(t), u_2(t))$ verifies the two-species system

$$\begin{cases} u_1'(t) = u_1(t)[a_1(t) - b_{11}(t)\,u_1(t) - b_{12}(t)\,u_2(t) - b_{13}(t)\,u_3(t)]\,, & t \neq t_k, \\ u_2'(t) = u_2(t)[a_2(t) - b_{21}(t)\,u_1(t) - b_{22}(t)\,u_2(t) - b_{23}(t)\,u_3(t)]\,, & t \neq t_k, \\ u_i(t_k^+) = (1 + c_{ik})u(t_k)\,, & i = 1, 2\,. \end{cases} \tag{8.19}$$

As $(\overset{\circ}{u}_1(t), \overset{\circ}{u}_2(t))$ satisfies system (7.1), the substitution

$$x(t) = \frac{u_1(t)}{\overset{\circ}{u}_1(t)} - 1\,, \quad y(t) = \frac{u_2(t)}{\overset{\circ}{u}_2(t)} - 1$$

turns differential system (8.19) into

$$\begin{cases} x' = (1 + x)(-a_{11}(t)\,x - a_{12}(t)\,y - (b_{13} \cdot u_3)(t)) \\ y' = (1 + y)(-a_{21}(t)\,x - a_{22}(t)\,y - (b_{23} \cdot u_3)(t)) \end{cases}$$

with coefficients $a_{ij}(t) \in PC_T$ defined by (7.9).

At this point, considering the Lyapunov function

$$V(t) = x(t) - \ln(1 + x(t)) + y(t) - \ln(1 + y(t))\,, \tag{8.20}$$

and calculating its derivative, we obtain

$$V'(t) = -a_{11}(t)\,x^2(t) - (a_{12} + a_{21})(t)\,x(t)y(t) - a_{22}(t)\,y^2(t) + g(t)\,, \quad t \neq t_k$$

where

$$g(t) = (-b_{13} \cdot x - b_{23} \cdot y)(t)\,u_3(t)\,.$$

Using the arguments employed in Theorem 7.4, we deduce

$$V'(t) \leq \alpha(t)(1 - e^{-V(t)}) + g(t)\,, \quad t \neq t_k\,.$$

Since

$$\lim_{t \to +\infty} g(t) = 0$$

and

$$m[\alpha] < 0\,,$$

using the arguments of Theorem 5.4, we get

$$\lim_{t \to +\infty} V(t) = 0\,.$$

Going back to $x(t)$ and $y(t)$ through (8.20), we obtain

$$\lim_{t \to +\infty} x(t) = 0 = \lim_{t \to +\infty} y(t)\,.$$

The relationship between $(x(t), y(t))$ and $(u_1(t), u_2(t))$ gives property (ii). □

Roughly speaking, we can conclude that the influence of impulsive perturbations on the Lotka–Volterra system (8.1) lies in constants ν_i. In fact, in the absence of impulses, we have $\nu_i = m[a_i]$ so that the results of this last section agree with those of Section 5.

Bibliography

[1] L. Ya. Adrianova, *Introduction to Linear Systems of Differential Equations*, American Mathematical Society, Providence, Rhode Island, 1995.

[2] S. Ahmad and A. C. Lazer, *Necessary and sufficient average growth in a Lotka–Volterra system*, Nonlinear Anal. **34** (1998), 191–228.

[3] S. Ahmad and A. C. Lazer, *Average conditions for global asymptotic stability in a nonautonomous Lotka–Volterra system*, Nonlinear Anal. **40** (2000), 37–49.

[4] S. Ahmad and I. M. Stamova, *Asymptotic stability of an N-dimensional impulsive competitive system*, Nonlinear Anal. Real World Appl. **8** (2007), 654-663.

[5] S. Ahmad and M. R. Mohana Rao, *Theory of Ordinary Differential Equations*, Affiliated East-West Press Private Limited, New Delhi, 1999.

[6] S. Ahmad and F. Montes de Oca, *Extinction in nonautonomous T-periodic competitive Lotka–Volterra system*, Appl. Math. Comput. **90** (1998), 155–166.

[7] S. Ahmad and A. Tineo, *Three-dimensional population systems*, Nonlinear Anal. Real World Appl. **9** (2008), 1607–1611.

[8] D. D. Bainov and P. S. Simeonov, *Impulsive Differential Equations: Periodic Solutions and Applications*, Longmann, New York, 1993.

[9] T. A. Burton, *Stability and Periodic Solutions of Ordinary and Functional Differential Equations*, Academic Press, 1985.

[10] P. de Mottoni and A. Schiaffino, *Competition systems with periodic coefficients: a geometric approach*, J. Math. Anal. Appl. **11** (1981), 319–335.

[11] J. C. Eilbeck and J. Lopez-Gomez, *On the periodic Lotka–Volterra competition model*, J. Math. Anal. Appl. **210** (1997), 58–87.

[12] T. G. Hallam, L. J. Svoboda and T. C. Gard, *Persistence and extinction in three species Lotka–Volterra systems*, Math. Biosci. **46** (1979), 117–124.

[13] J. Hou, Z. Teng and S. Gao, *Permanence and global stability for nonautonomous N-species Lotka–Volterra competitive system with impulses*, Nonlinear Anal. Real World Appl. **11** (2010), 1882–1896.

[14] B. Lisena, *Global stability in periodic competitive systems*, Nonlinear Anal. Real World Appl. **5** (2004), 613–627.

[15] B. Lisena, *Competitive exclusion in a periodic Lotka–Volterra system*, Appl. Math. Comput. **177** (2006), 761–768.

[16] B. Lisena, *Extinction in three species competitive systems with periodic coefficients*, Dynam. Systems Appl. **14** (2005), 393–406.

[17] B. Lisena, *Dynamic behaviour of a periodic competitive system with pulses*, Discrete Cont. Dynam. Syst. Ser. S **6** (2013), no. 3, 723–729.

[18] B. Lisena, *Stability and periodicity in competitive systems with impulses*, Mediterr. J. Math. **6** (2009), 291–302.

[19] B. Lisena, *Coexistence and extinction in competitive systems with impulses*, Dynam. Contin. Discrete Impuls. Systems Series A **17** (2010), 619–637.

[20] J. Lopez-Gomez, R. Ortega and A. Tineo, *The periodic predator–prey Lotka–Volterra model*, Adv. Differential Equations **1** (1996), 405–423.

[21] B. Liu, Z. Teng and W. Liu, *Dynamic behaviors of the periodic Lotka–Volterra competing systems with impulsive perturbations*, Chaos, Solutions Fractals **31** (2007), 356–370.

[22] I. M. Stamova, *Stability Analysis of Impulsive Functional Differential Equations*, Walter de Gruyter, Berlin, 2009.

[23] E. S. Stein and R. Shakarchi, *Real Analysis: Measure Theory, Integration, Hilbert Spaces*, Princeton University Press, Princeton and Oxford, 2005.

[24] A. Tineo, *Necessary and sufficient conditions for extinction of one species*, Adv. Nonlinear Studies **5** (2005), 57–71.

[25] A. Tineo and C. Alvarez, *A different consideration about the globally asymptotically stable solution of the periodic n-competing species problem*, J. Math. Anal. Appl. **159** (1991), 44–50.

[26] W. Wang, J. Shen and Z. Luo, *Partial survival and extinction in two competing species with impulses*, Nonlinear Anal. Real World Appl. **10** (2009), 1243–1254.

[27] J. Zhao, J. Jiang and A. C. Lazer, *The permanence and global attractivity in a nonautonomous Lotka–Volterra system*, Nonlinear Anal. Real World Appl. **5** (2004), 265–276.

Marina Pireddu and Fabio Zanolin

Fixed points, periodic points and chaotic dynamics for continuous maps with applications to population dynamics

Marina Pireddu: Università degli Studi di Milano-Bicocca, Dipartimento di Matematica e Applicazioni, Milano, Italy
Fabio Zanolin: Università degli Studi di Udine, Dipartimento di Matematica e Informatica, Udine, Italy

1 Introduction

In this chapter, we introduce a topological method for the search of fixed points and periodic points for continuous maps defined on rectangular domains in finite-dimensional Euclidean spaces. We name our technique the "Stretching Along the Paths" (SAP) method since we deal with maps that expand the arcs (paths) along one direction. This theory was developed in the planar case by Papini and Zanolin in [116–118] and motivated by a previous work concerning the study of superlinear equations with sign-indefinite weight [115]. It has been extended to the N-dimensional setting in [119] and by the present authors in [125]. Further extensions and applications to three-dimensional Lotka–Volterra systems have been recently proposed in [144, 145]. In order to develop our method and to prove the main results, we observe that, in the two-dimensional setting, elementary theorems from plane topology suffice. However, in a higher dimension, some results from topological degree theory are needed, leading to the study of the so-called "Cutting Surfaces" [125].

In the past decades, a big effort has been addressed towards the search of fixed and periodic points for maps in Euclidean spaces. In this respect, several authors [7, 8, 14, 15, 59, 107, 131, 156, 158, 163, 174, 182, 186] have studied maps that are expansive along some directions and compressive along other ones, obtaining results that are also significant from a dynamical point of view, in particular, in relation to Markov partitions. Most of these achievements usually require sophisticated algebraic tools, for example, the Conley index or the Lefschetz number. Our approach, even if confined to a special configuration, instead looks more elementary. The description of the Stretching Along the Paths method and suitable variants of it can be found in Section 4. In Section 7, we discuss the chaotic features that can be obtained for a given map when our technique applies. In particular, we are able to prove the semiconjugacy to the Bernoulli shift and thus the positivity of the topological entropy, the presence of topological transitivity and sensitivity with respect to initial conditions, and the density of periodic points. Moreover, we show the mutual relationships among various classical notions of chaos (such as those by Devaney, Li–Yorke, etc.). We also introduce an alternative geometrical framework related to the so-called "linked twist maps," where it is possible to employ our method in order to detect complex dynamics. Sections 6–9 are as self-contained as possible and hence accessible also to the reader who is not familiar with the concept of chaos. Due to the large quantity of definitions available in the literature, our exposition is thought of as a brief survey where a comparison with the notion of chaos in the sense of coin-tossing is presented as well. Such characterization is indeed natural when considered in relation with the approach in Section 3. The reader who is not strictly interested in topics concerning "chaotic dynamics" and the mutual relationships between different definitions and concepts of "chaos" can skip Sections 8 and 9 without harm.

The theoretical results obtained so far find an application to discrete- and continuous-time systems. In [105], in collaboration with Professor Alfredo Medio, applications have been proposed to some one-dimensional and planar discrete economic models, both of the Overlapping Generation and Duopoly Game classes. These economic models are taken from [104, 136] and [2], respectively. On the other hand, in Section 12, we analyze some nonlinear ODEs with periodic coefficients, with specific emphasis placed on the applications for population dynamics. In more detail, we consider a modified version of the Volterra predator–prey model in which a periodic harvesting is included. When dealing with ODEs with periodic coefficients, our method is applied to the associated Poincaré map and thus we are led back to work with discrete dynamical systems.

The above summary is just meant to mention the main aspects of this work. Each section is indeed equipped with a more detailed introduction, including the corresponding bibliography.

The contents of the present chapter (except for Sections 8 and 9) are based on the papers [105, 120, 125–127] and partially on [124], where maps which are expansive along several directions were considered. The results collected in Sections 8 and 9 are due to the first co-author (M. Pireddu) alone and taken from her Ph.D. thesis.

The so-called "SAP method" has already been applied in several different situations, mainly in connection with the search of periodic solutions and chaotic dynamics for nonlinear ordinary differential equations with periodic coefficients. Recent results in this area can be found in [27, 44, 115, 117]. Specific applications to nonlinear ODEs have been produced in [31, 32, 99, 123] for nonlinear pendulum-type equations, in [121, 122] for the study of a simplified version of the Lazer–McKenna suspension bridges model ([89, 90]), and in [98] for planar Hamiltonian systems. Applications to mathematical models like the nonlinear Schrödinger equation and to some Nagumo type equations for the nerve fiber can be found in [180] and [178, 179], respectively. In [143], Alfonso Ruiz-Herrera has recently applied these techniques to the study of some Lotka–Volterra equations with impulsive effects. Extensions to three-dimensional systems arising in ecology have been achieved in [144, 145]. The abstract framework for the SAP method has been developed in a series of papers in the past ten years [105, 115, 117–119, 121, 125], sometimes with different perspectives. We hope that the first sections of this chapter, collecting a selection of the main results taken from some of the above quoted papers, including proofs with complete technical details, may provide the interested reader with a comprehensive treatment of the subject.

Acknowledgments

As mentioned above, this work is mainly based on recent articles by the authors and their collaborators. Among the collaborators who were directly involved in the research results collected in the present chapter, we wish to deeply thank Professor Alfredo Medio for [105] and Dr Anna Pascoletti for [120]. In particular, the profound knowledge in Mathematical Economics of Professor Alfredo Medio was vital for the applications of the SAP method to discrete dynamical systems arising from the study of mathematical models in Economics ([105]). Discussions with Professor Medio also largely contributed to enhance the presentation and the clarification of some aspects of the theory. The second co-author (Fabio Zanolin) has a long list of friends, collaborators and students that he needs to thank on this occasion. First of all, a hearty thanks to Professor Duccio Papini, who initiated the research program regarding this topic in 1998. With his collaboration, the first articles [115–118] (from which all the subsequent developments of the theory originated) were produced. The start of the research with Duccio Papini in this area was strongly influenced by the study of the work of G. J. Butler [35] and by discussions with Professors Susanna Terracini, Gianmaria Verzini, Anna Capietto, Walter Dambrosio, Rafael Ortega, Tongren Ding, Bin Liu and Zhong Li. Special thanks are due to Professor Jean Mawhin for his constant encouragement and support. Further applications of the theory in the area of nonlinear ODEs with periodic coefficients have been obtained with Alberto Boscaggin [27], Lakshmi Burra [31, 32], Alessandro Margheri and Carlota Rebelo [98, 99], Alfonso Ruiz-Herrera [144, 145], and Chiara Zanini [178–180]. To all of them, a sincere thanks! Among the students (or former students) who were involved in this project and obtained some interesting results, Diego Covolan, Elena Bosa, Elisa Sovrano, Manuela Mazzariol, Gugliemo Feltrin, and Giuseppe Cian of the University of Udine must be mentioned for their stimulating discussions. Last, but not the least, a special and deep thanks to Professor Shair Ahmad. Without his strong support, it would have been impossible to conclude this project.

2 Notation

For the reader's convenience, we introduce some basic notation that will be used throughout this work.

We denote by $\mathbb{N}$, $\mathbb{Z}$, $\mathbb{Q}$, $\mathbb{R}$ and $\mathbb{C}$, the sets of *natural*, *integer*, *rational*, *real* and *complex numbers*, respectively. In particular, $\mathbb{N}$ is the set of nonnegative integers, while $\mathbb{N}_0$ denotes the set of the positive integers. The sets of nonnegative and positive real numbers will be indicated with $\mathbb{R}^+$ and $\mathbb{R}_0^+$. Accordingly, the first quadrant and the open first quadrant will be denoted by $(\mathbb{R}^+)^2$ and $(\mathbb{R}_0^+)^2$, respectively.

For a subset M of the topological space W, we denote by $\overline{M}$, $\mathrm{Int}(M)$ and ∂M the *closure* the *interior* and the *boundary* of M in W, respectively. The set $M \subseteq W$ is said to be *dense* in W if $\overline{M} = W$. If $M \subseteq W$, we indicate with $W \setminus M$ the *complement* of M in W. In the case that $A, B \subseteq W$, we denote by $A \setminus B$ the *relative complement* of B in A. By $|A|$, we mean the *cardinality* of the set A.

We denote by Id the *identity map* and by Id_W the identity on the space W when we need to specify it. Given a function f, we sometimes indicate its *domain* with D_f. By $f \restriction_M$, we mean the *restriction* of f to a subset M of its domain. For a function $f\colon W \supseteq D_f \to Z$ between the topological spaces W and Z, we define the *preimage* of $z \in Z$ as $f^{-1}(z) := \{w \in D_f\colon f(w) = z\}$.

A function $f\colon W \to W$ is called a *self-map* of the space W. The *iterates* of f are defined recursively with the convention $f^0 = \mathrm{Id}_W$, $f^1 = f$ and $f^n = f \circ f^{n-1}$, $\forall n \geq 2$. We say that $\overline{w} \in W$ is a *fixed point* for f if $f(\overline{w}) = \overline{w}$. We say that $\overline{w} \in W$ is a *periodic point* for f if there exists an integer $l \geq 1$ such that $f^l(\overline{w}) = \overline{w}$. The minimal l such that $f^l(\overline{w}) = \overline{w}$ is called the *period* of $\overline{w}$.

The N-dimensional Euclidean space $\mathbb{R}^N$ is endowed with the usual scalar product $\langle \cdot, \cdot \rangle$, norm $\|\cdot\|$ and distance $\mathrm{dist}(\cdot, \cdot)$. In the case of $\mathbb{R}$, the norm $\|\cdot\|$ will be replaced with the *absolute value* $|\cdot|$. If X is a metric space different from $\mathbb{R}^N$, we denote by d_X the distance defined on it. The distance between $M_1, M_2 \subseteq X$ is indicated with $d_X(M_1, M_2) := \inf\{d_X(x, y)\colon x \in M_1,\ y \in M_2\}$.

In a normed space $(X, \|\cdot\|_X)$, we denote by $B(x_0, r)$ and $B[x_0, r]$ the *open* and *closed balls* centered in $x_0 \in X$ with radius $r > 0$, i.e. $B(x_0, r) := \{x \in X\colon \|x - x_0\|_X < r\}$ and $B[x_0, r] := \{x \in X\colon \|x - x_0\|_X \leq r\}$. For $M \subseteq X$, we set $B(M, r) := \{x \in X\colon \exists w \in M \text{with} \|x - w\|_X < r\}$. The set $B[M, r]$ is defined accordingly. If we wish to specify the dimension m of the ball when it is contained in $\mathbb{R}^m$, we write $B_m(x_0, r)$ and $B_m[x_0, r]$ in place of $B(x_0, r)$ and $B[x_0, r]$. We denote by S^{N-1} the $N-1$-dimensional *unit sphere* embedded in $\mathbb{R}^N$, that is, $S^{N-1} := \partial B_N[0, 1]$.

Given an open bounded subset Ω of $\mathbb{R}^N$, an element p of $\mathbb{R}^N$ and a continuous function $f\colon \overline{\Omega} \to \mathbb{R}^N$, we indicate with $\deg(f, \Omega, p)$ the *topological degree* of the map f with respect to Ω and p. We say that $\deg(f, \Omega, p)$ is defined if $f(x) \neq p$, $\forall x \in \partial\Omega$ [95].

We denote by $C(D, \mathbb{R}^N)$ the set of the continuous maps $f\colon D \to \mathbb{R}^N$ on a compact set D endowed with the infinity norm $|f|_\infty := \max_{x \in D} \|f(x)\|$.

Given a compact interval $[a, b] \subset \mathbb{R}$, $L^1([a, b]) := \{f\colon [a, b] \to \mathbb{R}\colon f$ is Lebesgue measurable and $\int_a^b |f(t)| dt < +\infty\}$, where the integral is the Lebesgue integral. This space is endowed with the norm $||f||_1 := \int_a^b |f(t)| dt$.

3 Search of fixed points for maps expansive along one direction

One of the main areas in general topology concerns the search of fixed points for continuous maps defined on arbitrary topological spaces. Given a topological space $W \supseteq A \neq \emptyset$ and a continuous map $f\colon A \to W$, a *fixed point* for f is simply a point $\bar{x} \in A$ that is not moved by the map, i. e. $f(\bar{x}) = \bar{x}$. In spite of the simplicity of the definition, the concept of fixed points turns out to be central in the study of dynamical systems since, for instance, fixed points of suitable operators correspond to periodic solutions (cf. Section 11). A concept related to that of a fixed point is the *periodic point*. Namely, the periodic points of a continuous map f are the fixed points of the iterates of the map, that is, $\bar{x}$ is a periodic point for f if there exists an integer $l \geq 1$ such that $f^l(\bar{x}) = \bar{x}$.

As it often happens, the more general the spaces we consider, the more sophisticated the tools to be employed: the search of fixed points and periodic points can indeed require the use of advanced theories, like that of the Conley or fixed point index [107, 109, 139, 163, 174, 182], Lefschetz number [87, 155, 156, 158] and many other geometric or algebraic methods [18, 52, 159, 173]. In the past decades, several efforts have been made in order to find elementary tools to deal with such a problem. For example, the Poincaré–Miranda Theorem or some other equivalent versions of the Brouwer fixed point Theorem have turned out to be very useful when dealing with N-dimensional Euclidean spaces [13, 100].

It is among these not too sophisticated approaches that the theory of the Cutting Surfaces for the search of fixed point and periodic points, introduced by the authors in [125], has to be placed. More precisely, it concerns continuous maps defined on generalized N-dimensional rectangles and having an expansive direction. Our main tools are a modified version of the classical Hurewicz–Wallman Intersection Lemma [54, p. 72], [68, D), p. 40] and the Fundamental Theorem of Leray–Schauder [91, Théorème Fondamental]. The former result (also referred to Eilenberg–Otto [53, Th. 3], according to [51, Theorem on partitions, p. 100]) is one of the basic lemmas in dimension theory and it is known to be one of the equivalent versions of the Brouwer fixed point Theorem. It may be interesting to observe that extensions of the Intersection Lemma led to generalizations of the Brouwer fixed point Theorem to some classes of possibly noncontinuous functions [168]. On the other hand, the Leray–Schauder Continuation Theorem concerns topological degree theory and has found important applications in nonlinear analysis and differential equations ([101]). Actually, in Theorem 5.3 of Section 5, we will employ a more general version of such results due to Fitzpatrick, Massabó and Pejsachowicz [56]. However, we point out that for our applications regarding the study of periodic points and chaotic-like dynamics, we rely on the classical Leray–Schauder Fundamental Theorem. Another particular feature of our approach consists of a combination of Poincaré–Miranda Theorem [82, 102], which is an N-dimensional version of the intermediate value theorem, with the properties of topological surfaces cutting the arcs between two given sets. The theory of Cutting Surfaces is explained in Section 5.

When we confine ourselves to the bidimensional setting, suitable results from plane topology can be applied [120], like the Crossing Lemma 4.5, and no topological degree arguments are needed. In this case, we enter the context of the "Stretching Along the Paths" method developed by Papini and Zanolin in [117, 118]. The name comes from the fact that one deals with maps that expand the arcs, in a sense that will be clarified in Section 4, where the interested reader can find all the related details. We stress that, with respect to [117, 118], the terminology and notation are slightly different and some proofs are new.

In any case, our approach based on the Cutting Surfaces represents a possible extension of the planar theory in [117, 118] to the N-dimensional setting for any $N \geq 2$. By the similarity between the concepts employed and the results obtained in the two frameworks, we will keep the "stretching" terminology also for the higher dimensional case without distinguishing between them. Indeed, the main difference between the planar setting and the N-dimensional one resides in the tools to be employed in order to get some results, for instance, the fixed point Theorems 4.1 and 5.6, on which the two theories are respectively based. Namely, as already mentioned, in the plane elementary, results from general topology suffices, while in the higher dimension, one needs more sophisticated tools, such as topological degree.

4 The planar case

4.1 Stretching along the paths and variants

Before introducing the main concepts of the "Stretching Along the Paths" method, let us recall some facts about paths, arcs and continua.

Let W be a topological space. By a *path* γ in W, we mean a continuous map $\gamma\colon \mathbb{R} \supseteq [a,b] \to W$. Its range will be denoted by $\overline{\gamma}$, that is, $\overline{\gamma} := \gamma([a,b])$. A *subpath* ω of γ is the restriction of γ to a closed subinterval of its domain and hence it is defined as $\omega := \gamma \restriction_{[c,d]}$ for some $[c,d] \subseteq [a,b]$. If W, Z are topological spaces and $\psi\colon W \supseteq D_\psi \to Z$ is a map which is continuous on a set $\mathcal{M} \subseteq D_\psi$, then for any path γ in W with $\overline{\gamma} \subseteq \mathcal{M}$, it follows that $\psi \circ \gamma$ is a path in Z with range $\psi(\overline{\gamma})$. Notice that there is no loss of generality in assuming the paths to be defined on $[0,1]$. Indeed, if $\theta_1\colon [a_1,b_1] \to W$ and $\theta_2\colon [a_2,b_2] \to W$, with $a_i < b_i$, $i = 1,2$ are two paths in W, we define the equivalence relation "$\sim$" between θ_1 and θ_2 by setting $\theta_1 \sim \theta_2$ if there is a homeomorphism h of $[a_1,b_1]$ onto $[a_2,b_2]$ such that $\theta_2(h(t)) = \theta_1(t)$, $\forall t \in [a_1,b_1]$. It is easy to check that if $\theta_1 \sim \theta_2$, then the ranges of θ_1 and θ_2 coincide. Hence, for any path γ, there exists an equivalent path defined on $[0,1]$. In view of this fact, we will usually deal with paths defined on $[0,1]$, though sometimes we will also consider paths defined on an arbitrary interval $[a, b]$ when this can help to simplify the exposition. A concept similar to the one of a path is that of an arc. More precisely, an *arc* is the homeomorphic image of the compact interval $[0, 1]$, while an *open arc* is an arc without its end-points. A *continuum* of W is a compact connected subset of W and a *subcontinuum* is a subset of a continuum which is itself a continuum.

Now, we can start with our definitions. Given a metric space X, we call a *generalized rectangle* any set $\mathcal{R} \subseteq X$ homeomorphic to the unit square $\mathcal{Q} := [0, 1]^2$ of $\mathbb{R}^2$. If $\mathcal{R}$ is a generalized rectangle and $h\colon \mathcal{Q} \to h(\mathcal{Q}) = \mathcal{R}$ is a homeomorphism defining it, we call a *contour* $\vartheta\mathcal{R}$ of $\mathcal{R}$ the set

$$\vartheta\mathcal{R} := h(\partial\mathcal{Q}) ,$$

where $\partial\mathcal{Q}$ is the usual boundary of the unit square. Notice that the contour $\vartheta\mathcal{R}$ is well-defined as it does not depend on the choice of the homeomorphism h. In fact, $\vartheta\mathcal{R}$ is also a homeomorphic image of S^1, that is, a Jordan curve.

By an *oriented rectangle*, we mean a pair

$$\widetilde{\mathcal{R}} := (\mathcal{R}, \mathcal{R}^-) ,$$

where $\mathcal{R} \subseteq X$ is a generalized rectangle and

$$\mathcal{R}^- := \mathcal{R}^-_\ell \cup \mathcal{R}^-_r$$

is the union of two disjoint arcs $\mathcal{R}^-_\ell, \mathcal{R}^-_r \subseteq \vartheta\mathcal{R}$ which we call the *left* and the *right sides* of $\mathcal{R}^-$. Since $\vartheta\mathcal{R}$ is a Jordan curve, it follows that $\vartheta\mathcal{R} \setminus (\mathcal{R}^-_\ell \cup \mathcal{R}^-_r)$ consists of two open arcs. We denote by $\mathcal{R}^+$ the closure of such open arcs that we name $\mathcal{R}^+_d$ and $\mathcal{R}^+_u$ (the *down* and *up sides* of $\mathcal{R}^+$). It is important to notice that we can always label the arcs $\mathcal{R}^-_\ell$, $\mathcal{R}^+_d$, $\mathcal{R}^-_r$ and $\mathcal{R}^+_u$, following the cyclic order $\ell - d - r - u - \ell$, and take a homeomorphism $g\colon \mathcal{Q} \to g(\mathcal{Q}) = \mathcal{R}$ so that

$$\begin{aligned} g(\{0\} \times [0,1]) &= \mathcal{R}^-_\ell , \quad & g(\{1\} \times [0,1]) &= \mathcal{R}^-_r , \\ g([0,1] \times \{0\}) &= \mathcal{R}^+_d , \quad & g([0,1] \times \{1\}) &= \mathcal{R}^+_u . \end{aligned} \tag{4.1}$$

Both the term "generalized rectangle" for $\mathcal{R}$ and the decomposition of the contour $\vartheta\mathcal{R}$ into $\mathcal{R}^-$ and $\mathcal{R}^+$ are somehow inspired to the construction of rectangular domains around hyperbolic sets arising in the theory of Markov partitions [62, p. 291], as well as by the Conley–Ważewski theory [42, 157]. Roughly speaking, in such frameworks, the sets labeled as $[\cdot]^-$, or as $[\cdot]^+$, are made by those points which are moved by the flow outward, respectively inward, with respect to $\mathcal{R}$. As we shall see in the next definition of the *stretching along the paths* property, also in our case, the $[\cdot]^-$-set is loosely related to the expansive direction. Indeed, we have:

Definition 4.1. Let X be a metric space and let $\psi\colon X \supseteq D_\psi \to X$ be a map defined on a set D_ψ. Assume that $\widetilde{\mathcal{A}} := (\mathcal{A}, \mathcal{A}^-)$ and $\widetilde{\mathcal{B}} := (\mathcal{B}, \mathcal{B}^-)$ are oriented rectangles of X and let $\mathcal{K} \subseteq \mathcal{A} \cap D_\psi$ be a compact set. We say that $(\mathcal{K}, \psi)$ *stretches* $\widetilde{\mathcal{A}}$ *to* $\widetilde{\mathcal{B}}$ *along the paths* and write

$$(\mathcal{K}, \psi)\colon \widetilde{\mathcal{A}} \Leftrightarrow\!\!\!\!\to \widetilde{\mathcal{B}} \tag{4.2}$$

if the following conditions hold:
- ψ is continuous on $\mathcal{K}$;
- For every path $\gamma\colon [0,1] \to \mathcal{A}$ such that $\gamma(0) \in \mathcal{A}_\ell^-$ and $\gamma(1) \in \mathcal{A}_r^-$ (or $\gamma(0) \in \mathcal{A}_r^-$ and $\gamma(1) \in \mathcal{A}_\ell^-$), there exists a subinterval $[t',t''] \subseteq [0,1]$ with

  $$\gamma(t) \in \mathcal{K}\,, \quad \psi(\gamma(t)) \in \mathcal{B}\,, \quad \forall t \in [t',t'']$$

  and, moreover, $\psi(\gamma(t'))$ and $\psi(\gamma(t''))$ belong to different sides of $\mathcal{B}^-$.
  In the special case in which $\mathcal{K} = \mathcal{A}$, we simply write

  $$\psi\colon \widetilde{\mathcal{A}} \,\Longleftrightarrow\!\!\!\!\to\, \widetilde{\mathcal{B}}\,.$$

The role of the compact set $\mathcal{K}$ is crucial in the results which use Definition 4.1. For instance, if (4.2) is satisfied with $\widetilde{\mathcal{B}} = \widetilde{\mathcal{A}}$, that is,

$$(\mathcal{K},\psi)\colon \widetilde{\mathcal{A}} \,\Longleftrightarrow\!\!\!\!\to\, \widetilde{\mathcal{A}}\,, \tag{4.3}$$

we are able to prove the existence of a fixed point for ψ in $\mathcal{K}$ (Theorem 4.1).

Notice that when (4.2) holds, the stretching condition

$$(\mathcal{K}',\psi)\colon \widetilde{\mathcal{A}} \,\Longleftrightarrow\!\!\!\!\to\, \widetilde{\mathcal{B}}$$

is fulfilled for any compact set $\mathcal{K}'$, with $\mathcal{K} \subseteq \mathcal{K}' \subseteq \mathcal{A} \cap D_\psi$, on which ψ is continuous. On the other hand, since $\mathcal{K}$ provides a localization of the fixed point, it is convenient to choose it as small as possible.

We point out that in Definition 4.1 (as well as in its variant in Definition 4.2 below), we don't require $\psi(\mathcal{A}) \subseteq \mathcal{B}$. In fact, we will refer to $\mathcal{B}$ as a "target set" and not as a codomain. For a geometrical interpretation of Definition 4.1 (and Definition 4.2 below), see Figure 1.

The relation introduced in Definition 4.1 is fundamental in our "stretching along the paths" method. Indeed, such strategy consists of checking property (4.3) so that, in view of Theorem 4.1, there exists at least a fixed point for ψ in $\mathcal{K}$. If (4.3) is particularly satisfied with respect to two or more pairwise disjoint compact sets $\mathcal{K}_i$'s, we get a multiplicity of fixed points. On the other hand, when (4.3) holds for some iterate of ψ, the existence of periodic points is ensured. Since the stretching property in (4.2) is preserved under composition of mappings (cf. Lemma 4.1), combining such arguments, the presence of chaotic dynamics also follows. A detailed treatment of such a topic can be found in Section 7.

Definition 4.2. Let X be a metric space and let $\psi\colon X \supseteq D_\psi \to X$ be a map defined on a set D_ψ. Assume that $\widetilde{\mathcal{A}} := (\mathcal{A},\mathcal{A}^-)$ and $\widetilde{\mathcal{B}} := (\mathcal{B},\mathcal{B}^-)$ are oriented rectangles of X and let $\mathcal{D} \subseteq \mathcal{A} \cap D_\psi$ be a compact set. Also, let $m \geq 2$ be an integer. We say that $(\mathcal{D},\psi)$ *stretches* $\widetilde{\mathcal{A}}$ *to* $\widetilde{\mathcal{B}}$ *along the paths with crossing number* m and write

$$(\mathcal{D},\psi)\colon \widetilde{\mathcal{A}} \,\Longleftrightarrow\!\!\!\!\to^{m}\, \widetilde{\mathcal{B}}\,,$$

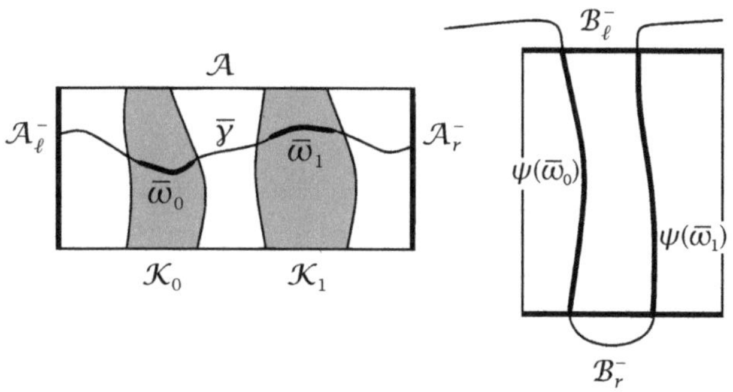

Fig. 1: A pictorial comment to Definitions 4.1 and 4.2. The rectangles $\mathcal{A}$ and $\mathcal{B}$ have been oriented by selecting the sets $\mathcal{A}^-$ and $\mathcal{B}^-$ (drawn with thicker lines), respectively. We represent a case in which the relation $(\mathcal{K}_i, \psi)\colon \widetilde{\mathcal{A}} \leftrightarrow\!\!\!\rightarrow \widetilde{\mathcal{B}}$, $i = 0, 1$ is satisfied for a map $\psi\colon \mathbb{R}^2 \supseteq \mathcal{A} \to \mathbb{R}^2$ and for the two darker compact subsets $\mathcal{K}_0$ and $\mathcal{K}_1$ of $\mathcal{A}$, on which ψ is continuous. For a generic path $\gamma\colon [0,1] \to \mathcal{A}$ with $\gamma(0)$ and $\gamma(1)$ belonging to different components of $\mathcal{A}^-$, we have highlighted two subpaths ω_0 and ω_1 with range in $\mathcal{K}_0$ and $\mathcal{K}_1$, respectively, such that their composition with ψ determines two new paths (drawn by bolder vertical lines) with values in $\mathcal{B}$ and joining the two sides of $\mathcal{B}^-$. In this framework, according to Definition 4.2, we could also write $\psi\colon \widetilde{\mathcal{A}} \overset{2}{\leftrightarrow\!\!\!\rightarrow} \widetilde{\mathcal{B}}$.

if there exist m pairwise disjoint compact sets

$$\mathcal{K}_0, \ldots, \mathcal{K}_{m-1} \subseteq \mathcal{D}$$

such that

$$(\mathcal{K}_i, \psi)\colon \widetilde{\mathcal{A}} \leftrightarrow\!\!\!\rightarrow \widetilde{\mathcal{B}}, \quad i = 0, \ldots, m-1 .$$

When $\mathcal{D} = \mathcal{A}$, we simply write

$$\psi\colon \widetilde{\mathcal{A}} \overset{m}{\leftrightarrow\!\!\!\rightarrow} \widetilde{\mathcal{B}} .$$

Definition 4.2 is an extension of Definition 4.1 as they are coincident when $m = 1$. The concept of a "crossing number" is borrowed from Kennedy and Yorke [73] and adapted to our framework. Indeed, in [73], the authors deal with a very general setting that looks closely related to ours when restricted to the planar case. More precisely, in [73, horseshoe hypotheses Ω], a locally connected and compact subset Q of the separable metric space X is considered, on which two disjoint and compact sets end_0, $end_1 \subseteq Q$ are selected, so that any component of Q intersects both of them. On Q, a continuous map $f\colon Q \to X$ is defined in such a way that every continuum $\Gamma \subseteq Q$ joining end_0 and end_1 (i. e. a *connection* according to [73]) admits at least $m \geq 2$ pairwise disjoint compact and connected subsets whose images under f are again connections. Such subcontinua are named *preconnections* and m is the so-called *crossing number*.

Clearly, the arcs $\mathcal{R}_\ell^-$ and $\mathcal{R}_r^-$ in our definition of oriented rectangles are a particular case of the sets end_0 and end_1 considered by Kennedy and Yorke. Moreover, any path γ with values in $\mathcal{R}$ and joining $\mathcal{R}_\ell^-$ with $\mathcal{R}_r^-$ determines a connection (according to [73]) via its image $\overline{\gamma}$. Similarly, any subpath ω of γ, with $\omega = \gamma \restriction_{[t',t'']}$ as in Definition 4.1, makes $\overline{\omega}$ a preconnection according to Kennedy and Yorke. In fact, the idea of considering paths to detect a sort of expansion was independently developed by these authors in [74, 75], where a terminology analogous to ours was employed. A special request in our approach, not explicitly assumed in [73], is the one concerning the compact sets like $\mathcal{K}$. Indeed, in [73], there are no sets playing the role of the $\mathcal{K}_i$'s in Definition 4.2.

On the other hand, our stretching condition is strong enough to ensure the existence of fixed points and periodic points for the map ψ inside the $\mathcal{K}_i$'s as we shall see in Theorems 4.1 and 4.3 below. However, in [73, Example 10], a fixed point free map defined on $\mathbb{R}^2 \times S^1$ and satisfying the horseshoe hypotheses Ω is presented ([39]). Indeed, we point out that Theorem 4.1 not only allows one to infer the existence of fixed points, but also to localize them. This turns out to be of particular importance when multiple coverings in the sense of Definition 4.2 occur.

In the proof of Theorems 4.1 and 4.3, we employ a classical result from plane topology (cf. Crossing Lemma 4.5) that we recall in Section 4.2 for the reader's convenience. See also [117, 118].

Theorem 4.1. *Let X be a metric space and let $\psi\colon X \supseteq D_\psi \to X$ be a map defined on a set D_ψ. Assume that $\widetilde{\mathcal{R}} := (\mathcal{R}, \mathcal{R}^-)$ is an oriented rectangle of X. If $\mathcal{K} \subseteq \mathcal{R} \cap D_\psi$ is a compact set for which it holds that*

$$(\mathcal{K}, \psi)\colon \widetilde{\mathcal{R}} \Leftrightarrow\!\!\!\rightarrow \widetilde{\mathcal{R}}\,, \tag{4.4}$$

then there exists at least one point $z \in \mathcal{K}$ with $\psi(z) = z$.

Proof. By the definition of an oriented rectangle, there exists a homeomorphism $h\colon \mathbb{R}^2 \supseteq \mathcal{Q} \to h(\mathcal{Q}) = \mathcal{R} \subseteq X$, mapping in a correct way (i. e. as in (4.1)) the sides of $\mathcal{Q} = [0,1]^2$ into the arcs that compose the sets $\mathcal{R}^-$ and $\mathcal{R}^+$. Then, passing to the planar map $\phi := h^{-1} \circ \psi \circ h$ defined on $D_\phi := h^{-1}(D_\psi) \subseteq \mathcal{Q}$, we can confine ourselves to the search of a fixed point for ϕ in the compact set $\mathcal{H} := h^{-1}(\mathcal{K}) \subseteq \mathcal{Q}$. The stretching assumption on ψ is now translated to

$$(\mathcal{H}, \phi)\colon \widetilde{\mathcal{Q}} \Leftrightarrow\!\!\!\rightarrow \widetilde{\mathcal{Q}}\,.$$

On $\widetilde{\mathcal{Q}}$, we consider the natural "left-right" orientation, choosing

$$\mathcal{Q}^- = (\{0\} \times [0,1]) \cup (\{1\} \times [0,1])\,.$$

A fixed point for ϕ in $\mathcal{H}$ corresponds to a fixed point for ψ in $\mathcal{K}$.

For $\phi = (\phi_1, \phi_2)$ and $x = (x_1, x_2)$, we define the compact set

$$V := \{x \in \mathcal{H}\colon 0 \le \phi_2(x) \le 1, x_1 - \phi_1(x) = 0\}\,.$$

The proof consists of showing that V contains a continuum C which joins in $\mathcal{Q}$ the lower side $[0,1] \times \{0\}$ to the upper side $[0,1] \times \{1\}$. To this end, in view of Lemma 4.5, it is sufficient to prove that V acts as a "cutting surface" (in the sense of Definition 5.1) between the left and the right sides of $\mathcal{Q}$, that is, V intersects any path in $\mathcal{Q}$ joining the left side $\{0\} \times [0,1]$ to the right side $\{1\} \times [0,1]$. Such a cutting property can be checked via the intermediate value theorem by observing that if $\gamma = (\gamma_1, \gamma_2)\colon [0,1] \to \mathcal{Q}$ is a continuous map with $\gamma(0) \in \{0\} \times [0,1]$ and $\gamma(1) \in \{1\}\times[0,1]$, then the stretching hypothesis $(\mathcal{H}, \phi)\colon \widetilde{\mathcal{Q}} \leftrightarrow\!\!\!\!\!\rightarrow \widetilde{\mathcal{Q}}$ implies that there exists an interval $[t', t''] \subseteq [0,1]$ such that $\gamma(t) \in \mathcal{H}$, $\phi(\gamma(t)) \in \mathcal{Q}$, $\forall t \in [t', t'']$ and $\gamma_1(t') - \phi_1(\gamma(t')) \geq 0 \geq \gamma_1(t'') - \phi_1(\gamma(t''))$ or $\gamma_1(t') - \phi_1(\gamma(t')) \leq 0 \leq \gamma_1(t'') - \phi_1(\gamma(t''))$. Notice that, by the definition of V, it follows that $\phi_2(z) \in [0,1], \forall z \in C$. Hence, for every point $p = (p_1, p_2) \in C \cap ([0,1] \times \{0\})$, we have $p_2 - \phi_2(p) \leq 0$ and, similarly, $p_2 - \phi_2(p) \geq 0$ for every $p = (p_1, p_2) \in C \cap ([0,1] \times \{1\})$. Applying the Bolzano Theorem, we obtain the existence of at least a point $v = (v_1, v_2) \in C \subseteq V \subseteq \mathcal{H}$ such that $v_2 - \phi_2(v) = 0$. Hence, v is a fixed point of ϕ in $\mathcal{H}$ and $z := h(v)$ is a fixed point for ψ in $\mathcal{K} \subseteq \mathcal{R}$. □

Remark 4.1. Notice that for the validity of Theorem 4.1, it is fundamental that the orientation of the generalized rectangle $\mathcal{R}$ in (4.4) remains the same for $\mathcal{R}$ considered as the "starting set" and "target set" of the map ψ. Indeed, if one chooses two different orientations for $\mathcal{R}$, in general, the above result no longer holds and the existence of fixed points for ψ is no longer ensured, not only in $\mathcal{K}$, but even in $\mathcal{R}$, as shown by the example depicted in Figure 2.

As a comment to Theorem 4.1, we review its main differences with respect to the more classical Brouwer fixed point Theorem (recalled in the two-dimensional case as Theorem 4.5 in Section 4.2). It is a well-known fact that the fixed point property for continuous maps is preserved by homeomorphisms. Therefore, it is straightforward to prove the existence of a fixed point for a continuous map ψ if $\psi(\mathcal{R}) \subseteq \mathcal{R}$, with $\mathcal{R}$ as a generalized rectangle of a metric space X. The framework depicted in Theorem 4.1 is quite different. Indeed, first of all, the stretching assumption $(\mathcal{K}, \psi)\colon \widetilde{\mathcal{R}} \leftrightarrow\!\!\!\!\!\rightarrow \widetilde{\mathcal{R}}$ does not imply $\psi(\mathcal{R}) \subseteq \mathcal{R}$ and, secondly, we need ψ to be continuous only on $\mathcal{K}$ and not on the whole set $\mathcal{R}$. Finally, as already noticed, we stress that our result also localizes the presence of a fixed point in the subset $\mathcal{K}$. From the point of view of the applications, this means that we are able to obtain a multiplicity of fixed points provided that the stretching property is fulfilled with respect to pairwise disjoint compact subsets of $\mathcal{R}$. Indeed, if the condition $(\mathcal{D}, \psi)\colon \widetilde{\mathcal{R}} \overset{m}{\leftrightarrow\!\!\!\!\!\rightarrow} \widetilde{\mathcal{R}}$ is satisfied with a crossing number $m \geq 2$, in view of Theorem 4.1, the map ψ has at least a fixed point in each of the compact sets $\mathcal{K}_i$'s, $i = 0, \ldots, m-1$ from Definition 4.2. Therefore, there are at least m fixed points for ψ in $\mathcal{R}$ (Theorem 4.3).

A different case related to the stretching relation from Definition 4.1 in which it is possible to find fixed points is when the special geometric configuration in Definition 4.3 gets realized, as stated in Theorem 4.2 below.

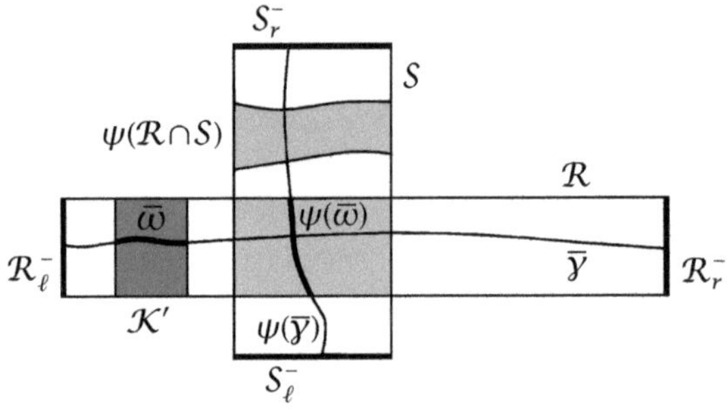

Fig. 2: The generalized rectangle $\mathcal{R}$ is transformed by a continuous planar map ψ onto the generalized rectangle $S = \psi(\mathcal{R})$ so that, in particular, $S_\ell^- = \psi(\mathcal{R}_\ell^-)$ and $S_r^- = \psi(\mathcal{R}_r^-)$. The boundary sets $\mathcal{R}^- = \mathcal{R}_\ell^- \cup \mathcal{R}_r^-$ and $S^- = S_\ell^- \cup S_r^-$ are drawn with thicker lines. As it is immediate to verify, for $\widetilde{\mathcal{R}} = (\mathcal{R}, \mathcal{R}^-)$ and $\widetilde{S} = (S, S^-)$, it holds that $\psi\colon \widetilde{\mathcal{R}} \Leftrightarrow\!\!\rightarrow \widetilde{S}$. On the other hand, calling $\widetilde{\widetilde{\mathcal{R}}}$ the generalized rectangle $\mathcal{R}$ oriented by choosing $\overline{\vartheta\mathcal{R} \setminus \mathcal{R}^-}$ as $[\cdot]^-$-set, it also holds that $(\mathcal{K}', \psi)\colon \widetilde{\mathcal{R}} \Leftrightarrow\!\!\rightarrow \widetilde{\widetilde{\mathcal{R}}}$, where $\mathcal{K}'$ is the subset of $\mathcal{R}$ depicted with a darker color. However, since $\mathcal{R} \cap S$ is mapped by ψ outside $\mathcal{R}$ (both $\mathcal{R} \cap S$ and $\psi(\mathcal{R} \cap S)$ are drawn with the same light color), there cannot exist fixed points for ψ in $\mathcal{R}$ and, a fortiori, neither in $\mathcal{K}'$. Notice that Theorem 4.1 does not apply because we have taken two different orientations for $\mathcal{R}$. A similar geometrical framework was already considered in [71, Figure 8] and [124, Figure 4].

Definition 4.3. Let $\widetilde{\mathcal{A}} := (\mathcal{A}, \mathcal{A}^-)$ and $\widetilde{\mathcal{B}} := (\mathcal{B}, \mathcal{B}^-)$ be two oriented rectangles of a metric space X. We say that $\widetilde{\mathcal{A}}$ is a *horizontal slab* of $\widetilde{\mathcal{B}}$ and write

$$\widetilde{\mathcal{A}} \subseteq_h \widetilde{\mathcal{B}}$$

if $\mathcal{A} \subseteq \mathcal{B}$ and, either

$$\mathcal{A}_\ell^- \subseteq \mathcal{B}_\ell^- \quad \text{and} \quad \mathcal{A}_r^- \subseteq \mathcal{B}_r^- ,$$

or

$$\mathcal{A}_\ell^- \subseteq \mathcal{B}_r^- \quad \text{and} \quad \mathcal{A}_r^- \subseteq \mathcal{B}_\ell^- ,$$

so that any path in $\mathcal{A}$ joining the two sides of $\mathcal{A}^-$ is also a path in $\mathcal{B}$ and joins the two opposite sides of $\mathcal{B}^-$.

We say that $\widetilde{\mathcal{A}}$ is a *vertical slab* of $\widetilde{\mathcal{B}}$ and write

$$\widetilde{\mathcal{A}} \subseteq_v \widetilde{\mathcal{B}}$$

if $\mathcal{A} \subseteq \mathcal{B}$ and every path in $\mathcal{B}$ joining the two sides of $\mathcal{B}^-$ admits a subpath in $\mathcal{A}$ that joins the two opposite sides of $\mathcal{A}^-$.

Given three oriented rectangles $\widetilde{\mathcal{A}} := (\mathcal{A}, \mathcal{A}^-)$, $\widetilde{\mathcal{B}} := (\mathcal{B}, \mathcal{B}^-)$ and $\widetilde{\mathcal{E}} := (\mathcal{E}, \mathcal{E}^-)$ of the metric space X, with $\mathcal{E} \subseteq \mathcal{A} \cap \mathcal{B}$, we say that $\widetilde{\mathcal{B}}$ *crosses* $\widetilde{\mathcal{A}}$ *in* $\widetilde{\mathcal{E}}$ and write

$$\widetilde{\mathcal{E}} \in \{\widetilde{\mathcal{A}} \pitchfork \widetilde{\mathcal{B}}\}$$

if

$$\widetilde{\mathcal{E}} \subseteq_h \widetilde{\mathcal{A}} \quad \text{and} \quad \widetilde{\mathcal{E}} \subseteq_v \widetilde{\mathcal{B}}\,.$$

The above definitions, which are adapted from the concept of "slice" in [118, 119] and imitate the classical terminology in [169, Ch. 2.3], are topological in nature and therefore do not necessitate any metric assumption (like smoothness, Lipschitzeanity, or similar properties often required in the literature). We also notice that the terms "horizontal" and "vertical" are employed in a purely conventional manner, as it is clear from Figure 3: the horizontal is the expansive direction and the vertical is the contractive one (in a quite broad sense). For instance, in [125], the terms "vertical" and "horizontal" were interchanged regarding the N-dimensional setting, but this did not make any difference with respect to the meaning of the results obtained.

The next theorem depicts a situation where the "starting set" and the "target set" of the mapping ψ are two intersecting oriented rectangles. A graphical illustration of it can be found in Figure 3.

Theorem 4.2. *Let X be a metric space and let $\psi\colon X \supseteq D_\psi \to X$ be a map defined on a set D_ψ. Assume that $\widetilde{\mathcal{A}} := (\mathcal{A}, \mathcal{A}^-)$ and $\widetilde{\mathcal{B}} := (\mathcal{B}, \mathcal{B}^-)$ are oriented rectangles of X and let $\mathcal{K} \subseteq \mathcal{A} \cap D_\psi$ be a compact set such that*

$$(\mathcal{K}, \psi)\colon \widetilde{\mathcal{A}} \,\overset{\leftrightarrow}{\longrightarrow}\, \widetilde{\mathcal{B}}\,. \tag{4.5}$$

If there exists an oriented rectangle $\widetilde{\mathcal{E}} := (\mathcal{E}, \mathcal{E}^-)$ with $\widetilde{\mathcal{E}} \in \{\widetilde{\mathcal{A}} \pitchfork \widetilde{\mathcal{B}}\}$, then ψ has at least a fixed point in $\mathcal{K} \cap \mathcal{E}$.

Proof. In order to achieve the thesis, we show that

$$(\mathcal{K} \cap \mathcal{E}, \psi)\colon \widetilde{\mathcal{B}} \,\overset{\leftrightarrow}{\longrightarrow}\, \widetilde{\mathcal{B}}\,. \tag{4.6}$$

Indeed, let γ be a path with $\overline{\gamma} \subseteq \mathcal{B}$ and $\overline{\gamma} \cap \mathcal{B}_\ell^- \neq \emptyset$, $\overline{\gamma} \cap \mathcal{B}_r^- \neq \emptyset$. Then, since $\widetilde{\mathcal{E}} \subseteq_v \widetilde{\mathcal{B}}$, there exists a subpath ω of γ such that $\overline{\omega} \subseteq \mathcal{E}$ and $\overline{\omega} \cap \mathcal{E}_\ell^- \neq \emptyset$, $\overline{\omega} \cap \mathcal{E}_r^- \neq \emptyset$. Recalling now that $\widetilde{\mathcal{E}} \subseteq_h \widetilde{\mathcal{A}}$, it holds that $\overline{\omega} \subseteq \mathcal{E} \subseteq \mathcal{A}$ and $\overline{\omega} \cap \mathcal{A}_\ell^- \neq \emptyset$, $\overline{\omega} \cap \mathcal{A}_r^- \neq \emptyset$. Finally, since $(\mathcal{K}, \psi)\colon \widetilde{\mathcal{A}} \overset{\leftrightarrow}{\longrightarrow} \widetilde{\mathcal{B}}$, there is a subpath η of ω such that $\overline{\eta} \subseteq \mathcal{K} \cap \mathcal{E}$, $\psi(\overline{\eta}) \subseteq \mathcal{B}$, with $\psi(\overline{\eta}) \cap \mathcal{B}_\ell^- \neq \emptyset$, $\psi(\overline{\eta}) \cap \mathcal{B}_r^- \neq \emptyset$. In this way, we have proved that any path γ with $\overline{\gamma} \subseteq \mathcal{B}$ and $\overline{\gamma} \cap \mathcal{B}_\ell^- \neq \emptyset$, $\overline{\gamma} \cap \mathcal{B}_r^- \neq \emptyset$ admits a subpath η such that $\overline{\eta} \subseteq \mathcal{K} \cap \mathcal{E}$ and $\psi(\overline{\eta}) \subseteq \mathcal{B}$ with $\psi(\overline{\eta}) \cap \mathcal{B}_\ell^- \neq \emptyset$, $\psi(\overline{\eta}) \cap \mathcal{B}_r^- \neq \emptyset$. Therefore, condition (4.6) has been checked and the existence of at least a fixed point for ψ in $\mathcal{K} \cap \mathcal{E}$ follows from Theorem 4.1. Notice that ψ is continuous on $\mathcal{K} \cap \mathcal{E}$ as, by (4.5), it is continuous on $\mathcal{K}$. □

With regards to Theorem 4.2, we stress that if (4.5) holds and there exist $m \geq 2$ pairwise disjoint oriented rectangles

$$\widetilde{\mathcal{E}}_0, \ldots, \widetilde{\mathcal{E}}_{m-1} \in \{\widetilde{\mathcal{A}} \pitchfork \widetilde{\mathcal{B}}\}\,, \tag{4.7}$$

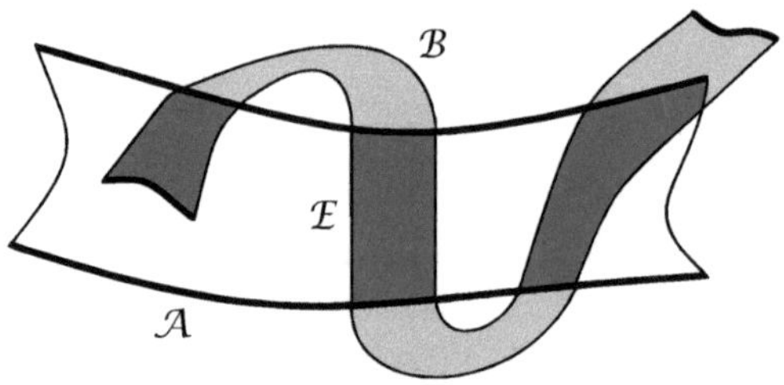

Fig. 3: The continuous planar map ψ transforms the generalized rectangle $\mathcal{A}$ onto the snake-like generalized rectangle $\mathcal{B} \supseteq \psi(\mathcal{A})$ so that $\mathcal{B}_\ell^- \supseteq \psi(\mathcal{A}_\ell^-)$ and $\mathcal{B}_r^- \supseteq \psi(\mathcal{A}_r^-)$. The boundary sets $\mathcal{A}^- = \mathcal{A}_\ell^- \cup \mathcal{A}_r^-$ and $\mathcal{B}^- = \mathcal{B}_\ell^- \cup \mathcal{B}_r^-$ have been drawn with thicker lines. Clearly, for $\widetilde{\mathcal{A}} = (\mathcal{A}, \mathcal{A}^-)$ and $\widetilde{\mathcal{B}} = (\mathcal{B}, \mathcal{B}^-)$, it holds that $\psi \colon \widetilde{\mathcal{A}} \Leftrightarrow\!\!\to \widetilde{\mathcal{B}}$. Notice that we do not require that the end sets $\mathcal{B}_\ell$ and $\mathcal{B}_r$ of $\mathcal{B}$ lie outside $\mathcal{A}$, differently from the approaches based on degree theory, as discussed in [126]. Among the three intersections between $\mathcal{A}$ and $\mathcal{B}$, only the central one, that we call $\mathcal{E}$, corresponds to a crossing in the sense of Definition 4.3 since $\widetilde{\mathcal{E}} \subseteq_h \widetilde{\mathcal{A}}$ and $\widetilde{\mathcal{E}} \subseteq_v \widetilde{\mathcal{B}}$. Therefore, Theorem 4.2 ensures the existence of at least a fixed point for ψ in $\mathcal{E}$. Even if the intersection between $\mathcal{A}$ and $\mathcal{B}$ on the right is "not far" from belonging to $\{\widetilde{\mathcal{A}} \pitchfork \widetilde{\mathcal{B}}\}$, it is possible to define ψ so that it has no fixed points therein. This shows that our fixed point theorems are, in some sense, "sharp." See [126] for more details and corresponding examples.

then the map ψ has at least a fixed point in $\mathcal{K} \cap \mathcal{E}_i$, $\quad \forall i = 0, \dots, m-1$. Hence, similarly to Definition 4.2, it is also possible in this framework to find a multiplicity of fixed points, obtaining conclusions analogous to the ones in Theorem 4.3 below.

If in place of fixed points we are concerned with the search of periodic points of any period (as in the following Theorems 4.3–4.4), then, in order to apply our stretching along the paths method, we need to check the stretching relation to be preserved under composition of maps. This fact can be easily proved by induction: the basic step is the content of the next lemma.

Lemma 4.1. *Let X be a metric space and let $\varphi \colon X \supseteq D_\varphi \to X$ and $\psi \colon X \supseteq D_\psi \to X$ be maps defined on the sets D_φ and D_ψ, respectively. Assume that $\widetilde{\mathcal{A}} := (\mathcal{A}, \mathcal{A}^-)$, $\widetilde{\mathcal{B}} := (\mathcal{B}, \mathcal{B}^-)$ and $\widetilde{\mathcal{C}} := (\mathcal{C}, \mathcal{C}^-)$ are oriented rectangles of X. If $\mathcal{H} \subseteq \mathcal{A} \cap D_\varphi$ and $\mathcal{K} \subseteq \mathcal{B} \cap D_\psi$ are compact sets such that*

$$(\mathcal{H}, \varphi) \colon \widetilde{\mathcal{A}} \Leftrightarrow\!\!\to \widetilde{\mathcal{B}} \quad \text{and} \quad (\mathcal{K}, \psi) \colon \widetilde{\mathcal{B}} \Leftrightarrow\!\!\to \widetilde{\mathcal{C}},$$

then it follows that

$$\left(\mathcal{H} \cap \varphi^{-1}(\mathcal{K}), \psi \circ \varphi\right) \colon \widetilde{\mathcal{A}} \Leftrightarrow\!\!\to \widetilde{\mathcal{C}}.$$

Proof. Let $\gamma \colon [0,1] \to \mathcal{A}$ be a path such that $\gamma(0)$ and $\gamma(1)$ belong to the different sides of $\mathcal{A}^-$. Then, since $(\mathcal{H}, \varphi) \colon \widetilde{\mathcal{A}} \Leftrightarrow\!\!\to \widetilde{\mathcal{B}}$, there exists a subinterval $[t', t''] \subseteq [0,1]$ such that

$$\gamma(t) \in \mathcal{H}, \quad \varphi(\gamma(t)) \in \mathcal{B}, \quad \forall t \in [t', t'']$$

and, moreover, $\varphi(\gamma(t'))$ and $\varphi(\gamma(t''))$ belong to different components of $\mathcal{B}^-$. Let us call ω the restriction of γ to $[t', t'']$ and define $\nu \colon [t', t''] \to \mathcal{B}$ as $\nu := \varphi \circ \omega$. Notice that $\nu(t')$ and $\nu(t'')$ belong to the different sides of $\mathcal{B}^-$ and so, by the stretching hypothesis $(\mathcal{K}, \psi): \widetilde{\mathcal{B}} \Leftrightarrow \widetilde{C}$, there is a subinterval $[s', s''] \subseteq [t', t'']$ such that

$$\nu(t) \in \mathcal{K}, \quad \psi(\nu(t)) \in C, \quad \forall t \in [s', s'']$$

with $\psi(\nu(s'))$ and $\psi(\nu(s''))$ belonging to different components of C^-. Rewriting all in terms of γ, this means that we have found a subinterval $[s', s''] \subseteq [0, 1]$ such that

$$\gamma(t) \in \mathcal{H} \cap \varphi^{-1}(\mathcal{K}), \quad \psi(\varphi(\gamma(t))) \in C, \quad \forall t \in [s', s'']$$

and $\psi(\varphi(\gamma(s')))$ and $\psi(\varphi(\gamma(s'')))$ belong to the different sides of C^-. By the arbitrariness of the path γ, the stretching property

$$\left(\mathcal{H} \cap \varphi^{-1}(\mathcal{K}), \psi \circ \varphi\right) : \widetilde{\mathcal{A}} \Leftrightarrow \widetilde{C}$$

is thus fulfilled. We just pointed out that the continuity of the composite mapping $\psi \circ \varphi$ on the compact set $\mathcal{H} \cap \varphi^{-1}(\mathcal{K})$ follows from the continuity of φ on $\mathcal{H}$ and of ψ on $\mathcal{K}$, respectively. □

Theorem 4.3. *Let X be a metric space and $\psi: X \supseteq D_\psi \to X$ be a map defined on a set D_ψ. Assume that $\widetilde{\mathcal{R}} := (\mathcal{R}, \mathcal{R}^-)$ is an oriented rectangle X. If $\mathcal{K}_0, \ldots, \mathcal{K}_{m-1}$ are $m \geq 2$ pairwise disjoint compact subsets of $\mathcal{R} \cap D_\psi$ and*

$$(\mathcal{K}_i, \psi): \widetilde{\mathcal{R}} \Leftrightarrow \widetilde{\mathcal{R}}, \quad \text{for } i = 0, \ldots, m-1,$$

then the following conclusions hold:
- *The map ψ has at least a fixed point in $\mathcal{K}_i$, $i = 0, \ldots, m-1$;*
- *For each two-sided sequence $(s_h)_{h\in\mathbb{Z}} \in \{0, \ldots, m-1\}^{\mathbb{Z}}$, there exists a sequence of points $(x_h)_{h\in\mathbb{Z}}$ such that $\psi(x_{h-1}) = x_h \in \mathcal{K}_{s_h}, \quad \forall h \in \mathbb{Z}$;*
- *For each sequence $\mathbf{s} = (s_n)_n \in \{0, \ldots, m-1\}^{\mathbb{N}}$, there exists a compact connected set $C_\mathbf{s} \subseteq \mathcal{K}_{s_0}$, satisfying*

$$C_\mathbf{s} \cap \mathcal{R}_d^+ \neq \emptyset, \quad C_\mathbf{s} \cap \mathcal{R}_u^+ \neq \emptyset$$

  and such that $\psi^i(x) \in \mathcal{K}_{s_i}, \quad \forall i \geq 1, \quad \forall x \in C_\mathbf{s}$;
- *Given an integer $j \geq 2$ and a $j+1$-uple $(s_0, \ldots, s_j)$, $s_i \in \{0, \ldots, m-1\}$, for $i = 0, \ldots, j$, and $s_0 = s_j$, then there exists a point $w \in \mathcal{K}_{s_0}$ such that*

$$\psi^i(w) \in \mathcal{K}_{s_i}, \forall i = 1, \ldots, j \quad \text{and} \quad \psi^j(w) = w.$$

As we shall see in Section 7, the previous result turns out to be our fundamental tool in the proof of Theorem 7.1 about chaotic dynamics. On the other hand, Theorem 4.3 can be viewed as a particular case of Theorem 4.4 below. The proof of the former result is thus postponed since it comes as a corollary of the latter more general theorem.

Before stating Theorem 4.4, we simply make an observation that will reveal its significance in Section 8 when dealing with symbolic dynamics.

Remark 4.2. We observe that in the hypotheses of Theorem 4.3, or equivalently when we enter the framework of Definition 4.2 with $\widetilde{\mathcal{A}} = \widetilde{\mathcal{B}}$, i. e. when $m \geq 2$ pairwise disjoint compact subsets $\mathcal{K}_0, \dots, \mathcal{K}_{m-1}$ of an oriented rectangle $\mathcal{A} \subseteq X$ exist, for which $(\mathcal{K}_i, \psi) : \widetilde{\mathcal{A}} \Leftrightarrow \widetilde{\mathcal{A}}$ holds, then it is possible to find m pairwise disjoint vertical slabs $\widetilde{\mathcal{R}}_i$ of $\widetilde{\mathcal{A}}$ such that $\mathcal{R}_i \supseteq \mathcal{K}_i$, for $i = 0, \dots, m-1$. Indeed, by Theorem 4.3, we know that any $\mathcal{K}_i$ contains a compact connected set C_i (actually, infinitely many) joining $\mathcal{A}_d^+$ and $\mathcal{A}_u^+$. Thus, the idea is to "fill" and "fatten" each C_i up in order to obtain a compact set $\mathcal{R}_i$ that still joins $\mathcal{A}_d^+$ and $\mathcal{A}_u^+$, but homeomorphic to the unit square of $\mathbb{R}^2$ and containing only $\mathcal{K}_i$ among all the $\mathcal{K}_j$'s. Moreover, we require $\mathcal{A}_d^+ \cap \mathcal{R}_i$ and $\mathcal{A}_d^+ \cap \mathcal{R}_i$ to be arcs. Notice that this is possible because the $\mathcal{K}_i$'s are compact and disjoint. To such generalized rectangles $\mathcal{R}_i$'s, we give the orientation "inherited" from $\mathcal{A}$, that is, we set $\mathcal{R}_d^i := \mathcal{A}_d^+ \cap \mathcal{R}_i$ and $\mathcal{R}_u^i := \mathcal{A}_u^+ \cap \mathcal{R}_i$, for $i = 0, \dots, m-1$, where we have denoted by $\mathcal{R}_d^i$ and $\mathcal{R}_u^i$, the two sides of $\mathcal{R}_i^+$. Indicating with $\mathcal{R}_\ell^i$ and $\mathcal{R}_r^i$ the two parts of $\mathcal{R}_i^-$, we find that they coincide with the two components of $\overline{\vartheta \mathcal{R}_i \setminus (\mathcal{R}_d^i \cup \mathcal{R}_u^i)}$. In particular, we can name such sets following the cyclic order $\ell - d - r - u - \ell$. As usual, we put $\widetilde{\mathcal{R}}_i = (\mathcal{R}_i, \mathcal{R}_i^-)$ and, by construction, these are the desired vertical slabs of $\widetilde{\mathcal{A}}$. See Figure 4 for a graphical illustration.

If, in addition, the map ψ is continuous on $\mathcal{A}$, we claim that

$$\psi : \widetilde{\mathcal{R}}_i \Leftrightarrow \widetilde{\mathcal{R}}_j\,, \quad \forall i, j \in \{0, \dots, m-1\}\,.$$

At first, we notice that ψ is continuous on each $\mathcal{R}_i$ because $\bigcup_{i=0}^{m-1} \mathcal{R}_i \subseteq \mathcal{A}$. Moreover, any $i = 0, \dots, m-1$ and for every path $\gamma : [a, b] \to \mathcal{R}_i$, with $\gamma(a)$ and $\gamma(b)$ belonging to different components of $\mathcal{R}_i^-$, there exists a subinterval $[t', t''] \subseteq [a, b]$ such that $\gamma(t) \in \mathcal{K}_i$ and $\psi(\gamma(t)) \in \mathcal{A}$, $\forall t \in [t', t'']$, with $\psi(\gamma(t'))$ and $\psi(\gamma(t''))$ belonging to different sides of $\mathcal{A}^-$. This follows from the fact that γ can be extended to a path $\gamma^* : [a', b'] \to \mathcal{A}$, with $[a', b'] \supseteq [a, b]$, such that $\gamma^* \restriction_{[a,b]} = \gamma$ and $\gamma^*(a')$, $\gamma^*(b')$ belong to different sides of $\mathcal{A}^-$, and by recalling that $(\mathcal{K}_i, \psi) : \widetilde{\mathcal{A}} \Leftrightarrow \widetilde{\mathcal{A}}$. Then, since $\bigcup_{i=0}^{m-1} \mathcal{R}_i \subseteq \mathcal{A}$, for any fixed $j = 0, \dots, m-1$, there exists a subinterval $[s', s''] \subseteq [t', t'']$ such that $\psi(\gamma(t)) \in \mathcal{R}_j$, $\forall t \in [s', s'']$, with $\psi(\gamma(s'))$ and $\psi(\gamma(s''))$ belonging to different sides of $\mathcal{R}_j^-$. However, this means that $\psi : \widetilde{\mathcal{R}}_i \Leftrightarrow \widetilde{\mathcal{R}}_j$. By the arbitrariness of $i, j \in \{0, \dots, m-1\}$, the proof of our claim is complete.

Theorem 4.4. *Assume there is a double sequence of oriented rectangles $(\widetilde{\mathcal{R}}_i)_{i \in \mathbb{Z}}$ (with $\widetilde{\mathcal{R}}_i = (\mathcal{R}_i, \mathcal{R}_i^-)$) of a metric space X and a sequence $((\mathcal{K}_i, \psi_i))_{i \in \mathbb{Z}}$, with $\mathcal{K}_i \subseteq \mathcal{R}_i$ compact sets, such that*

$$(\mathcal{K}_i, \psi_i) : \widetilde{\mathcal{R}}_i \Leftrightarrow \widetilde{\mathcal{R}}_{i+1}\,, \quad \forall i \in \mathbb{Z}\,.$$

Let us denote by $\mathcal{R}_\ell^i$ and $\mathcal{R}_r^i$ the two components of $\mathcal{R}_i^-$ and by $\mathcal{R}_d^i$ and $\mathcal{R}_u^i$ the two components of $\mathcal{R}_i^+$. Then, the following conclusions hold:

- *There is a sequence $(w_k)_{k \in \mathbb{Z}}$ such that $w_k \in \mathcal{K}_k$ and $\psi_k(w_k) = w_{k+1}$ for all $k \in \mathbb{Z}$;*

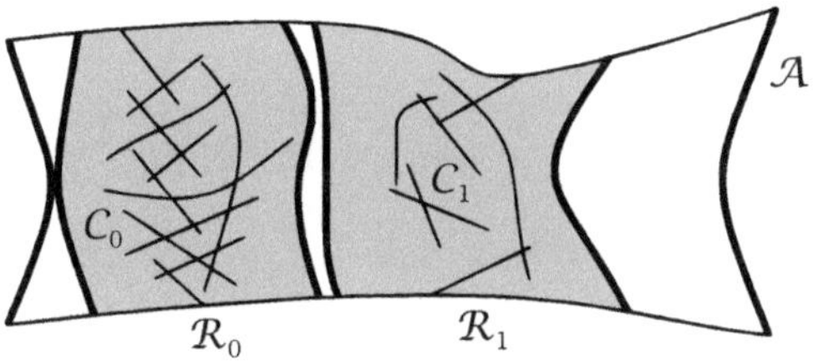

Fig. 4: With reference to the situation described in Remark 4.2 for $m = 2$, we have depicted the generalized rectangle $\mathcal{A}$ that we orientate by choosing as $\mathcal{A}^-$ the boundary set drawn with thicker lines. The continua C_0 and C_1, contained respectively in the disjoint compact sets $\mathcal{K}_0$ and $\mathcal{K}_1$ (not represented in the picture in order not to overburden it) and joining the two sides of $\mathcal{A}^+$, are embedded (together with the $\mathcal{K}_i$'s) in the disjoint generalized rectangles $\mathcal{R}_0$ and $\mathcal{R}_1$, whose $[\cdot]^-$-sets have been indicated with thicker lines. With this choice, $\widetilde{\mathcal{R}}_0$ and $\widetilde{\mathcal{R}}_1$ are vertical slabs of $\widetilde{\mathcal{A}}$.

- *For each $j \in \mathbb{Z}$, there exists a compact connected set $C_j \subseteq \mathcal{K}_j$, satisfying*

$$C_j \cap \mathcal{R}_d^j \neq \emptyset\,, \quad C_j \cap \mathcal{R}_u^j \neq \emptyset$$

  and such that, for every $w \in C_j$, there is a sequence $(y_i)_{i \geq j}$ with $y_j = w$ and

$$y_i \in \mathcal{K}_i\,, \quad \psi_i(y_i) = y_{i+1}\,, \quad \forall i \geq j\,;$$

- *If there are integers h and l with $h < l$ such that $\widetilde{\mathcal{R}}_h = \widetilde{\mathcal{R}}_l$, then there exists a finite sequence $(z_i)_{h \leq i \leq l-1}$, with $z_i \in \mathcal{K}_i$ and $\psi_i(z_i) = z_{i+1}$ for each $i = h, \dots, l-1$, such that $z_l = z_h$, that is, z_h is a fixed point of $\psi_{l-1} \circ \dots \circ \psi_h$ in $\mathcal{K}_h$.*

Proof. We prove the conclusions of the theorem in the reverse order. So, let us start with the verification of the last assertion. By the assumptions and by Definition 4.1, it is easy to check that

$$(\mathcal{H}, \psi_{l-1} \circ \dots \circ \psi_h)\colon \widetilde{\mathcal{R}}_h \Leftrightarrow\!\!\!\to \widetilde{\mathcal{R}}_l\,, \tag{4.8}$$

where

$$\mathcal{H} := \{z \in \mathcal{K}_h \colon \psi_i \circ \dots \circ \psi_h(z) \in \mathcal{K}_{i+1},\ \forall i = h, \dots, l-1\}\,.$$

With the positions $\widetilde{\mathcal{R}} = \widetilde{\mathcal{R}}_h = \widetilde{\mathcal{R}}_l$ and $\phi = \psi_{l-1} \circ \dots \circ \psi_h$, we read condition (4.8) as $(\mathcal{H}, \phi)\colon \widetilde{\mathcal{R}} \Leftrightarrow\!\!\!\to \widetilde{\mathcal{R}}$ and therefore the thesis follows immediately by Theorem 4.1.

Regarding the second conclusion, without loss of generality, we can assume $j = 0$. Let us define the closed set

$$S := \{z \in \mathcal{K}_0 \colon \psi_j \circ \dots \circ \psi_0(z) \in \mathcal{K}_{j+1},\ \forall j \geq 0\}$$

and fix a path $\gamma_0\colon [0,1] \to \mathcal{R}_0$ such that $\gamma_0(0)$ and $\gamma_0(1)$ belong to the different components of $\mathcal{R}_0^-$. Then, since $(\mathcal{K}_0, \psi_0)\colon \widetilde{\mathcal{R}}_0 \Leftrightarrow\!\!\!\to \widetilde{\mathcal{R}}_1$, there exists a subinterval

$$[t_1', t_1''] \subseteq [t_0', t_0''] := [0,1]$$

such that

$$\gamma_0(t) \in \mathcal{K}_0 \quad \text{and}\ \gamma_1(t) := \psi_0(\gamma_0(t)) \in \mathcal{R}_1\,, \quad \forall t \in [t_1', t_1'']\,.$$

By the same assumption, we also have that $\psi_0(\gamma_0(t_1'))$ and $\psi_0(\gamma_0(t_1''))$ belong to different components of $\mathcal{R}_1^-$. Similarly, there exists a subinterval

$$[t_2', t_2''] \subseteq [t_1', t_1'']$$

such that

$$\gamma_1(t) \in \mathcal{K}_1 \quad \text{and}\ \gamma_2(t) := \psi_1(\gamma_1(t)) \in \mathcal{R}_2\,, \quad \forall t \in [t_2', t_2'']\,,$$

with $\psi_1(\gamma_1(t_2'))$ and $\psi_1(\gamma_1(t_2''))$ belonging to the different components of $\mathcal{R}_2^-$. Defining

$$\Gamma_2 := \{x \in \gamma_0([t_1', t_1'']) \colon \psi_0(x) \in \gamma_1([t_2', t_2''])\} \subseteq \{z \in \mathcal{K}_0 \colon \psi_0(z) \in \mathcal{K}_1\}$$

and proceeding by induction, we can find a decreasing sequence of nonempty compact sets

$$\Gamma_0 := \gamma_0([t_0', t_0'']) \supseteq \Gamma_1 := \gamma_0([t_1', t_1'']) \supseteq \Gamma_2 \supseteq \cdots \supseteq \Gamma_n \supseteq \Gamma_{n+1} \supseteq \ldots$$

such that $\psi_j \circ \cdots \circ \psi_0(\Gamma_{j+1}) \subseteq \mathcal{R}_{j+1}$, with $\psi_j \circ \cdots \circ \psi_0(\Gamma_{j+1}) \cap \mathcal{R}_\ell^{j+1} \neq \emptyset$ and $\psi_j \circ \cdots \circ \psi_0(\Gamma_{j+1}) \cap \mathcal{R}_r^{j+1} \neq \emptyset$, for $j \geq 0$. Moreover, for every $i \geq 1$, we have that

$$\Gamma_{i+1} \subseteq \{z \in \mathcal{K}_0 \colon \psi_{j-1} \circ \cdots \circ \psi_0(z) \in \mathcal{K}_j,\ \forall\, j = 1, \ldots, i\}\,.$$

As it is straightforward to see, $\cap_{j=0}^{+\infty} \Gamma_j \neq \emptyset$ and for any $z \in \cap_{j=0}^{+\infty} \Gamma_j$, it holds that $\psi_n \circ \cdots \circ \psi_0(z) \in \mathcal{K}_{n+1}$, $\forall n \in \mathbb{N}$. In this way, we have shown that any path γ_0 joining in $\mathcal{R}_0$, the two sides of $\mathcal{R}_0^-$ intersects S. The existence of the connected compact set $C_0 \subseteq S \subseteq \mathcal{K}_0$ joining the two components of $\mathcal{R}_0^+$ comes from Lemma 4.5. By the definition of S, it is obvious that any point of $C_0 \subseteq S$ generates a sequence as required in the statement of the theorem.

The first conclusion follows now by a standard diagonal argument ([71, Proposition 5] and [117, Theorem 2.2]), which allows one to extend the result to bi-infinite sequences once it has been proved for one-sided sequences. □

Proof of Theorem 4.3. The first conclusion easily follows from Theorem 4.1. Regarding the remaining ones, given a two-sided sequence $(s_h)_{h\in\mathbb{Z}} \in \{0, \ldots, m-1\}^{\mathbb{Z}}$, they can be achieved by applying Theorem 4.4 with the positions $\mathcal{R}_i = \mathcal{R}$, $\psi_i = \psi$ and $\mathcal{K}_i = \mathcal{K}_{s_i}$, $\forall i \in \mathbb{Z}$. The details are omitted since they are a straightforward verification. □

We end this subsection with the presentation and discussion of some stretching relations alternative to the one in Definition 4.1. In particular, we will ask ourselves if the results on the existence and localization of fixed points are still valid with respect to these new concepts.

For the next definitions, the basic setting concerns the following framework: *Let X be a metric space. Assume $\psi\colon X \supseteq D_\psi \to X$ is a map defined on a set D_ψ and let $\widetilde{\mathcal{A}} := (\mathcal{A}, \mathcal{A}^-)$ and $\widetilde{\mathcal{B}} := (\mathcal{B}, \mathcal{B}^-)$ be oriented rectangles of X. Also, let $\mathcal{K} \subseteq \mathcal{A} \cap D_\psi$ be a compact set.*

Definition 4.4. We say that $(\mathcal{K}, \psi)$ *stretches $\widetilde{\mathcal{A}}$ to $\widetilde{\mathcal{B}}$ along the continua* and write

$$(\mathcal{K}, \psi)\colon \widetilde{\mathcal{A}} \mathrel{\stackrel{\frown}{\underset{\smile}{\rightarrow}}} \widetilde{\mathcal{B}}$$

if the following conditions hold:
- ψ is continuous on $\mathcal{K}$;
- For every continuum $\Gamma \subseteq \mathcal{A}$ with $\Gamma \cap \mathcal{A}_\ell^- \neq \emptyset$ and $\Gamma \cap \mathcal{A}_r^- \neq \emptyset$, there exists a continuum $\Gamma' \subseteq \Gamma \cap \mathcal{K}$ such that $\psi(\Gamma') \subseteq \mathcal{B}$ and

$$\psi(\Gamma') \cap \mathcal{B}_\ell^- \neq \emptyset, \quad \psi(\Gamma') \cap \mathcal{B}_r^- \neq \emptyset .$$

While the above definition is based on the one given by Kennedy and Yorke in [73], the following bears some resemblances to that of the "family of expanders" considered in [71].

Definition 4.5. We say that $(\mathcal{K}, \psi)$ *expands $\widetilde{\mathcal{A}}$ across $\widetilde{\mathcal{B}}$* and write

$$(\mathcal{K}, \psi)\colon \widetilde{\mathcal{A}} \mathrel{\stackrel{\smile}{\underset{\frown}{\rightarrow}}} \widetilde{\mathcal{B}}$$

if the following conditions hold:
- ψ is continuous on $\mathcal{K}$;
- For every continuum $\Gamma \subseteq \mathcal{A}$ with $\Gamma \cap \mathcal{A}_\ell^- \neq \emptyset$ and $\Gamma \cap \mathcal{A}_r^- \neq \emptyset$, there exists a nonempty compact set $P \subseteq \Gamma \cap \mathcal{K}$ such that $\psi(P)$ is a continuum contained in $\mathcal{B}$ and

$$\psi(P) \cap \mathcal{B}_\ell^- \neq \emptyset, \quad \psi(P) \cap \mathcal{B}_r^- \neq \emptyset .$$

When it is possible to take $\mathcal{K} = \mathcal{A}$, we simply write $\psi\colon \widetilde{\mathcal{A}} \mathrel{\stackrel{\frown}{\underset{\smile}{\rightarrow}}} \widetilde{\mathcal{B}}$ and $\psi\colon \widetilde{\mathcal{A}} \mathrel{\stackrel{\smile}{\underset{\frown}{\rightarrow}}} \widetilde{\mathcal{B}}$ in place of $(\mathcal{A}, \psi)\colon \widetilde{\mathcal{A}} \mathrel{\stackrel{\frown}{\underset{\smile}{\rightarrow}}} \widetilde{\mathcal{B}}$ and $(\mathcal{A}, \psi)\colon \widetilde{\mathcal{A}} \mathrel{\stackrel{\smile}{\underset{\frown}{\rightarrow}}} \widetilde{\mathcal{B}}$, respectively.

In analogy to Definition 4.2, one can define the variants $\mathrel{\stackrel{\frown}{\underset{\smile}{\rightarrow}}}^m$ and $\mathrel{\stackrel{\smile}{\underset{\frown}{\rightarrow}}}^m$ when multiple coverings occur.

As we shall see in a moment, the above relations behave in a different way with respect to the possibility of detecting fixed points. Indeed, since the ranges of paths are a particular kind of continua, one can follow the same steps as in the proof of Theorem 4.1 and show that an analogous result still holds when the property of stretching along the paths is replaced with the one of stretching along the continua (cf. [126, Theorem 2.10]). On the other hand, this is no more true when the relation in Definition 4.5 is fulfilled since it generally guarantees neither the existence of fixed points nor their localization (in case those fixed points do exist).

Concerning the existence of fixed points, a possible counterexample is described in Figure 5 and is inspired by the bulging horseshoe in [71, Fig. 4]. For sake of conciseness, we prefer to present it by means of a series of graphical illustrations in Figures 5–8: we point out, however, that it is based on a concrete definition of a planar map (whose form, although complicated, can be explicitly given in analytical terms).

With respect to the localization of fixed points, a counterexample can instead be obtained by suitably adapting a one-dimensional map to the planar case. Indeed, if $f\colon \mathbb{R} \supseteq [0,1] \to \mathbb{R}$ is any continuous function, we can set

$$\psi(x_1, x_2) := (f(x_1), x_2) \tag{4.9}$$

and have a continuous planar map defined on the unit square $[0,1]^2$ of $\mathbb{R}^2$, inheriting all the interesting properties of f. Notice that in this special case, any fixed point x^* for f generates a vertical line (x^*, s) (with $s \in [0,1]$) of fixed points for ψ. The more general framework of a map ψ defined as

$$\psi(x_1, x_2) := (f(x_1), g(x_2)),$$

for $g\colon [0,1] \to [0,1]$, a continuous function could be considered as well.

In view of the above discussion, we define a continuous map $f\colon [0,1] \to [0,1]$ of the form

$$f(s) := \begin{cases} \frac{1-c}{a}\, s + c & 0 \le s < a\,, \\ \frac{1}{a-b}(s-b) & a \le s \le b\,, \\ \frac{d}{1-b}(s-b) & b < s \le 1\,, \end{cases} \tag{4.10}$$

where a, b, c, d are fixed constants such that $0 < a < b < 1$ and $0 < c < d < 1$.

For $\widetilde{\mathcal{A}}$ the unit square oriented in the standard left-right manner and ψ as in (4.9), it holds that $\psi\colon \widetilde{\mathcal{A}} \Bumpeq\!\!\!\!\longrightarrow \widetilde{\mathcal{A}}$. In particular, we can write both

$$(\mathcal{K}_0, \psi)\colon \widetilde{\mathcal{A}} \Bumpeq\!\!\!\!\longrightarrow \widetilde{\mathcal{A}}, \quad \text{for } \mathcal{K}_0 = [a,b]$$

and

$$(\mathcal{K}_1, \psi)\colon \widetilde{\mathcal{A}} \Bumpeq\!\!\!\!\longrightarrow \widetilde{\mathcal{A}}, \quad \text{for } \mathcal{K}_1 = [0,a] \cup [b,1]\,.$$

In the former case, we also have

$$(\mathcal{K}_0, \psi)\colon \widetilde{\mathcal{A}} \Leftarrow\!\!\!\!\!\!\longrightarrow \widetilde{\mathcal{A}}$$

and therefore, consistently with Theorem 4.1, there exists at least a fixed point for ψ in $\mathcal{K}_0$. On the other hand, as it is clear from Figure 9, there are no fixed points for ψ in $\mathcal{K}_1$. Thus, the localization of the fixed points is not guaranteed when only the relation $\Bumpeq\!\!\!\!\longrightarrow$ is satisfied. The same example could be slightly modified in order to have $\psi\colon \widetilde{\mathcal{A}} \Bumpeq\!\!\!\!\longrightarrow^{m} \widetilde{\mathcal{A}}$ for an arbitrary $m \ge 2$, but just one fixed point does exist.

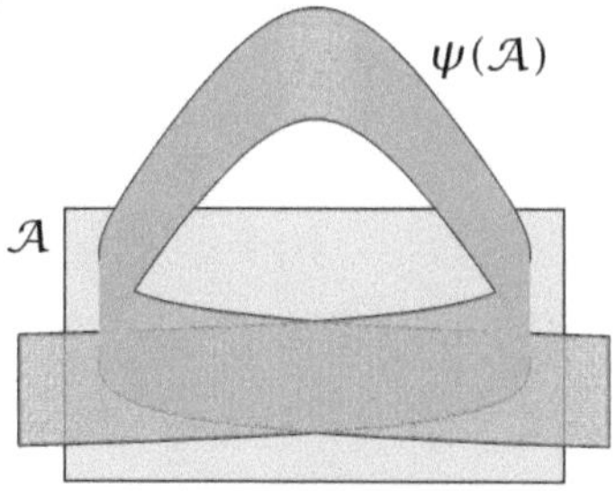

Fig. 5: A rectangle $\mathcal{A}$, oriented by taking as $[\cdot]^-$-set the union of its left and right vertical segments, is deformed by a continuous planar map ψ onto $\psi(\mathcal{A})$, a ribbon bent across the rectangle itself. The left and right sides of the domain are homeomorphically transformed onto the two endings of the ribbon. This is an example of $\psi\colon \widetilde{\mathcal{A}} \rightsquigarrow \widetilde{\mathcal{A}}$.

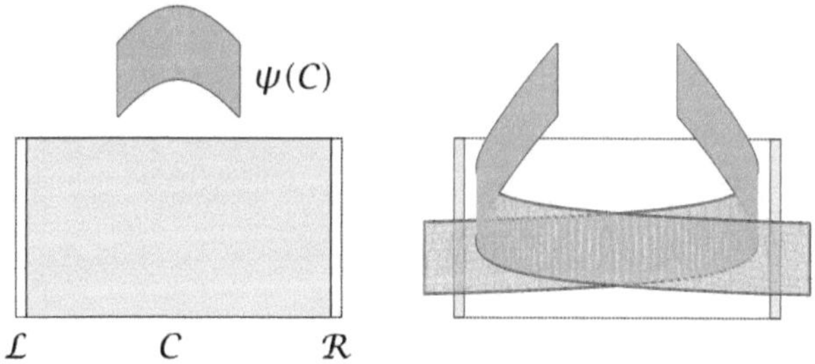

Fig. 6: With regards to the example portrayed in the previous picture, we emphasize how some subrectangles of $\mathcal{A}$ are transformed by the map ψ. In particular, the main central part C of the rectangle is pushed out of $\mathcal{A}$ (left), while the two narrow rectangles $\mathcal{L}$ and $\mathcal{R}$ near the left and right sides of it, respectively, are stirred onto two overlapping bent strips (right). This can happen in two different ways, as shown in Figures 7–8. One can pass from a configuration to the other by considering, instead of the map ψ, the related map $(x_1, x_2) \mapsto \psi(-x_1, x_2)$.

We finally observe that by playing with the coefficients a, b, c, d and choosing a suitable compact set $\mathcal{K}_2 \subset [0, 1]$, it is possible to have

$$(\mathcal{K}_2, \psi)\colon \widetilde{\mathcal{A}} \rightsquigarrow \widetilde{\mathcal{A}}$$

for a case in which neither ψ nor ψ^2 possess fixed points in $\mathcal{K}_2$. The set $\mathcal{K}_2$ will consist of the union of some compact subintervals of $\mathcal{K}_0$ and $\mathcal{K}_1$. A graphical representation of f and f^2 is given in Figure 9.

4.2 The Crossing Lemma

The fundamental result from plane topology that we have employed in the proof of the main theorems of Section 4.1 is the Crossing Lemma (cf. Lemma 4.5 below), which

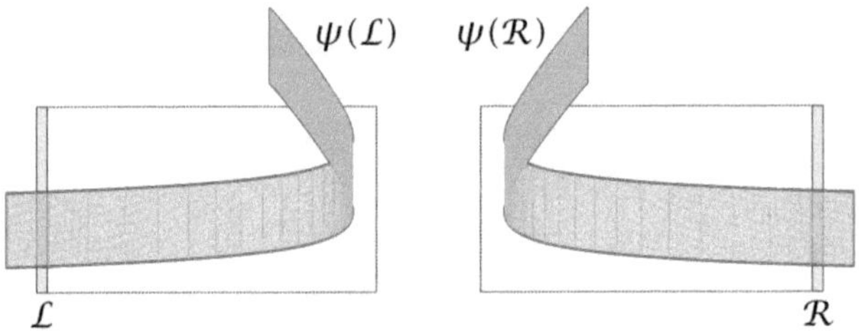

Fig. 7: In this case, we enter the setting of Theorem 4.1. By applying our result to the narrow rectangles $\mathcal{L}$ and $\mathcal{R}$ (provided they are suitably oriented in an obvious left-right manner), we can prove the existence of at least two fixed points for the planar map ψ in $\mathcal{A}$ (one fixed point lies in $\mathcal{L}$ and the other in $\mathcal{R}$).

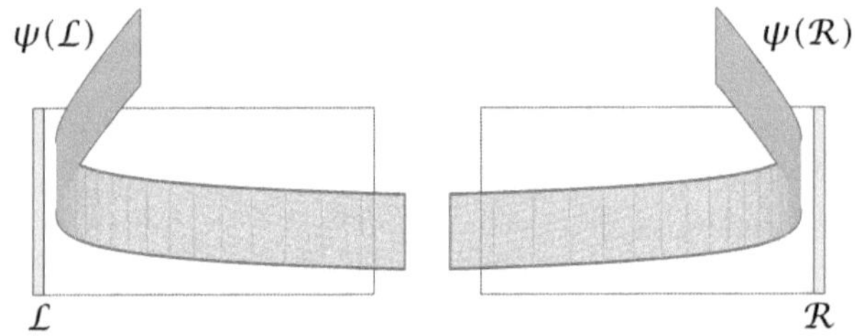

Fig. 8: In the framework depicted above, we do not have any fixed point. In fact, the central main part C of the rectangle $\mathcal{A}$ is mapped by ψ onto a hat-like figure outside the domain (see the left of Figure 6) and, at the same time, each of the two narrow rectangles $\mathcal{L}$ and $\mathcal{R}$, which constitute the remaining part of the domain, is mapped onto a set which is disjoint from itself.

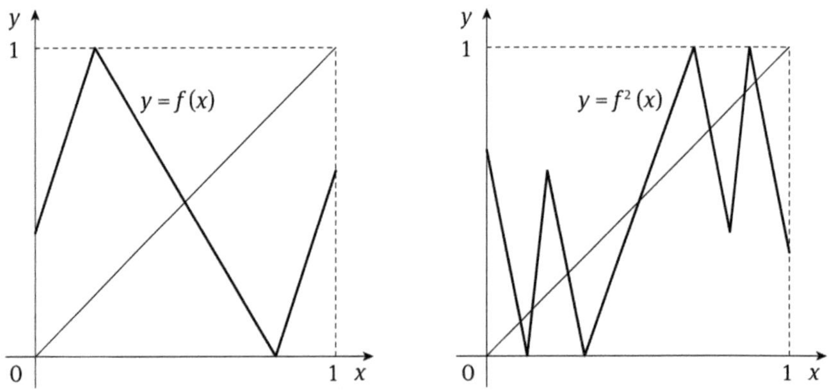

Fig. 9: Example of the graph of a function f (left) and of its iterate f^2 (right), with f defined as in (4.10). We have drawn the line $y = x$ in order to show the fixed points and the points of period 2.

establishes the existence of a continuum C joining two opposite sides of a rectangle and contained in a given compact set S. More precisely, the existence of C is guaranteed by the *cutting property* (CP) for S in Lemma 4.5, that is, a condition which asserts that S intersects any path contained in the rectangle and joining the other two sides. In the two-dimensional case, what we have called the "Crossing Lemma" is a classical result which has been used, more or less explicitly, by various authors in their proofs ([35, 41]). The interested reader will find in [123] more information connecting this lemma to other classical results of plane topology. In the higher dimensional case, the set S looks like a *cutting surface* (Section 5).

On the next pages, we provide a verification of the Crossing Lemma, presenting a few topological facts that allow one to achieve it. Even if, for the reader's convenience, we try to make the treatment as self-contained as possible, some theorems are stated below without proof.

The first among such results is the Whyburn Lemma (Lemma 4.2). It is a central tool in bifurcation theory and in the application of continuation methods to nonlinear equations [3, 101, 134]. We quote here one of its most used versions, as taken from [134, Lemma 1.33]. The interested reader can find a proof as well as further details in Section V of Kuratowski's book [85]. We refer also to [3] and [101] for interesting surveys about the role of connectedness in the study of fixed point theory and operator equations.

Lemma 4.2 (Whyburn Lemma). *If A and B are (nonempty) disjoint closed subsets of a compact metric space K such that no connected components of K intersect both A and B, then $K = K_A \cup K_B$, where K_A and K_B are disjoint compact sets containing A and B, respectively.*

In the applications, the Whyburn Lemma is often employed in the form of the following alternative:

Either there exists a continuum $C \subseteq K$ with $C \cap A \neq \emptyset$ and $C \cap B \neq \emptyset$, or A and B are separated, in the sense that there exists K_A and K_B as in Lemma 4.2.

The other result that we state without proof is a two-dimensional version of the well-known Brouwer fixed point Theorem. In such a simplified form, it can be proved by using the winding number [40].

Theorem 4.5. *Let $\phi\colon D \to D$ be a continuous self-mapping of a set $D \subseteq \mathbb{R}^2$, where D is homeomorphic to a closed disc. Then, ϕ has at least one fixed point in D.*

In the sequel, the set D will be the product of two nondegenerate compact intervals $[a, b]$ and $[c, d]$.

The next lemma is a particular case of the celebrated Leray–Schauder Continuation Principle [91, Théorème Fondamental]. This result plays a useful role in various contexts: for a different application, see, for instance, [83], while the interested reader can find more corresponding details in [101, 181] and in the bibliography therein. By its significance in our approach, we present two proofs of Lemma 4.3: the first one is less direct, but elementary, while the second one is based on degree theory and thus it is immediately extendable to a higher dimensional setting [125].

Introducing the projections of the plane onto the coordinate axes

$$\pi_1, \pi_2 : \mathbb{R}^2 \to \mathbb{R}, \quad \pi_1(x,y): = x, \quad \pi_2(x,y) := y, \quad \forall (x,y) \in \mathbb{R}^2,$$

it can be stated as follows:

Lemma 4.3. *Let $f\colon [a,b] \times [c,d] \to \mathbb{R}$ be a continuous map such that*

$$f(x,y) \le 0, \quad \forall (x,y) \in [a,b] \times \{c\},$$
$$f(x,y) \ge 0, \quad \forall (x,y) \in [a,b] \times \{d\}.$$

Then, there exists a compact connected set

$$S \subseteq \{(x,y) \in [a,b] \times [c,d] : f(x,y) = 0\},$$

such that

$$\pi_1(S) = [a,b].$$

Proof. Let

$$K := \{(x,y) \in [a,b] \times [c,d] : f(x,y) = 0\}$$

be the zero-set for the map f and let

$$A := K \cap (\{a\} \times [c,d]), \quad B := K \cap (\{b\} \times [c,d]).$$

By the intermediate value theorem, A and B are both nonempty. They are also disjoint compact subsets of the compact set K.

Assume, by contradiction, that there are no compact connected subsets of K intersecting both A and B. Then, according to the Whyburn Lemma 4.2, there exist compact sets K_A and K_B with

$$K = K_A \cup K_B, \quad K_A \supseteq A, \quad K_B \supseteq B, \quad K_A \cap K_B = \emptyset.$$

We can thus define the continuous function ([113, Lemma on Interpolation])

$$g(x,y) := \frac{\operatorname{dist}(p,K_A) - \operatorname{dist}(p,K_B)}{\operatorname{dist}(p,K_A) + \operatorname{dist}(p,K_B)} \quad \text{for } p = (x,y) \in [a,b] \times [c,d],$$

which has $[-1,1]$ as range and attains the value -1 on K_A and the value 1 on K_B. Notice that the denominator never vanishes as the compact sets K_A and K_B are disjoint.

We also introduce the projections of the real line onto the intervals $[a,b]$ and $[c,d]$, respectively. For $s \in \mathbb{R}$, they are defined as

$$P_{[a,b]}(s) := \max\{a, \min\{s,b\}\}, \quad P_{[c,d]}(s) := \max\{c, \min\{s,d\}\}. \tag{4.11}$$

Finally, we consider the continuous map

$$\phi = (\phi_1, \phi_2) \colon [a,b] \times [c,d] \to [a,b] \times [c,d]\,,$$

with components

$$\phi_1(x,y) := P_{[a,b]}(x - g(x,y))\,, \quad \phi_2(x,y) := P_{[c,d]}(y - f(x,y))\,.$$

Since $D := [a,b] \times [c,d]$ is homeomorphic to the closed unit disc, the Brouwer Theorem 4.5 ensures the existence of a fixed point $z = (z_1, z_2) \in D$ for the map ϕ, that is,

$$z_1 = P_{[a,b]}(z_1 - g(z_1,z_2))\,, \quad z_2 = P_{[c,d]}(z_2 - f(z_1,z_2))\,. \tag{4.12}$$

We claim that

$$z_2 - f(z_1,z_2) \in [c,d]\,. \tag{4.13}$$

Indeed, if $z_2 - f(z_1,z_2) \notin [c,d]$, then either $z_2 - f(z_1,z_2) < c$ or $z_2 - f(z_1,z_2) > d$. In the former case, $P_{[c,d]}(z_2 - f(z_1,z_2)) = c$ and thus, from the second relation in (4.12), we find $z_2 = c$. By the sign condition on $f(x,y)$ for $(x,y) \in [a,b] \times \{c\}$, it follows that $f(z) = f(z_1,c) \le 0$ and therefore $z_2 - f(z_1,z_2) = c - f(z_1,c) \ge c$. This contradicts the assumption $z_2 - f(z_1,z_2) < c$. On the other hand, if $z_2 - f(z_1,z_2) > d$, then $P_{[c,d]}(z_2 - f(z_1,z_2)) = d$ and thus the second relation in (4.12) implies $z_2 = d$. By the sign condition on $f(x,y)$ for $(x,y) \in [a,b] \times \{d\}$, it follows that $f(z) = f(z_1,d) \ge 0$ and therefore $z_2 - f(z_1,z_2) = d - f(z_1,d) \le d$. This contradicts the assumption $z_2 - f(z_1,z_2) > d$. Condition (4.13) is thus proved.

From (4.13) and the second relation in (4.12), we obtain $z_2 = P_{[c,d]}(z_2 - f(z_1,z_2)) = z_2 - f(z_1,z_2)$. Therefore, $f(z) = 0$, that is, $z \in K = K_A \cup K_B$ and, recalling the definition of the interpolation map g, we conclude that

$$g(z) = \pm 1\,. \tag{4.14}$$

Arguing as in the proof of (4.13), we can also show that

$$z_1 - g(z_1,z_2) \in [a,b]\,.$$

Then, the first relation in (4.12) implies that $z_1 = P_{[a,b]}(z_1 - g(z_1,z_2)) = z_1 - g(z_1,z_2)$. Hence,

$$g(z) = 0\,,$$

in contradiction with (4.14). Thus, by the Whyburn Lemma 4.2, the existence of a continuum $S \subseteq K = \{(x,y) \in [a,b] \times [c,d] : f(x,y) = 0\}$, with $S \cap A \neq \emptyset$ and $S \cap B \neq \emptyset$, follows. This means that the set S contains points of the form (a,u) and (b,v), with $u, v \in [c,d]$, and consequently the image of S under π_1 covers the interval $[a,b]$. The verification of the lemma is complete.

We present also an alternative proof via degree theory.

Consider the auxiliary function $\widehat{f}\colon [a,b] \times \mathbb{R} \to \mathbb{R}$ defined by

$$\widehat{f}(x,y) := f(x, P_{[c,d]}(y)) + \min\{y - c, \max\{0, y - d\}\}\,.$$

The map $\widehat{f}$ is continuous such that

$$\widehat{f}(x,y) = f(x,y)\,, \quad \forall (x,y) \in [a,b] \times [c,d]\,.$$

Moreover, it satisfies the following inequalities:

$$\widehat{f}(x,y) < 0\,, \quad \forall (x,y) \in [a,b] \times (-\infty, c\,[\,,$$
$$\widehat{f}(x,y) > 0\,, \quad \forall (x,y) \in [a,b] \times \,]\,d, +\infty)\,.$$

If we treat the variable $x \in [a,b]$ as a parameter and consider the open set $\Omega := \,]c-1, d+1[$, we have that $\widehat{f}(x,y) \neq 0$, for every $x \in [a,b]$ and $y \in \partial\Omega$. The Brouwer degree $\deg(\widehat{f}(a,\cdot), \Omega, 0)$ is well-defined and nontrivial since, by the sign condition $\widehat{f}(a, c-1) < 0 < \widehat{f}(a, d+1)$, it holds that

$$\deg\left(\widehat{f}(a,\cdot), \Omega, 0\right) = 1\,.$$

The Leray–Schauder Continuation Theorem ([91, 101, 181]) ensures that the set of solution pairs

$$\widehat{K} := \left\{(x,y) \in [a,b] \times \Omega \colon \widehat{f}(x,y) = 0\right\}$$

contains a continuum along which x assumes all the values in $[a,b]$. By the definition of $\widehat{f}$, it is clear that $\widehat{K} = K$ and this gives another proof of the existence of S. □

The next result shows that the continuum S found in Lemma 4.3 can be ε-approximated by paths.

Lemma 4.4. *Let $f\colon [a,b] \times [c,d] \to \mathbb{R}$ be a continuous map such that*

$$f(x,y) \le -1\,, \quad \forall (x,y) \in [a,b] \times \{c\},$$
$$f(x,y) \ge 1\,, \quad \forall (x,y) \in [a,b] \times \{d\}\,.$$

Then, for each $\varepsilon > 0$, there exists a continuous map

$$\gamma = \gamma_\varepsilon : [0,1] \to [a,b] \times [c,d]$$

such that

$$\gamma(0) \in \{a\} \times [c,d]\,, \quad \gamma(1) \in \{b\} \times [c,d]$$

and

$$|f(\gamma(t))| < \varepsilon\,, \quad \forall t \in [0,1]\,.$$

Proof. Without loss of generality, we can assume $0 < \varepsilon < 1$.

First of all, we introduce the continuous function

$$\tilde{f}\colon \mathbb{R}^2 \to \mathbb{R}, \quad \tilde{f}(x,y) := f(P_{[a,b]}(x), P_{[c,d]}(y)),$$

which extends f to the whole plane by means of the projections in (4.11).

From the sign conditions on f, we can easily see that

$$\tilde{f}(x,y) \le -1, \quad \forall (x,y) \in \mathbb{R}^2\colon y \le c,$$
$$\tilde{f}(x,y) \ge 1, \quad \forall (x,y) \in \mathbb{R}^2\colon y \ge d.$$

We also define the set

$$T_\varepsilon := \left\{(x,y) \in \mathbb{R}^2\colon\ |\tilde{f}(x,y)| < \varepsilon\right\},$$

which is an open subset of the strip $\mathbb{R}\times\,]c,d[$.

By Lemma 4.3, there exists a compact connected set

$$S \subseteq \{(x,y) \in [a,b] \times [c,d] : f(x,y) = 0\},$$

such that $\pi_1(S) = [a,b]$. Notice that

$$S \subseteq T_\varepsilon.$$

For every $p \in S$, there exists an open disc $B(p,\delta_p)$ with a center in p and radius $\delta_p > 0$ such that $B(p,\delta_p) \subseteq T_\varepsilon$. By compactness, we can find a finite number of points $p_1, \dots, p_n \in S$ such that

$$S \subseteq B := \bigcup_{i=1}^{n} B(p_i,\delta_i) \subseteq T_\varepsilon,$$

where we have set $\delta_i := \delta_{p_i}$. Without loss of generality (adding two further discs, if necessary), we can suppose that

$$p_1 \in (\{a\} \times [c,d]) \cap T_\varepsilon, \quad \text{and} \quad p_n \in (\{b\} \times [c,d]) \cap T_\varepsilon.$$

The open set B is connected. Indeed, it is a union of connected sets (the open discs $B(p_i,\delta_i)$) and each of such connected sets has nonempty intersection with the connected set S.

Since every open connected set in the plane is also arcwise connected, there exists a continuous map $\omega_\varepsilon\colon [s_0,s_1] \to B$ with $\omega_\varepsilon(s_0) = p_1$ and $\omega_\varepsilon(s_1) = p_n$. Then, we define

$$s' := \max\{s \in [s_0,s_1] : \omega_\varepsilon(s) \in \{a\} \times [c,d]\}$$

and

$$s'' := \min\{s \in [s',s_1] : \omega_\varepsilon(s) \in \{b\} \times [c,d]\}.$$

Hence, the path

$$\gamma_\varepsilon(t) := \omega_\varepsilon(s' + t \cdot (s'' - s')) , \quad \text{for } t \in [0,1]$$

is continuous and has the desired properties. Indeed, $\gamma_\varepsilon(t) \in ([a,b] \times \mathbb{R}) \cap T_\varepsilon \subseteq [a,b] \times]c,d[$ and therefore $|\tilde{f}(\gamma_\varepsilon(t))| = |f(\gamma_\varepsilon(t))| < \varepsilon$, for all $t \in [0,1]$. Moreover, $\gamma_\varepsilon(0) = \omega_\varepsilon(s') \in \{a\} \times [c,d]$ and $\gamma_\varepsilon(1) = \omega_\varepsilon(s'') \in \{b\} \times [c,d]$. The proof is complete. □

We are now in a position to prove the following:

Lemma 4.5 (Crossing Lemma). *Let $K \subseteq [0,1]^2$ be a compact set which satisfies the cutting property*[1]*:*

$$\begin{gathered} K \cap \gamma([0,1]) \neq \emptyset, \\ \text{for each continuous map } \gamma\colon [0,1] \to [0,1]^2, \\ \text{with } \gamma(0) \in \{0\} \times [0,1] \text{ and } \gamma(1) \in \{1\} \times [0,1]. \end{gathered} \tag{CP}$$

Then, there exists a continuum $C \subseteq K$ with

$$C \cap ([0,1] \times \{0\}) \neq \emptyset \quad \text{and} \quad C \cap ([0,1] \times \{1\}) \neq \emptyset .$$

Proof. By the assumptions, the sets

$$A := K \cap ([0,1] \times \{0\}) , \quad B := K \cap ([0,1] \times \{1\})$$

are compact, nonempty and disjoint. The thesis is achieved if we prove that there exists a continuum $C \subseteq K$ with $C \cap A \neq \emptyset$ and $C \cap B \neq \emptyset$. If, by contradiction, there is no continuum of this kind, the Whyburn Lemma 4.2 implies that the set K can be decomposed as

$$K = K_A \cup K_B , \quad K_A \cap K_B = \emptyset , \quad K_A \supseteq A , \quad K_B \supseteq B ,$$

with K_A and K_B compact sets. Let us then define the compact sets

$$K'_A := K_A \cup ([0,1] \times \{0\}) , \quad K'_B := K_B \cup ([0,1] \times \{1\}) .$$

We claim that they are disjoint. Indeed, since $K_A \cap K_B = \emptyset$, if there exists $\bar{x} \in K'_A \cap K'_B$, then $\bar{x} \in K_A \cap ([0,1] \times \{1\})$ or $\bar{x} \in K_B \cap ([0,1] \times \{0\})$. In the former case, recalling that $K_A \subseteq K$, we find that $\bar{x} \in B \subseteq K_B$ and thus $\bar{x} \in K_A \cap K_B = \emptyset$. The same contradiction can be achieved by assuming $\bar{x} \in K_B \cap ([0,1] \times \{0\})$. The claim on the disjointness of K'_A and K'_B is thus checked.

1 In view of Definition 5.1 in Section 5, we could also say that K *cuts the arcs between* $\{0\} \times [0,1]$ *and* $\{1\} \times [0,1]$.

Arguing like in the proof of Lemma 4.3, we introduce the continuous interpolation function

$$f(x,y) := \frac{\mathrm{dist}(p,K'_A) - \mathrm{dist}(p,K'_B)}{\mathrm{dist}(p,K'_A) + \mathrm{dist}(p,K'_B)} \quad \text{for } p = (x,y) \in [0,1]^2$$

which has $[-1,1]$ as range and attains the value -1 on K'_A and the value 1 on K'_B. By Lemma 4.4, there exists a continuous map $\gamma\colon [0,1] \to [0,1]^2$ with $\gamma(0) \in \{0\} \times [0,1]$ and $\gamma(1) \in \{1\} \times [0,1]$ such that

$$|f(\gamma(t))| < \frac{1}{2}, \quad \forall t \in [0,1]. \tag{4.15}$$

The cutting property (CP) ensures the existence of $t^* \in [0,1]$ such that

$$\gamma(t^*) \in K.$$

By the splitting of $K = K_A \cup K_B$ and the definition of f, we find that

$$f(\gamma(t^*)) = \pm 1,$$

in contradiction with (4.15). This concludes the proof. □

The above Crossing Lemma may be used to give a simple proof of the Poincaré–Miranda Theorem in dimension two. We recall that the Poincaré–Miranda Theorem asserts the existence of a zero for a continuous vector field $F = (F_1, F_2)$ defined on a rectangle $[a_1,b_1] \times [a_2,b_2] \subseteq \mathbb{R}^2$ such that $F_1(a_1,x_2) \cdot F_1(b_1,x_2) \le 0$ for every $x_2 \in [a_2,b_2]$ and $F_2(x_1,a_2) \cdot F_2(x_1,b_2) \le 0$ for every $x_1 \in [a_1,b_1]$. The result holds also for the standard hypercube of $\mathbb{R}^N$ and for N-dimensional rectangles, as we shall see in Section 5 (cf. Theorem 5.1). Such a theorem is usually referred to Carlo Miranda, who in 1940, noticed its equivalence to the Brouwer fixed point Theorem [106]. On the other hand, as remarked in [102], Henry Poincaré in [128, 129] had already announced this result with a suggestion of a correct proof using the Kronecker's index. In [129], with regard to the two-dimensional case, Poincaré (assuming strict inequalities for the components of the vector field on the boundary of the rectangle) also described a heuristic proof[2] as follows: the "curve" $F_2 = 0$ departs from a point of the side $x_1 = b_1$ and ends at some point of $x_1 = a_1$; in the same manner, the curve $F_1 = 0$, departing from a point of $x_2 = b_2$ and ending at some point of $x_2 = a_2$, must necessarily meet the first "curve" in the interior of the rectangle. Further information and historical remarks about Poincaré–Miranda Theorem can be found in [82, 102].

Following Poincaré's heuristic argument, one could adapt his proof in the following manner: if $\gamma(t)$ is a path contained in the rectangle $[a_1,b_1] \times [a_2,b_2]$ and

---

2 *"pour faire comprendre comment on peut démontrer ce théorème"*, quoting H. Poincaré's own words [129].

joining the left and the right sides, by the Bolzano Theorem, there exists at least a zero of $F_1(\gamma(t))$ and this, in turns, means that any path as above meets the set $S := F_1^{-1}(0)$. The Crossing Lemma then implies the existence of a compact connected set $C_1 \subseteq F_1^{-1}(0)$ which intersects the lower and the upper sides of the rectangle. At this point, one can easily achieve the conclusion in various ways. For instance, one could just repeat the same argument on F_2 in order to obtain a compact connected set $C_2 \subseteq F_2^{-1}(0)$ which intersects the left and the right sides of the rectangle and thus prove the existence of a zero of the vector field F using the fact that $C_1 \cap C_2 \neq \emptyset$. Alternatively, one could apply the Bolzano Theorem and find a zero for F_2 restricted to C_1 (see also [135] for a similar use of a variant of the Crossing Lemma and [125] for extensions to the N-dimensional setting). Conversely, it is possible to provide a proof of the Crossing Lemma via the Poincaré–Miranda Theorem [118].

5 The N-dimensional setting: Intersection Lemma

In addition to the notions on paths introduced in Section 4, we need some further concepts in order to present a generalization to higher dimensional settings of the planar theory previously explained.

If $\gamma_1, \gamma_2 \colon [0,1] \to W$ are paths in the topological space W with $\gamma_1(1) = \gamma_2(0)$, we define the *gluing of* γ_1 *with* γ_2 as the path $\gamma_1 \star \gamma_2 \colon [0,1] \to W$ such that

$$\gamma_1 \star \gamma_2(t) := \begin{cases} \gamma_1(2t) & \text{for } 0 \le t \le \frac{1}{2}, \\ \gamma_2(2t-1) & \text{for } \frac{1}{2} \le t \le 1. \end{cases}$$

Moreover, given a path $\gamma\colon [0,1] \to W$, we denote by $\gamma^- \colon [0,1] \to W$ the path having $\overline{\gamma}$ as support, but run with reverse orientation with respect to γ, i. e. $\gamma^-(t) := \gamma(1-t)$, for all $t \in [0,1]$. We also recall that a topological space W is said to be *arcwise connected* if, for any couple of distinct points $p, q \in W$, there is a path $\gamma\colon [0,1] \to W$ such that $\gamma(0) = p$ and $\gamma(1) = q$. In the case of a Hausdorff topological space W, the range $\overline{\gamma}$ of γ turns out to be a locally connected metric continuum (a Peano space according to [66]). Thus, if W is a metric space, the above definition of arcwise connectedness is equivalent to the fact that, given any two points $p, q \in W$ with $p \neq q$, there exists an arc (i. e. the homeomorphic image of [0,1]) contained in W and having p and q as extreme points ([66, pp. 115–131]).

We start our explanation by presenting some topological lemmas concerning the relationship between particular surfaces and zero-sets of continuous real valued functions. Analogous results can be found, often in a more implicit form, in different contexts. However, since for our applications we need a specific version of the statements, we give an independent proof with all the details.

At first, let us introduce the central concept of “cutting set” .

Definition 5.1. Let X be an arcwise connected metric space and let A, B, C be closed nonempty subsets of X with $A \cap B = \emptyset$. We say that C *cuts the arcs between* A *and* B if for any path $\gamma\colon [0,1] \to X$, with $\overline{\gamma} \cap A \neq \emptyset$ and $\overline{\gamma} \cap B \neq \emptyset$, it follows that $\overline{\gamma} \cap C \neq \emptyset$. In the sequel, if X is a subspace of a larger metric space Z and we wish to stress that we consider only paths contained in X, we make our definition more precise by saying that C *cuts the arcs between* A *and* B *in* X.

Such a definition is a modification of the classical one regarding the cutting of a space between two points in [85]. See [20] for a more general concept concerning a set C that intersects every connected set meeting two nonempty sets A and B. In the special case that A and B are the opposite faces of an N-dimensional cube, J. Kampen [69, p. 512] says that C separates A and B. We prefer to use the “cutting” terminology in order to avoid any misunderstanding with other notions of separation which are more common in topology. In particular, we remark that our definition agrees with the usual one of cut when A, B, C are pairwise disjointed [55].

In the sequel, even when not explicitly mentioned, we assume the basic space X to be arcwise connected. In some of the following results, the local arcwise connectedness of X is required as well, i. e. for any $p \in X$, it holds that each neighborhood of p contains an arcwise connected neighborhood of p. In this respect, we recall that any connected and locally arcwise connected metric space is arcwise connected ([85, Th. 2, p. 253]).

Lemma 5.1. *Let X be a connected and locally arcwise connected metric space and let $A, B, C \subseteq X$ be closed and nonempty sets with $A \cap B = \emptyset$. Then, C cuts the arcs between A and B if and only if there exists a continuous function $f\colon X \to \mathbb{R}$ such that*

$$f(x) \le 0\,, \quad \forall x \in A\,, \quad f(x) \ge 0\,, \quad \forall x \in B \tag{5.1}$$

and

$$C = \{x \in X\colon f(x) = 0\}\,. \tag{5.2}$$

Proof. Assume there exists a continuous function $f\colon X \to \mathbb{R}$, satisfying (5.1) and (5.2). Let $\gamma\colon [0,1] \to X$ be a path such that $\gamma(0) \in A$ and $\gamma(1) \in B$. We want to prove that $\overline{\gamma} \cap C \neq \emptyset$. Indeed, for the composite continuous function $\theta := f \circ \gamma\colon [0,1] \to \mathbb{R}$, we have that $\theta(0) \le 0 \le \theta(1)$ and so the Bolzano Theorem ensures the existence of $t^* \in [0,1]$ with $\theta(t^*) = 0$. This means that $\gamma(t^*) \in C$ and therefore $\overline{\gamma} \cap C \neq \emptyset$. Hence, we have proved that C cuts the arcs between A and B.

Conversely, let us assume that C cuts the arcs between A and B. We introduce the auxiliary functions

$$\begin{aligned} &\rho\colon X \to \mathbb{R}^+\,, \\ &\rho(x) := \operatorname{dist}(x, C)\,, \quad \forall x \in X \end{aligned} \tag{5.3}$$

and

$$\mu\colon X \to \{-1,0,1\}\,,$$

$$\mu(x) := \begin{cases} 0 & \text{if } x \in C\,, \\ -1 & \text{if } x \notin C \text{ and } \exists \text{ a path } \gamma_x\colon [0,1] \to X \setminus C \\ & \text{such that } \gamma_x(0) \in A \text{ and } \gamma_x(1) = x\,, \\ 1 & \text{elsewhere.} \end{cases} \tag{5.4}$$

Observe that ρ is a continuous function with $\rho(x) = 0$ if and only if $x \in C$, and also $\mu(x) = 0$ if and only if $x \in C$. Moreover, μ is bounded.

Let $x_0 \notin C$. We claim that μ is continuous in x_0. Actually, μ is locally constant on $X \setminus C$. Indeed, since $x_0 \in X \setminus C$ (an open set) and X is locally arcwise connected, there is a neighborhood U_{x_0} of x_0 with $U_{x_0} \subseteq X \setminus C$ such that for each $x \in U_{x_0}$, there exists a path $\omega_{x_0,x}$ joining x_0 to x in U_{x_0}. Clearly, if there is a path γ_{a,x_0} in $X \setminus C$ joining some point $a \in A$ with x_0, then the path $\gamma_{a,x_0} \star \omega_{x_0,x}$ connects a to x in $X \setminus C$. This proves that if $\mu(x_0) = -1$, then $\mu(x) = -1$ for every $x \in U_{x_0}$. On the other hand, if there is a path $\gamma_{a,x}$ in $X \setminus C$ which connects some point $a \in A$ to $x \in U_{x_0}$, then the path $\gamma_{a,x} \star \omega^{-}_{x_0,x}$ connects a to x_0 in $X \setminus C$. This shows that if $\mu(x_0) = 1$ (that is, it is not possible to connect x_0 to any point of A in $X \setminus C$ using a path), then $\mu(x) = 1$ for every $x \in U_{x_0}$ (that is, it is not possible to connect any point $x \in U_{x_0}$ to any point of A in $X \setminus C$ using a path).

We can now define

$$f\colon X \to \mathbb{R}\,, \qquad f(x) := \rho(x)\mu(x)\,. \tag{5.5}$$

Clearly, $f(x) = 0$ if and only if $x \in C$ and, moreover, f is continuous. Indeed, if $x_0 \notin C$, we have that f is continuous in x_0 because both ρ and μ are continuous in x_0. If $x_0 \in C$ and $x_n \to x_0$ (as $n \to \infty$), then $\rho(x_n) \to 0$ and $|\mu(x_n)| \le 1$, so that $f(x_n) \to 0 = f(x_0)$. Finally, by the definition of μ in (5.4), it holds that $\mu(a) = -1$ for every $a \in A \setminus C$, and hence, for such an a, it holds that $f(a) < 0$. On the contrary, if we suppose that $b \in B \setminus C$, we must have $\mu(b) = 1$. In fact, by the cutting condition, there is no path connecting in $X \setminus C$ the point b to any point of A. Therefore, in this case, we have $f(b) > 0$. The proof is complete. □

The following lemma will find an application in Theorem 5.6.

Lemma 5.2. *Let X be a connected and locally arcwise connected metric space and let $A, B \subseteq X$ be closed and nonempty sets with $A \cap B = \emptyset$. Let $\Gamma \subseteq X$ be a compact connected set such that*

$$\Gamma \cap A \neq \emptyset\,, \quad \Gamma \cap B \neq \emptyset\,.$$

Then, for every $\varepsilon > 0$, there exists a path $\gamma = \gamma_\varepsilon : [0,1] \to X$ with $\gamma(0) \in A$, $\gamma(1) \in B$ and

$$\overline{\gamma} \subseteq B(\Gamma, \varepsilon)\,.$$

Moreover, if X is locally compact and $C \subseteq X$ is a closed set which cuts the arcs between A and B, then

$$\Gamma \cap C \neq \emptyset .$$

Proof. Let $\varepsilon > 0$ be fixed and consider, for every $p \in \Gamma$, a radius $\delta_p \in]0, \varepsilon[$ such that any two points in $B(p, \delta_p)$ can be joined by a path in $B(p, \varepsilon)$. Since Γ is compact, we can find a finite number of points

$$p_1, p_2, \ldots, p_k \in \Gamma$$

such that

$$\Gamma \subseteq \mathcal{B} := \bigcup_{i=1}^{k} B(p_i, \delta_i) \subseteq B(\Gamma, \varepsilon), \quad \text{where } \delta_i := \delta_{p_i} .$$

By the connectedness of Γ, for any partition of $\{1, \ldots, k\}$ into two nonempty disjoint subsets J_1 and J_2, there exist $i \in J_1$ and $j \in J_2$ such that $B(p_i, \delta_i) \cap B(p_j, \delta_j) \neq \emptyset$. This, in turn, implies that we can rearrange the p_i's (possibly changing their order in the labeling) so that

$$B(p_i, \delta_i) \cap B(p_{i+1}, \delta_{i+1}) \neq \emptyset, \quad \forall i = 1, \ldots, k-1 .$$

Hence, we can conclude that for any pair of points $w, z \in \mathcal{B}$ with $w \neq z$, there is a path $\gamma = \gamma_{w,z}$ joining w with z and such that $\overline{\gamma} \subseteq B(\Gamma, \varepsilon)$. In particular, taking $a \in A \cap \Gamma$ and $b \in B \cap \Gamma$, we have that there exists a path $\gamma = \gamma_\varepsilon : [0,1] \to B(\Gamma, \varepsilon)$, with $\gamma(0) = a$ and $\gamma(1) = b$ and this proves the first part of the statement.

Assume now that X is locally compact (i. e. for any $p \in X$ and $\eta > 0$, there exists $0 < \mu_p \leq \eta$ such that $\overline{B(p, \mu_p)}$ is compact). By the compactness of Γ, we can find a finite number of points $q_1, q_2, \ldots, q_l \in \Gamma$ and corresponding radii $\mu_i := \mu_{q_i}$ such that

$$\Gamma \subseteq \mathcal{A} := \bigcup_{i=1}^{l} B(q_i, \mu_i)$$

and $\overline{\mathcal{A}} = \bigcup_{i=1}^{l} \overline{B(q_i, \mu_i)}$ is compact. Since $\mathcal{A}$ is an open neighborhood of the compact set Γ, there exists $\varepsilon_0 > 0$ such that $B(\Gamma, \varepsilon_0) \subseteq \mathcal{A}$. Hence, for each $0 < \varepsilon \leq \varepsilon_0$, the set $\overline{B(\Gamma, \varepsilon)}$ is compact. Taking $\varepsilon = \frac{1}{n}$, we know that for every $n \in \mathbb{N}_0$, there exists a path $\gamma_n : [0,1] \to X$, with $\gamma_n(0) \in A$, $\gamma_n(1) \in B$ and $\overline{\gamma_n} \subseteq B(\Gamma, \frac{1}{n})$. However, since C cuts the arcs between A and B, it follows that for every $n \in \mathbb{N}_0$, there is a point $c_n \in C \cap B(\Gamma, \frac{1}{n})$. For $n \geq \hat{n}$ that is large enough ($\hat{n} > 1/\varepsilon_0$), the sequence $(c_n)_{n \geq \hat{n}}$ is contained in the compact set $\overline{B(\Gamma, \varepsilon_0)}$ and therefore it admits a converging subsequence $c_{n_k} \to c^* \in \overline{B(\Gamma, \varepsilon_0)}$. Since $d_X(c_{n_k}, \Gamma) < \frac{1}{n_k}$ and the sets C, Γ are closed, the limit point $c^* \in \Gamma \cap C$. This concludes the proof. □

Now, we are ready to apply the previous results to the intersection of generalized surfaces which separate the opposite edges of an N-dimensional cube. Such generalized surfaces (cf. Definition 5.2) will be described as zero-sets of continuous scalar functions and therefore a nonempty intersection will be obtained as a zero of a suitably defined vector field. To this aim, we recall a classical result about the existence of zeros for continuous maps in $\mathbb{R}^N$, the Poincaré–Miranda Theorem.

Theorem 5.1. *Let $I^N := [0,1]^N$ be the N-dimensional unit cube of $\mathbb{R}^N$, for which we denote by $[x_i = k] := \{x = (x_1,\dots,x_N) \in I^N \colon x_i = k\}$. Let $F = (F_1,\dots,F_N)\colon I^N \to \mathbb{R}^N$ be a continuous mapping such that, for each $i \in \{1,\dots,N\}$,*

$$F_i(x) \le 0\,, \quad \forall x \in [x_i = 0] \text{ and } F_i(x) \ge 0\,, \quad \forall x \in [x_i = 1]$$

or

$$F_i(x) \ge 0\,, \quad \forall x \in [x_i = 0] \text{ and } F_i(x) \le 0\,, \quad \forall x \in [x_i = 1]\,.$$

Then, there exists $\bar{x} \in I^N$ such that $F(\bar{x}) = 0$.

Let us now introduce the spaces we are going to consider.

Definition 5.2. Let Z be a metric space and

$$h\colon \mathbb{R}^N \supseteq I^N \to X \subseteq Z$$

be a homeomorphism of $I^N := [0,1]^N$ onto its image X. We call the pair

$$\widehat{X} := (X, h)$$

a *generalized N-dimensional rectangle of Z*. We also set

$$X_i^{\ell} := h([x_i = 0])\,, \quad X_i^{r} := h([x_i = 1])$$

and name them the *left* and the *right* i-faces of X.

Finally, we define

$$\vartheta X := h\left(\partial I^N\right)$$

and call it the *contour* of X.

As it is immediate to see, the generalized N-dimensional rectangles are a natural extension of the higher dimension of the generalized rectangles in Section 4. In a similar way, also the concept of an oriented rectangle will be transposed to the N-dimensional framework (cf. Definition 5.4).

Our main result on the intersection of generalized N-dimensional rectangles is Theorem 5.2 below, which can be considered as a variant of the Hurewicz–Wallman Lemma regarding dimension [68]. The statements of the two results are in fact very similar, but the lemma in [68] concerns, instead of our concept of cutting, the stronger notion of separation and requires the sets A, B, C in Definition 5.1 to be pairwise disjoint ([85]). Furthermore, with reference to Definition 5.2, in the statement of the Hurewicz–Wallman Lemma only, the very special case in which $Z = \mathbb{R}^N$, $X = I^N$ and $h = \mathrm{Id}_{\mathbb{R}^N}$ is considered. For such reasons, we have chosen to provide all the details.

Theorem 5.2 (Intersection Lemma). *Let $\widehat{X} := (X, h)$ be a generalized N-dimensional rectangle of a metric space Z. Assume that, $\forall\, i \in \{1, \dots, N\}$, there exists a compact set $S_i \subseteq X$ that cuts the arcs between X_i^{ℓ} and X_i^{r} in X. Then,*

$$\bigcap_{i=1}^{N} S_i \neq \emptyset\,.$$

Proof. Through the inverse of the homeomorphism $h\colon \mathbb{R}^N \supseteq I^N \to X \subseteq Z$, we can define the compact sets

$$C_i := h^{-1}(S_i)\,,$$

which cut the arcs between $[x_i = 0]$ and $[x_i = 1]$ in I^N (for $i = 1, \dots, N$). Clearly, it is sufficient to prove that

$$\bigcap_{i=1}^{N} C_i \neq \emptyset\,.$$

By Lemma 5.1, for every $i = 1, \dots, N$, there exists a continuous function $f_i\colon I^N \to \mathbb{R}$ such that $f_i \le 0$ on $[x_i = 0]$ and $f_i \ge 0$ on $[x_i = 1]$. Moreover,

$$C_i = \left\{x \in I^N\colon\, f_i(x) = 0\right\}\,.$$

The continuous vector field $\vec{f} := (f_1, \dots, f_N)\colon I^N \to \mathbb{R}^N$ satisfies the assumptions of the Poincaré–Miranda Theorem 5.1 and therefore there exists $\bar{x} \in I^N$ such that

$$f_i(\bar{x}) = 0\,, \quad \forall\, i = 1, \dots, N\,.$$

Hence, $\bar{x} \in \bigcap_{i=1}^{N} C_i$ and the proof is complete. □

We recall that the previous result was applied in [125] in order to extend some recent theorems about fixed points and periodic points for continuous mappings in Euclidean spaces. In particular, in [125, Corollary 3.5], we generalized, via a simplified proof, a theorem by Kampen [69], while in [125, Theorem 3.8], we obtained an extension of a result by Zgliczyński [184] about periodic points associated to Markov partitions.

5.1 Zero-sets of maps depending on parameters

As a next step, we deal with the intersection of generalized surfaces which separate the opposite edges of an N-dimensional cube in the case that the number of cutting surfaces is smaller than the dimension of the space. Our main tool is a result by Fitzpatrick, Massabó and Pejsachowicz ([56, Theorem 1.1]) on the covering dimension of the zero-set of an operator depending on parameters. For the reader's convenience, we recall the concept of covering dimension [54] in Definition 5.3 below, where by *order* of a family $\mathfrak{A}$ of subsets of the metric space Z, we mean the largest integer n such that the family $\mathfrak{A}$ contains $n+1$ sets with a nonempty intersection; if no such integer exists, the family $\mathfrak{A}$ has order infinity.

Definition 5.3 ([54, p. 54, p. 208]). Let Z be a metric space. We say that $\dim Z \leq n$ if every finite open cover of the space Z has a finite open [closed] refinement of order $\leq n$. The object $\dim Z \in \mathbb{N} \cup \{\infty\}$ is called the *covering dimension* or the *Čech–Lebesgue dimension* of the metric space Z. According to [56], if $z_0 \in Z$, we also say that $\dim Z \geq j$ *at* z_0 if each neighborhood of z_0 has a dimension of at least j.

By a classical result from topology (cf. [54, The coincidence theorem]), in separable metric spaces, the covering dimension coincides with the inductive dimension [54, p. 3].

In view of the following results, we also recall that given an open bounded set $\mathcal{O} \subseteq \mathbb{R}^N$ and $n \in \{1,\dots,N-1\}$, a continuous map $\pi\colon \overline{\mathcal{O}} \to \mathbb{R}^{N-n}$ is a *complement* for the continuous map $F\colon \overline{\mathcal{O}} \to \mathbb{R}^n$ if the topological degree $\deg((\pi,F),\mathcal{O},0)$ is defined and nonzero [56]. According to [66], a mapping f of a space X into a space Y is said to be *inessential* if f is homotopic to a constant; otherwise, f is *essential.*

At last, we introduce a further notation. Given an N-dimensional rectangle $\mathcal{R} := \prod_{i=1}^N [a_i,b_i]$, we denote its opposite i-faces by

$$\mathcal{R}_i^\ell := \{x \in \mathcal{R}\colon x_i = a_i\} \quad \text{and} \quad \mathcal{R}_i^r := \{x \in \mathcal{R}\colon x_i = b_i\}\,.$$

Theorem 5.3. *Let $\mathcal{R} := \prod_{i=1}^N [a_i,b_i]$ be an N-dimensional rectangle and let $P = (p_1,\dots,p_N)$ be any interior point of $\mathcal{R}$. Let $n \in \{1,\dots,N-1\}$ be fixed. Suppose that $F = (F_1,\dots,F_n)\colon \mathcal{R} \to \mathbb{R}^n$ is a continuous mapping such that for each $i \in \{1,\dots,n\}$,*

$$F_i(x) < 0\,, \quad \forall x \in \mathcal{R}_i^\ell \quad \textit{and} \quad F_i(x) > 0\,, \quad \forall x \in \mathcal{R}_i^r$$

or

$$F_i(x) > 0\,, \quad \forall x \in \mathcal{R}_i^\ell \quad \textit{and} \quad F_i(x) < 0\,, \quad \forall x \in \mathcal{R}_i^r\,.$$

Also, define the affine map

$$\pi\colon \mathbb{R}^N \to \mathbb{R}^{N-n}\,, \quad \pi_j(x_1,\dots,x_N) := x_j - p_j\,, \quad j = n+1,\dots,N\,. \tag{5.6}$$

Then, there exists a connected subset $\mathcal{Z}$ of

$$F^{-1}(0) = \{x \in \mathcal{R}\colon F_i(x) = 0,\ \forall i = 1,\dots,n\}$$

whose dimension at each point is at least $N-n$. Moreover,

$$\dim(\mathcal{Z} \cap \partial\mathcal{R}) \geq N-n-1$$

and

$$\pi\colon \mathcal{Z} \cap \partial\mathcal{R} \to \mathbb{R}^{N-n} \setminus \{0\}$$

is essential.

Proof. We define the continuous mapping

$$H := (F, \pi)\colon \mathcal{R} \to \mathbb{R}^N .$$

By the assumptions on F and π, we have

$$\deg(H, \mathrm{Int}(\mathcal{R}), 0) = (-1)^d \neq 0 ,$$

where d is the number of components $i \in \{1, \dots, n\}$ such that $F_i(x) > 0$ for $x \in \mathcal{R}_i^\ell$ and $F_i(x) < 0$ for $x \in \mathcal{R}_i^r$. Hence, π turns out to be a complementing map for F (according to [56]). A direct application of [56, Theorem 1.1] gives the thesis (observe that the dimension m in [56, Theorem 1.1] corresponds to our $N - n$). □

Notice that if $a_i < 0 < b_i$ for $i = 1, \dots, N$, then we can take $P = 0$, so that the complementing map is just the projection $\pi\colon \mathbb{R}^N \to \mathbb{R}^{N-n}$.

A more elementary version of Theorem 5.3 can be given for the zero-set of a vector field with range in $\mathbb{R}^{N-1}$. In this case, it is possible to achieve the thesis by a direct use of the classical Leray–Schauder Continuation Theorem [91] instead of the more sophisticated tools in [56]. Namely, we have:

Theorem 5.4. *Let $\mathcal{R} := \prod_{i=1}^N [a_i, b_i]$ be an N-dimensional rectangle and let $F = (F_1, \dots, F_{N-1})\colon \mathcal{R} \to \mathbb{R}^{N-1}$ be a continuous mapping such that for each $i \in \{1, \dots, N-1\}$,*

$$F_i(x) < 0 , \quad \forall x \in \mathcal{R}_i^\ell \quad \text{and} \quad F_i(x) > 0 , \quad \forall x \in \mathcal{R}_i^r$$

or

$$F_i(x) > 0 , \quad \forall x \in \mathcal{R}_i^\ell \quad \text{and} \quad F_i(x) < 0 , \quad \forall x \in \mathcal{R}_i^r .$$

Then there exists a closed connected subset $\mathcal{Z}$ of

$$F^{-1}(0) = \{x \in \mathcal{R}\colon F_i(x) = 0, \forall i = 1, \dots, N-1\}$$

such that

$$\mathcal{Z} \cap \mathcal{R}_N^\ell \neq \emptyset , \quad \mathcal{Z} \cap \mathcal{R}_N^r \neq \emptyset .$$

Proof. We split $x = (x_1, \dots, x_{N-1}, x_N) \in \mathcal{R} \subseteq \mathbb{R}^N$ as $x = (y, \lambda)$ with

$$y = (x_1, \dots, x_{N-1}) \in \mathcal{M} := \prod_{i=1}^{N-1} [a_i, b_i] , \quad \lambda = x_N \in [a_N, b_N]$$

and define

$$f = f(y, \lambda)\colon \mathcal{M} \times [a_N, b_N] \to \mathbb{R}^{N-1} , \quad f(y, \lambda) := F(x_1, \dots, x_{N-1}, \lambda) ,$$

treating the variable $x_N = \lambda$ as a parameter for the $(N-1)$-dimensional vector field

$$f_\lambda(\cdot) = f(\cdot, \lambda) .$$

By the assumptions on F, we have

$$\deg(f_\lambda, \operatorname{Int}(\mathcal{M}), 0) = (-1)^d \neq 0\,, \quad \forall \lambda \in [a_N, b_N]\,,$$

where d is the number of components $i \in \{1, \ldots, N-1\}$ such that $F_i(x) > 0$ for $x \in \mathcal{R}_i^\ell$ and $F_i(x) < 0$ for $x \in \mathcal{R}_i^r$. The Leray–Schauder Continuation Theorem [91, Théorème Fondamental] ensures the existence of a closed connected set

$$\mathcal{Z} \subseteq \left\{(y, \lambda) \in \mathcal{M} \times [a_N, b_N] : f(y, \lambda) = 0 \in \mathbb{R}^{N-1}\right\},$$

whose projection onto the λ-component covers the interval $[a_N, b_N]$. By the above positions, the thesis immediately follows. □

For the interested reader, we mention that Theorem 5.4 can be found in [84] and that it was then applied in [83].

In the next lemma, we take the unit cube $I^N := [0,1]^N$ as a N-dimensional rectangle and choose the interior point $P = (\frac{1}{2}, \frac{1}{2}, \ldots, \frac{1}{2})$ in order to apply Theorem 5.3. Obviously, any other point interior to I^N could be chosen as well.

Lemma 5.3. *Let $n \in \{1, \ldots, N-1\}$ be fixed. Assume that, $\forall\, i \in \{1, \ldots, n\}$, there is a compact set $S_i \subseteq I^N$ that cuts the arcs between $[x_i = 0]$ and $[x_i = 1]$ in I^N. Then, there exists a connected subset $\mathcal{Z}$ of $\bigcap_{i=1}^{n} S_i \neq \emptyset$ whose dimension at each point is at least $N - n$. Moreover,*

$$\dim\left(\mathcal{Z} \cap \partial I^N\right) \geq N - n - 1$$

and

$$\pi \colon \mathcal{Z} \cap \partial I^N \to \mathbb{R}^{N-n} \setminus \{0\}$$

is essential, where π is the affine map defined in (5.6).

Proof. For any fixed index $i^* \in \{1, \ldots, n\}$, we introduce the tunnel set

$$T_{i^*} := \prod_{i=1}^{i^*-1} [0,1] \times \mathbb{R} \times \prod_{i=i^*+1}^{N} [0,1]\,.$$

It is immediate to check that S_{i^*} cuts the arcs between $[x_{i^*} = 0]$ and $[x_{i^*} = 1]$ in T_{i^*}.

By Lemma 5.1, there exists a continuous function $f_{i^*} : T_{i^*} \to \mathbb{R}$ such that

$$\begin{aligned} &f_{i^*}(x) \leq 0\,, \quad \forall x \in T_{i^*}\,, \quad \text{with } x_{i^*} \leq 0 \quad \text{and} \\ &f_{i^*}(x) \geq 0\,, \quad \forall x \in T_{i^*}\,, \quad \text{with } x_{i^*} \geq 1\,, \end{aligned}$$

and moreover

$$S_{i^*} = \{x \in T_{i^*} : f_{i^*}(x) = 0\}\,.$$

(See also [125].) By this latter property and the fact that $S_{i^*} \subseteq I^N$, it follows that

$$f_{i^*}(x) < 0\,, \quad \forall x \in T_{i^*}\,, \quad \text{with } x_{i^*} < 0 \quad \text{and}$$
$$f_{i^*}(x) > 0\,, \quad \forall x \in T_{i^*}\,, \quad \text{with } x_{i^*} > 1\,.$$

Now, we define, for $x = (x_1, \dots, x_{i^*-1}, x_{i^*}, x_{i^*+1}, \dots, x_N) \in \mathbb{R}^N$, the continuous function

$$F_{i^*}(x) := f_{i^*}(\eta_{[0,1]}(x_1), \dots, \eta_{[0,1]}(x_{i^*-1}), x_{i^*}, \eta_{[0,1]}(x_{i^*+1}), \dots, \eta_{[0,1]}(x_N))\,,$$

where

$$\eta\colon \mathbb{R} \to [0,1]\,, \quad \eta_{[0,1]}(s) := \max\{0, \min\{s,1\}\} \tag{5.7}$$

is the projection of $\mathbb{R}$ onto the interval $[0,1]$. As a consequence of the above positions, we find that

$$F_{i^*}(x) < 0\,, \quad \forall x \in \mathbb{R}^N\colon x_{i^*} < 0 \quad \text{and} \quad F_{i^*}(x) > 0\,, \quad \forall x \in \mathbb{R}^N\colon x_{i^*} > 1\,.$$

We can thus apply Theorem 5.3 to the map $F = (F_1, \dots, F_n)$ restricted to the N-dimensional rectangle

$$\mathcal{R} := \prod_{i=1}^{n} [-1,2] \times \prod_{i=n+1}^{N} [0,1]\,.$$

Clearly,

$$(F\restriction_{\mathcal{R}})^{-1}(0) = \bigcap_{i=1}^{n} S_i \subseteq I^N$$

and the proof is complete. □

Remark 5.1. Both in Theorem 5.3 and in Lemma 5.3, the fact that we have privileged the first n components is purely conventional. It is evident that the results are valid for any finite sequence of indexes $i_1 < i_2 < \cdots < i_n$ in $\{1, \dots, N\}$. The same observation applies systematically to all the other results (preceding and subsequent) in which some directions are conventionally chosen. Moreover, notice that Lemma 5.3 is invariant under homeomorphisms in a sense that is described in Theorem 5.5 below.

Theorem 5.5. *Let $\widehat{X} := (X, h)$ be a generalized N-dimensional rectangle of a metric space Z. Let $i_1 < i_2 < \cdots < i_n$ be a finite sequence of $n \geq 1$ indexes in $\{1, \dots, N\}$. Assume that for each $j \in \{i_1, \dots, i_n\}$, there is a compact set $S_j \subseteq X$ that cuts the arcs between X_j^ℓ and X_j^r in X. Then there exists a compact connected subset $\mathcal{Z}$ of $\bigcap_{k=1}^{n} S_{i_k} \neq \emptyset$ whose dimension at each point is at least $N - n$. Moreover,*

$$\dim(\mathcal{Z} \cap \vartheta X) \geq N - n - 1$$

and

$$\pi\colon h^{-1}(\mathcal{Z}) \cap \partial I^N \to \mathbb{R}^{N-n} \setminus \{0\}$$

is essential, where π is defined as in (5.6) for $P = (\frac{1}{2}, \frac{1}{2}, \dots, \frac{1}{2})$.

Proof. The result easily follows by moving to the setting of Lemma 5.3 through the homeomorphism h^{-1} and repeating the arguments used therein. □

At last, we present a result (Corollary 5.1) which plays a crucial role in the subsequent proofs. It concerns the case $n = N - 1$ and could be obtained by suitably adapting the arguments employed in Lemma 5.3. However, due to its significance for our applications, we wish to provide a detailed proof using Theorem 5.4 (which only requires the knowledge of the Leray–Schauder principle and therefore, in some sense, is more elementary). Corollary 5.1 extends, to an arbitrary dimension, some results in [135, Appendix] which were there proved only for $N = 2$ using [146].

Corollary 5.1. *Let $\widehat{X} := (X, h)$ be a generalized N-dimensional rectangle of a metric space Z. Let $k \in \{1, \ldots, N\}$ be fixed. Assume that, for each $j \in \{1, \ldots, N\}$ with $j \neq k$, there exists a compact set $S_j \subseteq X$ that cuts the arcs between X_j^{ℓ} and X_j^{r} in X. Then, there exists a compact connected subset C of $\bigcap_{i \neq k} S_i \neq \emptyset$ such that*

$$C \cap X_k^{\ell} \neq \emptyset, \quad C \cap X_k^{r} \neq \emptyset .$$

Proof. Without loss of generality (if necessary, by a permutation of the coordinates), we assume $k = N$. In this manner, using the homeomorphism $h^{-1}\colon Z \supseteq X = h(I^N) \to I^N$, we can confine ourselves to the following framework:

For each $j \in \{1, \ldots, N-1\}$, there exists a compact set

$$S_j' := h^{-1}(S_j) \subseteq I^N$$

that cuts the arcs between $[x_j = 0]$ and $[x_j = 1]$ in I^N.

Proceeding as in the proof of Lemma 5.3, for any fixed $i^* \in \{1, \ldots, N-1\}$, we introduce the tunnel set

$$T_{i^*} := \prod_{i=1}^{i^*-1} [0,1] \times \mathbb{R} \times \prod_{i=i^*+1}^{N} [0,1]$$

and find that S_{i^*}' cuts the arcs between $[x_{i^*} = 0]$ and $[x_{i^*} = 1]$ in T_{i^*}.

Hence, by Lemma 5.1, there exists a continuous function $f_{i^*} : T_{i^*} \to \mathbb{R}$ such that

$$f_{i^*}(x) \le 0, \quad \forall x \in T_{i^*}, \quad \text{with } x_{i^*} \le 0 \quad \text{and}$$
$$f_{i^*}(x) \ge 0, \quad \forall x \in T_{i^*}, \quad \text{with } x_{i^*} \ge 1,$$

as well as

$$S_{i^*}' = \{x \in T_{i^*} : f_{i^*}(x) = 0\} .$$

(See also [125].) More precisely, it holds that

$$f_{i^*}(x) < 0, \quad \forall x \in T_{i^*}, \quad \text{with } x_{i^*} < 0 \quad \text{and}$$
$$f_{i^*}(x) > 0, \quad \forall x \in T_{i^*}, \quad \text{with } x_{i^*} > 1 .$$

We define, for $x = (x_1, \dots, x_{i^*-1}, x_{i^*}, x_{i^*+1}, \dots, x_N) \in \mathbb{R}^N$, the continuous function

$$F_{i^*}(x) := f_{i^*}(\eta_{[0,1]}(x_1), \dots, \eta_{[0,1]}(x_{i^*-1}), x_{i^*}, \eta_{[0,1]}(x_{i^*+1}), \dots, \eta_{[0,1]}(x_N))\,,$$

where $\eta_{[0,1]}$ is the projection of $\mathbb{R}$ onto the interval $[0,1]$ defined as in (5.7). Then, we have

$$F_{i^*}(x) < 0\,, \quad \forall x \in \mathbb{R}^N : x_{i^*} < 0 \quad \text{and } F_{i^*}(x) > 0\,, \quad \forall x \in \mathbb{R}^N : x_{i^*} > 1\,.$$

Now, we consider the map $F = (F_1, \dots, F_{N-1})$ restricted to the N-dimensional rectangle

$$\mathcal{R} := \prod_{i=1}^{N-1} [-\varepsilon, 1+\varepsilon] \times [0,1]\,,$$

for any fixed $\varepsilon > 0$. Since

$$(F \restriction_{\mathcal{R}})^{-1}(0) = \bigcap_{i=1}^{N-1} S'_i \subseteq I^N\,,$$

we can set

$$C := h(\mathcal{Z})\,,$$

where $\mathcal{Z} \subseteq (F \restriction_{\mathcal{R}})^{-1}(0)$ comes from the statement of Theorem 5.4, and this concludes the proof. □

As we shall see, the latter result turns out to be our main ingredient in the next subsection for the study of the dynamics of continuous maps defined on topological N-dimensional rectangles which possess, in a very broad sense, a one-dimensional expansive direction. Indeed, it allows one to prove the crucial fixed point Theorem 5.6 (similar to Theorem 4.1) and the subsequent results on the existence of multiple periodic points. Notice that the role of Corollary 5.1 in the proof of Theorem 5.6 is analogous to the one of the Crossing Lemma 4.5 in the verification of Theorem 4.1.

5.2 Stretching along the paths in the N-dimensional case

Finally, we are in position to provide an extension to N-dimensional spaces of the results obtained in Section 4 for the planar maps which expand the paths along a certain direction. To this aim, we reconsider Definition 5.2 in order to focus our attention on generalized N-dimensional rectangles in which we have fixed once and for all the left and right sides. In the applications, these opposite sides give a sort of orientation to the generalized N-dimensional rectangles and, in fact, they are usually related to the expansive direction.

Definition 5.4. Let Z be a metric space and let $\widehat{X} := (X, h)$ be a generalized N-dimensional rectangle of Z. We set

$$X_\ell := h([x_N = 0])\,, \quad X_r := h([x_N = 1])$$

and

$$X^- := X_\ell \cup X_r .$$

The pair

$$\widetilde{X} := (X, X^-)$$

is called an *oriented N-dimensional rectangle.*

Remark 5.2. A comparison between Definitions 5.2 and 5.4 shows that an oriented N-dimensional rectangle is just a generalized N-dimensional rectangle in which we privilege the two subsets of its contour which correspond to the opposite faces for some fixed component (namely, the x_N-component). In analogy with Remark 5.1, we point out that the choice of the N-th component is purely conventional. For example, in some other papers ([59, 124, 186]) and also in the planar examples from Section 4, the first component was selected. Clearly, there is no substantial difference as the homeomorphism $h\colon \mathbb{R}^N \supseteq I^N \to X \subseteq Z$ could be composed with a permutation matrix, yielding to a new homeomorphism with the same image set. From this point of view, our definition fits to the one of h-set of $(1, N-1)$-type, given by Zgliczyński and Gidea in [186] for a subset of $\mathbb{R}^N$ which is obtained as the inverse image of the unit cube through a homeomorphism of $\mathbb{R}^N$ onto itself. The similar concept of $(1, N-1)$-window is then considered by Gidea and Robinson in [59]: it is defined as a homeomorphic copy of the unit cube I^N of $\mathbb{R}^N$ through a homeomorphism whose domain is an open neighborhood of I^N.

We now adapt to maps between oriented N-dimensional rectangles the concept of "stretching along the paths", already considered in Definition 4.1 with regards to the planar case.

Definition 5.5. Let Z be a metric space and let $\psi\colon Z \supseteq D_\psi \to Z$ be a map defined on a set D_ψ. Assume that $\widetilde{X} := (X, X^-)$ and $\widetilde{Y} := (Y, Y^-)$ are oriented rectangles of Z and let $\mathcal{K} \subseteq X \cap D_\psi$ be a compact set. We say that $(\mathcal{K}, \psi)$ *stretches $\widetilde{X}$ to $\widetilde{Y}$ along the paths* and write

$$(\mathcal{K}, \psi)\colon \widetilde{X} \Leftrightarrow\!\!\!\to \widetilde{Y}$$

if the following conditions hold:

- ψ is continuous on $\mathcal{K}$;
- For every path $\gamma\colon [0,1] \to X$ such that $\gamma(0) \in X_\ell^-$ and $\gamma(1) \in X_r^-$ (or $\gamma(0) \in X_r^-$ and $\gamma(1) \in X_\ell^-$), there exists a subinterval $[t', t''] \subseteq [0,1]$ such that
$$\gamma(t) \in \mathcal{K}, \quad \psi(\gamma(t)) \in Y, \quad \forall t \in [t', t'']$$
and, moreover, $\psi(\gamma(t'))$ and $\psi(\gamma(t''))$ belong to different sides of Y^-.

By the similarity between Definitions 4.1 and 5.5, it is easy to see that the remarks in Section 4 regarding the stretching along the path relation for planar maps remain valid also in the higher dimensional setting. In particular, the comments on the relationship with the Brouwer fixed point Theorem or with the theory of the topological horseshoes in [71, 73] hold true. Suitable modifications of Definitions 4.2–4.5 could be presented as well. For the sake of conciseness, we prefer to omit them. In view of its significance for the following treatment, we present, however, the next fixed point theorem which shows that the set $\mathcal{K}$ still plays a crucial role.

Theorem 5.6. *Let Z be a metric space and let $\psi\colon Z \supseteq D_\psi \to Z$ be a map defined on a set D_ψ. Assume that $\widetilde{X} := (X, X^-)$ is an oriented N-dimensional rectangle of Z. If $\mathcal{K} \subseteq X \cap D_\psi$ is a compact set for which it holds that*

$$(\mathcal{K}, \psi)\colon \widetilde{X} \Longleftrightarrow\!\!\!\!\to \widetilde{X}\,, \tag{5.8}$$

then there exists at least one point $z \in \mathcal{K}$ with $\psi(z) = z$.

Proof. Let $h\colon I^N = [0,1]^N \to h(I^N) = X \subseteq Z$ be a homeomorphism such that $X_\ell = h([x_N = 0])$, $X_r = h([x_N = 1])$ and consider the compact subset of I^N

$$\mathcal{W} := h^{-1}(\mathcal{K} \cap \psi^{-1}(X))$$

as well as the continuous mapping $\phi = (\phi_1, \dots, \phi_N)\colon \mathcal{W} \to I^N$ defined by

$$\phi(x) := h^{-1}(\psi(h(x)))\,, \quad \forall x \in \mathcal{W}\,.$$

By the Tietze–Urysohn Theorem [54, p. 87], there exists a continuous map

$$\varphi = (\varphi_1, \dots, \varphi_N)\colon I^N \to I^N\,, \quad \varphi \restriction_{\mathcal{W}} = \phi\,.$$

Let us introduce, for every $i = 1, \dots, N-1$, the closed sets

$$S_i := \left\{x = (x_1, \dots, x_{N-1}, x_N) \in I^N : x_i = \varphi_i(x)\right\} \subseteq I^N\,.$$

Since $\varphi(I^N) \subseteq I^N$, by the continuity of the φ_i's, it is straightforward to check that S_i cuts the arcs between $[x_i = 0]$ and $[x_i = 1]$ in I^N for $i = 1, \dots, N-1$. Indeed, if $\gamma\colon [0,1] \to I^N$ is a path with $\gamma_i(0) = 0$ and $\gamma_i(1) = 1$, then, for the auxiliary function $g\colon [0,1] \ni t \mapsto \gamma_i(t) - \varphi_i(\gamma(t))$, we have $g(0) \le 0 \le g(1)$ and therefore Bolzano Theorem implies the existence of $s \in [0,1]$ such that $\gamma_i(s) = \varphi_i(\gamma(s))$, that is, $\overline{\gamma} \cap S_i \ne \emptyset$. The cutting property is thus proved.

Now, Corollary 5.1 guarantees that there is a continuum

$$C \subseteq \bigcap_{i=1}^{N-1} S_i \tag{5.9}$$

such that

$$C \cap [x_N = 0] \ne \emptyset\,, \quad C \cap [x_N = 1] \ne \emptyset\,.$$

Lemma 5.2 implies that for every $\varepsilon > 0$, there exists a path $\gamma_\varepsilon\colon [0,1] \to I^N$ such that

$$\gamma_\varepsilon(0) \in [x_N = 0]\,, \quad \gamma_\varepsilon(1) \in [x_N = 1] \quad \text{and } \gamma_\varepsilon(t) \in B(C, \varepsilon) \cap I^N\,, \quad \forall t \in [0,1]\,.$$

By the stretching assumption (5.8) and the definition of $\mathcal{W}$ and ϕ, there is a subpath ω_ε of γ_ε such that

$$\overline{\omega_\varepsilon} \subseteq \mathcal{W} \text{ and } \phi(\overline{\omega_\varepsilon}) \subseteq I^N\,, \qquad \text{with } \phi(\overline{\omega_\varepsilon}) \cap [x_N = 0] \neq \emptyset\,, \quad \phi(\overline{\omega_\varepsilon}) \cap [x_N = 1] \neq \emptyset\,.$$

The Bolzano Theorem applied to the continuous mapping $x \mapsto x_N - \varphi_N(x)$ on $\overline{\omega_\varepsilon}$ ensures the existence of a point

$$\tilde{x}^{\varepsilon} = (\tilde{x}^{\varepsilon}_1, \ldots, \tilde{x}^{\varepsilon}_N) \in \overline{\omega_\varepsilon} \subseteq \mathcal{W}$$

such that

$$\tilde{x}^{\varepsilon}_N = \varphi_N(\tilde{x}^{\varepsilon})\,.$$

Taking $\varepsilon = \frac{1}{n}$ (for $n \in \mathbb{N}_0$) and letting $n \to \infty$, by a standard compactness argument, we find a point

$$\tilde{x} = (\tilde{x}_1, \ldots, \tilde{x}_N) \in C \cap \mathcal{W}$$

such that

$$\tilde{x}_N = \varphi_N(\tilde{x})\,.$$

By (5.9), recalling also the definition of the S_i's, we get

$$\tilde{x} = \varphi(\tilde{x}) \in \mathcal{W}\,.$$

Then, since $\varphi \restriction_{\mathcal{W}} = \phi$, by the relation

$$h(\phi(x)) = \psi(h(x))\,, \quad \forall x \in \mathcal{W}\,,$$

we have that $h(\tilde{x}) = \psi(h(\tilde{x})) \in h(\mathcal{W})$ and therefore

$$z := h(\tilde{x}) \in \mathcal{K} \cap \psi^{-1}(X)$$

is the desired fixed point for ψ. □

See Figure 10 for a geometrical description of Theorem 5.6.

Having proved Theorems 5.5 and 5.6, we have the tools available for extending the results obtained in Section 4 for the planar case [117, 118] to any finite dimension. For brevity's sake, we focus our attention only on a few of them presented below in the more general setting. Notice that now we can complement the previous results, adding some information on the dimension of the cutting surfaces.

Theorem 5.7. *Let Z be a metric space and $\psi\colon Z \supseteq D_\psi \to Z$ be a map defined on a set D_ψ. Assume that $\widetilde{X} := (X, X^-)$ is an oriented N-dimensional rectangles of Z. If $\mathcal{K}_0, \ldots, \mathcal{K}_{m-1}$ are $m \geq 2$ pairwise disjoint compact subsets of $X \cap D_\psi$ and*

$$(\mathcal{K}_i, \psi)\colon \widetilde{X} \Leftrightarrow\!\!\!\!\to \widetilde{X}\,, \quad \textit{for } i = 0, \ldots, m-1\,,$$

then the following conclusions hold:

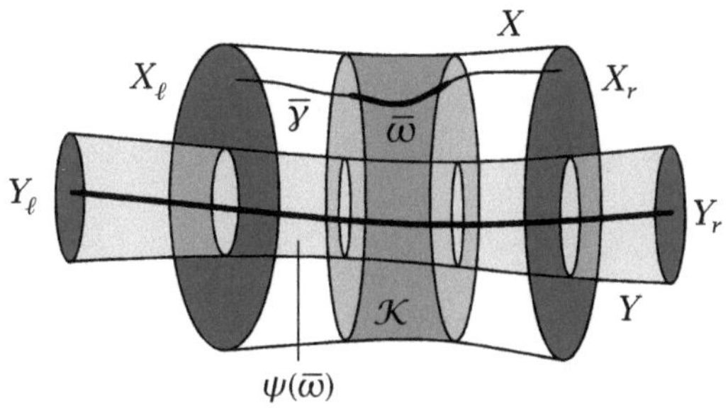

Fig. 10: The tubular sets X and Y represent two generalized three-dimensional rectangles in which we have indicated the compact set $\mathcal{K}$ and the boundary sets X_ℓ and X_r, as well as Y_ℓ and Y_r. The function ψ is continuous on $\mathcal{K}$ and maps X onto $Y = \psi(X)$ so that $Y_\ell = \psi(X_\ell)$ and $Y_r = \psi(X_r)$. However, in this particular case, the map ψ stretches the paths of X not only across Y, but also across X itself and therefore the existence of a fixed point for ψ inside $\mathcal{K}$ is ensured by Theorem 5.6.

- *The map ψ has at least a fixed point in $\mathcal{K}_i$, $\quad i = 0, \dots, m-1$;*
- *For each two-sided sequence $(s_h)_{h\in\mathbb{Z}} \in \{0,\dots,m-1\}^{\mathbb{Z}}$, there exists a sequence of points $(x_h)_{h\in\mathbb{Z}}$ such that $\psi(x_{h-1}) = x_h \in \mathcal{K}_{s_h}$, $\quad \forall h \in \mathbb{Z}$;*
- *For each sequence of m symbols $\mathbf{s} = (s_n)_n \in \{0,1,\dots,m-1\}^{\mathbb{N}}$, there exists a compact connected set $C_\mathbf{s} \subseteq \mathcal{K}_{s_0}$ which cuts the arcs between X_ℓ and X_r in X and such that for every $w \in C_\mathbf{s}$, there is a sequence $(y_n)_n$ with $y_0 = w$ and*

$$y_n \in \mathcal{K}_{s_n}, \quad \psi(y_n) = y_{n+1}, \quad \forall n \geq 0\,.$$

The dimension of $C_\mathbf{s}$ at any point is at least $N-1$. Moreover, $\dim(C_\mathbf{s} \cap \vartheta X) \geq N-2$ and $\pi\colon h^{-1}(C_\mathbf{s}) \cap \partial I^N \to \mathbb{R}^{N-1} \setminus \{0\}$ is essential (where π is defined as in (5.6) for $p_i = \frac{1}{2}$, $\forall i$ and h is the homeomorphism defining X);
- *Given an integer $j \geq 2$ and a $j+1$-uple $(s_0,\dots,s_j)$, $s_i \in \{0,\dots,m-1\}$, for $i = 0,\dots,j$, and $s_0 = s_j$, then there exists a point $w \in \mathcal{K}_{s_0}$ such that*

$$\psi^i(w) \in \mathcal{K}_{s_i}, \quad \forall i = 1,\dots,j, \quad \text{and} \quad \psi^j(w) = w\,.$$

As in the two-dimensional framework, such a result can be easily obtained as a corollary of the more general:

Theorem 5.8. *Let $(\widetilde{X}_i)_{i\in\mathbb{Z}}$ (with $\widetilde{X}_i = (X_i, X_i^-)$) be a double sequence of oriented N-dimensional rectangles of a metric space Z and let $(\mathcal{K}_i, \psi_i)_{i\in\mathbb{Z}}$, with $\mathcal{K}_i \subseteq X_i$, be a sequence such that*

$$(\mathcal{K}_i, \psi_i)\colon \widetilde{X}_i \Leftrightarrow\!\!\!\!\rightarrow \widetilde{X}_{i+1}, \quad \forall i \in \mathbb{Z}\,.$$

Let us denote by X_ℓ^i and X_r^i the two components of X_i^-. Then, the following conclusions hold:

- *There is a sequence $(w_k)_{k\in\mathbb{Z}}$ such that $w_k \in \mathcal{K}_k$ and $\psi_k(w_k) = w_{k+1}$, for all $k \in \mathbb{Z}$;*

- *For each* $j \in \mathbb{Z}$*, there exists a compact connected set* $C_j \subseteq \mathcal{K}_j$ *which cuts the arcs between* X_ℓ^j *and* X_r^j *in* X_j *and such that for every* $w \in C_j$*, there is a sequence* $(y_i)_{i \geq j}$ *with* $y_j = w$ *and*

$$y_i \in \mathcal{K}_i\,, \quad \psi_i(y_i) = y_{i+1}\,, \quad \forall\, i \geq j\,.$$

The dimension of C_j *at any point is at least* $N-1$*. Moreover,* $\dim(C_j \cap \vartheta X_j) \geq N-2$ *and* $\pi\colon h_j^{-1}(C_j) \cap \partial I^N \to \mathbb{R}^{N-1} \setminus \{0\}$ *is essential (where* π *is defined as in (5.6) for* $p_i = \frac{1}{2}$*,* $\forall\, i$ *and* h_j *is the homeomorphism defining* X_j*);*
- *If there are integers* h *and* l*, with* $h < l$ *such that* $\widetilde{X}_h = \widetilde{X}_l$*, then there exists a finite sequence* $(z_i)_{h \leq i \leq l-1}$*, with* $z_i \in \mathcal{K}_i$ *and* $\psi_i(z_i) = z_{i+1}$ *for each* $i = h, \ldots, l-1$*, such that* $z_l = z_h$*, that is,* z_h *is a fixed point of* $\psi_{l-1} \circ \cdots \circ \psi_h$*.*

Proof. The thesis follows by steps analogous to the ones in the proof of Theorem 4.4, using Theorem 5.6 in place of Theorem 4.1 and recalling Definition 5.1. The estimates on the dimension of the continuum C_j come from Theorem 5.5. □

As a final remark, we observe that instead of maps expansive just along one direction, it is possible to deal with the more general case of functions defined on N-dimensional rectangles (or homeomorphic images of them) with u unstable directions (along which an expansion occurs) and $s = N - u$ stable directions (along which the system is compressive). In this respect, many works are available in the literature, for instance, [14, 15, 59, 69, 131, 186]. In particular, we recall that a similar framework was analyzed in [124], where the present authors introduced a "deformation relation" among generalized N-dimensional rectangles with some similarities to the stretching property in Definition 5.5. Indeed, that relation could also be viewed (cf. [124, Remark 3.4]) as a higher dimensional counterpart of the planar stretching property in Definition 4.1 from [117, 118] and moreover it was suitable for the detection of fixed points, periodic points and chaotic dynamics as well. However, its definition required some assumptions on the topological degree of the map involved, while the conditions in Definition 5.5 do not need any sophisticated topological tool. Therefore, it is in this perspective of simplicity that the above results from [125] have to be considered.

6 Chaotic dynamics for continuous maps

In the present section, we show how to apply the method of "stretching along the paths" from Section 4 to prove the presence of chaotic dynamics in discrete-time dynamical systems. In order to do that, we first set our framework among more classical ones available in the literature concerning the concepts of "covering" and "crossing." Indeed, this preliminary introductory discussion is meant to give an outline of the reasons that led various authors to deal with similar settings as well as to quickly present the topics developed along Sections 7–10, where the undefined terms will be rigorously explained.

Generally speaking, in the investigation of a dynamical system defined by a map f on a metric space, it can be difficult or impossible to find a direct proof of the presence of chaos according to one or the other of the several, more or less equivalent, existing definitions. Hence, a canonical strategy is to establish a conjugacy or semiconjugacy between f, perhaps restricted to a suitable (positively) invariant subset of its domain, and the shift map $\sigma\colon (s_i)_{i\in\mathbb{I}} \mapsto (s_{i+1})_{i\in\mathbb{I}}$ on $\{0,\dots,m-1\}^{\mathbb{I}}$, for $\mathbb{I} = \mathbb{N}$ or $\mathbb{I} = \mathbb{Z}$, i. e. on the space of unilateral or bilateral sequences of $m \geq 2$ symbols, which displays many chaotic features, for example, transitivity, sensitivity, positive entropy, etc. In such an indirect manner, one can conclude that f possesses all the properties of σ that are preserved by the conjugacy/semiconjugacy relation. In particular, the known fact (cf. Section 8) that $h_{\text{top}}(f^n) = nh_{\text{top}}(f)$ for $n \geq 1$, where $h_{\text{top}}(f)$ denotes the topological entropy of f, guarantees that if a power of a map is conjugate or semiconjugate to σ, then such a map has positive entropy. Accordingly, to show the existence of chaos (positive entropy) of a map f, it is sufficient to look at any of its iterates.

A crucial role in this strategy is played by the study of a topological property for maps which are called “crossing” and, broadly speaking, refers to the way in which the image of suitable sets under the iterates of the maps intersect the original sets. In particular, the kind of crossing known as “horseshoe” has turned out to be a fundamental tool of analysis to prove the existence of chaotic dynamics in a well-defined manner. The name and idea of a horseshoe was derived from the celebrated work by Smale [149, 150] that provided a mathematically rigorous and geometrically fascinating proof of the existence of chaos for a special planar diffeomorphism.

More precisely (refer to [47, 111, 169] for the technical details), in the horseshoe model, a square S is first shrunk uniformly in one direction and expanded in the other one. Subsequently, the elongated rectangle obtained in the previous step is bent along the original square in order to cross it twice. The resulting map F is a diffeomorphism with F and F^{-1} transforming the vertical and horizontal lines of the square into similar lines crossing the domain. In Smale’s construction, the set $\mathcal{I}_S := \{q \in S\colon F^k(q) \in S\,, \quad \forall k \in \mathbb{Z}\}$, consisting of the points which remain in the square under all the iterates of F in both forward and backward time, is a compact invariant set for F which contains, as a dense subset, the periodic points of F such that F is sensitive on initial conditions and topologically transitive on $\mathcal{I}_S$. Actually, F acts on $\mathcal{I}_S$ like the shift map σ on two symbols since $F \restriction_{\mathcal{I}_S}$ and σ are conjugate. As mentioned above, the conjugacy, which is given by a homeomorphism $\pi\colon \mathcal{I}_S \to \{0,1\}^{\mathbb{Z}}$, with $\pi \circ F = \sigma \circ \pi$, allows one to transfer to $F \restriction_{\mathcal{I}_S}$ the well-known dynamical properties of the Bernoulli shift.

In the applications to concrete dynamical systems, the presence of a complex behavior for a given map F can be verified by proving the existence of a horseshoe structure either for the map itself or for one of its iterates. This led some authors ([30]) to define a *horseshoe* like a set Λ that is invariant under F^{n_0}, for some $n_0 \geq 1$, with the property that $F^{n_0} \restriction_\Lambda$ is topologically conjugate to σ (or, more generally, to a nontrivial subshift of finite type). As pointed out by Burns and Weiss in [30], the difficulty in finding a horseshoe lies in showing that the map $\pi\colon \Lambda \to \{0,1\}^{\mathbb{Z}}$ is injective. Such a step, in turn, often requires the verification of some assumptions on F, like being a diffeomorphism satisfying suitable hyperbolicity conditions, which either are not fulfilled or are hard to check for a given map. Hence, more general and less stringent definitions of a horseshoe have been suggested to reproduce some geometrical features typical of the Smale horseshoe while discarding the hyperbolicity conditions associated with it, which are difficult or impossible to prove in practical cases. This led to the study of the so-called "topological horseshoes" [30, 73, 186].

The core of such a field of research consists of providing an adequate notion of crossing property for higher dimensional dynamical systems so that a map F (or one of its iterates) satisfying certain geometrical conditions is proved to be semiconjugate to a full shift on $m \geq 2$ symbols. This is enough to conclude that F displays chaotic dynamics in the sense that F has, for instance, positive topological entropy. More generally, one could show that F is semiconjugate to a nontrivial subshift of finite type by employing some tools from symbolic dynamics [77, 94] as explained in Section 8.

Regarding the one-dimensional case, a classical example of a horseshoe-type crossing property relies on the definition of "covering." We recall that, according to [25], given a continuous map $f\colon \mathbb{R} \to \mathbb{R}$ and two intervals $I, J \subseteq \mathbb{R}$, we say that *$I$ f-covers J* if there exists a subinterval $I_0 \subseteq I$ such that $f(I_0) = J$. We also say that *I f-covers J m times* if there exist $m \geq 2$ subintervals $I_0, \dots, I_{m-1} \subseteq I$, with pairwise disjoint interiors, such that $f(I_k) = J$ for $k = 0, \dots, m-1$. Special significance from the point of view of complex dynamics is assigned to the case in which I f-covers I m times. Different authors have investigated possible variants of the above definitions dealing with the framework in which there are $m \geq 2$ intervals $I_0, \dots, I_{m-1}$ such that I_i f-covers I_j, for each $i, j \in \{0, \dots, m-1\}$, and some disjointness condition is imposed on them. For instance, the case $m = 2$ with I_0, I_1 open disjoint intervals was taken by Glendinning in [60] as a definition of a "horseshoe" for f, while Block and Coppel in [23, 24] define "turbulence" for f by the similar framework when I_0 and I_1 are compact intervals with disjoint interiors and "strict turbulence" when I_0, I_1 are compact and disjoint. In such settings, one obtains the typical features associated with the concept of chaos (e. g. existence of periodic points of each period, semiconjugacy to the Bernoulli shift for the map f or for some of its iterates and thus positive topological entropy, ergodicity with respect to some invariant measure, sensitivity to initial conditions, transitivity). Besides the works already quoted, see [10, 26, 88, 142].

An elementary introduction to these ideas is provided by the analysis of the logistic map $F\colon [0,1] \to \mathbb{R}$, $F(x) = \mu x(1-x)$, with μ a positive real parameter. It is easy to see that for $\mu \geq 4$, the interval $[0,1]$ F-covers $[0,1]$ twice, whereas for $\mu < 4$, the existence of multiple coverings can be established for some iterate of F, as shown in Section 8. The concepts of covering and horseshoe can also be used to revisit some classical results in the theory of one-dimensional unimodal maps, such as the celebrated Li–Yorke Theorem [93]. For instance, [60, Theorem 11.13] shows that whenever a unimodal map f has a point of period three, then its second iterate f^2 has a horseshoe. However, it is in the higher dimensional frameworks that the approach described above shows its full power.

The most natural way to extend the notion of multiple f-covering to the abstract setting of a continuous self-map f of a metric space X may be that of assuming the existence of $m \geq 2$ compact sets $K_0, \ldots, K_{m-1} \subseteq X$ such that $f(K_i) \supseteq \bigcup_{j=0}^{m-1} K_j$, $\forall i = 0, \ldots, m-1$. Moreover, in order to avoid trivial conclusions, one usually requires some disjointness conditions on the sets K_i's, such as $K_i \cap K_j = \emptyset$, $\forall i \neq j$ (or, more generally, $\bigcap_{i=0}^{m-1} K_i = \emptyset$). Blokh and Teoh [26] describe such framework by saying that $(f, K_0, \ldots, K_{m-1})$ form a m-horseshoe (or a weak m-horseshoe, respectively). Under these conditions, it is possible to show the existence of some chaotic behavior for the discrete dynamical system generated by f ([88, 142]). In particular, in [26], it is proved that if a power of f admits a weak m-horseshoe, then the topological entropy of f is positive and there exists a set B and a power g of f such that B is g-invariant and $g \restriction_B$ is semiconjugate to the Bernoulli shift on $\{0, \ldots, m-1\}^{\mathbb{N}}$.

In spite of the generality of the setting in which such results can be obtained and their effectiveness for the one-dimensional framework, when one tries to apply the theory to specific higher dimensional mathematical models arising in applications, it may be expedient to follow Burns and Weiss' suggestion [30] and replace the previous covering relation with a weaker condition of the form $f(K_i)$ "goes across" $\bigcup_{j=0}^{m-1} K_j$, $\forall i = 0, \ldots, m-1$, which does not require the map f to be surjective on the K_i's.

In our approach, the expression "goes across" (also called the "Markov property" [62, p. 291]) has to be understood in relation to the stretching along the paths effect of a map f from Definition 4.1 or Definition 5.5.[3] More precisely, we assume that each path γ in our generalized (N-dimensional) rectangle $\mathcal{R}$, joining the two sides of $\mathcal{R}^-$, intercepts every K_i and is then expanded to a path $f \circ \gamma$ which crosses all the K_j's. Further details can be found in Section 8.

Different characterizations of the concept of "crossing" have been suggested by various authors in order to establish the presence of complex dynamics for continuous maps in higher dimensional spaces (refer to [71, 73, 107, 156, 182, 184, 186] and the references therein). In this respect, the already discussed approach by Kennedy, Koçak and Yorke in [71], and Kennedy and Yorke in [73] is perhaps the most general, regarding the spaces considered and the restrictions on the maps involved. However, as we have seen, the great generality of such setting comes with a price in the

3 By the similarity between the results obtained in Section 4 and 5 for the planar and the N-dimensional settings, for the sake of generality, we will generally work in the higher dimensional framework, maybe not indicating the dimension so that no confusion may occur.

sense that the existence of periodic points is not ensured. Other authors (for example, in [107, 156, 158, 174, 182, 184, 186]) have developed theories of topological horseshoes more focused on the search of fixed and periodic points for maps defined on subsets of the N-dimensional Euclidean space. The tools employed in these and related works range from the Conley index [107] to the Lefschetz fixed point theory [156] and the topological degree [186].

Our framework may be looked at as an intermediate point of view between the theory of topological horseshoes developed by Kennedy and Yorke in [73] and the above-mentioned works based on some more or less sophisticated fixed point index theories. Indeed, compared to [73], the method of stretching along the paths, thanks to its specialized framework, allows one to obtain sharper results concerning the existence of periodic points. On the other hand, our approach, although mathematically rigorous, avoids the use of more advanced topological theories and it is relatively easy to apply to specific models arising in applications.

In more details, in Section 7, we describe the chaotic features that we are able to obtain with our method. Some tools from symbolic dynamics are introduced in Section 8 in order to correlate our results on chaotic dynamics with other related theorems obtained by means of more classical approaches. A discussion on the various notions of chaos is pursued further in Section 9. An alternative geometrical context for the applicability of the stretching along the paths method is presented in Section 10, where we deal with the so-called "linked twist maps."

7 Definitions and main results

In the literature, various notions of complex dynamics can be found, so that one could say "as many authors, as many definitions of 'chaos'" [22]. However, some common features are shared by several of these definitions, such as the unpredictability of the future behavior of the system under consideration. In particular, one of the most natural notions of chaos is related to the possibility of realizing a generic coin-flipping experiment [151]. The definition of chaos that we choose is adapted from that considered by Kirchgraber and Stoffer in [76] under the name of *chaos in the coin-tossing sense*. In fact, exactly as in [76], our definition concerns the possibility of reproducing, via the iterates of a given map ψ, any coin-flipping sequence of two symbols. On the other hand, our definition, when compared to [76], is enhanced by the possibility of realizing periodic sequences of two symbols by means of periodic points of ψ. More formally, we have:

Definition 7.1. Let X be a metric space, $\psi\colon X \supseteq D_\psi \to X$ be a map and let $\mathcal{D} \subseteq D_\psi$. We say that ψ *induces chaotic dynamics on two symbols on the set* $\mathcal{D}$ if there exist two nonempty disjoint compact sets

$$\mathcal{K}_0,\ \mathcal{K}_1 \subseteq \mathcal{D}$$

such that, for each two-sided sequence $(s_i)_{i\in\mathbb{Z}} \in \{0,1\}^{\mathbb{Z}}$, there exists a corresponding sequence $(w_i)_{i\in\mathbb{Z}} \in \mathcal{D}^{\mathbb{Z}}$ such that

$$w_i \in \mathcal{K}_{s_i} \quad \text{and} \quad w_{i+1} = \psi(w_i), \quad \forall i \in \mathbb{Z}, \tag{7.1}$$

and, whenever $(s_i)_{i\in\mathbb{Z}}$ is a k-periodic sequence (that is, $s_{i+k} = s_i$, $\forall i \in \mathbb{Z}$) for some $k \geq 1$, there exists a corresponding k-periodic sequence $(w_i)_{i\in\mathbb{Z}} \in \mathcal{D}^{\mathbb{Z}}$, satisfying (7.1). When we want to emphasize the role of the sets $\mathcal{K}_j$'s, we also say that ψ *induces chaotic dynamics on two symbols on the set $\mathcal{D}$ relative to $\mathcal{K}_0$ and $\mathcal{K}_1$*.

To get a feel of such a definition, we can associate the name "head" $= H$ to the set $\mathcal{K}_0$ and the name "tail" $= T$ to $\mathcal{K}_1$. If we consider any sequence of symbols

$$(s_i)_{i\in\mathbb{Z}} \in \{0,1\}^{\mathbb{Z}} \cong \{H,T\}^{\mathbb{Z}}$$

so that for each i, s_i is either a "head" or "tail," and then we have the same itinerary of heads and tails realized through the map ψ. Namely, there exists a sequence $(w_i)_{i\in\mathbb{Z}}$ of points of the metric space X which is a full orbit for ψ, i. e.

$$w_{i+1} = \psi(w_i), \quad \forall i \in \mathbb{Z},$$

and such that $w_i \in \mathcal{K}_0$ or $w_i \in \mathcal{K}_1$, according to the fact that the i-th term of the sequence $(s_i)_i$ is a "head" or "tail." Moreover, as already remarked, our definition extends that in [76] in the sense that any periodic sequence of heads and tails can be realized by suitable points which are periodic points for ψ. For instance, there exists a fixed point of ψ in the set $\mathcal{K}_1$ corresponding to the constant sequence of symbols $s_i =$ "tail", $\forall i \in \mathbb{Z}$. There is also a point $w \in \mathcal{K}_0$ of period three with $\psi(w) \in \mathcal{K}_0$ and $\psi^2(w) \in \mathcal{K}_1$, corresponding to the periodic sequence ... *HHT HHT HHT* ... and so on.

We stress that we have presented our definition of chaos with reference to just two symbols in order to make the connection with the coin-flipping process clearer. On the other hand, it is possible to define *chaotic dynamics on m symbols* for an arbitrary integer $m \geq 2$ in a completely analogous manner when there are m pairwise disjoint compact sets $\mathcal{K}_0, \ldots, \mathcal{K}_{m-1}$ acting as $\mathcal{K}_0$ and $\mathcal{K}_1$ in Definition 7.1. More precisely, we have:

Definition 7.2. Let X be a metric space, $\psi\colon X \supseteq D_\psi \to X$ be a map and let $\mathcal{D} \subseteq D_\psi$. Also, let $m \geq 2$ be an integer. We say that ψ *induces chaotic dynamics on m symbols on the set $\mathcal{D}$* if there exist m nonempty pairwise disjoint compact sets

$$\mathcal{K}_0, \ldots, \mathcal{K}_{m-1} \subseteq \mathcal{D},$$

such that, for each two-sided sequence $(s_i)_{i\in\mathbb{Z}} \in \{0,\ldots,m-1\}^{\mathbb{Z}}$, there exists a corresponding sequence $(w_i)_{i\in\mathbb{Z}} \in \mathcal{D}^{\mathbb{Z}}$ such that

$$w_i \in \mathcal{K}_{s_i} \quad \text{and} \quad w_{i+1} = \psi(w_i), \quad \forall i \in \mathbb{Z} \tag{7.2}$$

and, whenever $(s_i)_{i\in\mathbb{Z}}$ is a k-periodic sequence (that is, $s_{i+k} = s_i$, $\forall i \in \mathbb{Z}$) for some $k \geq 1$, there exists a corresponding k-periodic sequence $(w_i)_{i\in\mathbb{Z}} \in \mathcal{D}^{\mathbb{Z}}$, satisfying (7.2). When we want to emphasize the role of the sets $\mathcal{K}_j$'s, we also say that ψ *induces chaotic dynamics on m symbols on the set $\mathcal{D}$ relative to $\mathcal{K}_0, \dots, \mathcal{K}_{m-1}$.*

Such a variant of Definition 7.1 will be of particular importance when dealing with linked twist maps (Section 10). Moreover, the latter generalization agrees with the kind of chaotic behavior detected in various papers where the dynamical systems are investigated using some topological tools related to fixed point theory [158, 174]. Hence, in order to collate our notion of chaos to other ones available in the literature, from now on, we will consider Definition 7.2 instead of Definition 7.1.

With regards to the results in [71, 73], we recall that the authors therein also deal with chaotic dynamics meant as the possibility of realizing any sequence of symbols. The difference, as it should be clear from the discussion in Section 4, resides in the possibility of realizing periodic coin-flipping sequences through periodic itineraries. Nonetheless, the kind of chaos detected in [71, 73] for a map f satisfying the "horseshoe hypotheses Ω" in [73] with crossing number $m \geq 2$ allows one to prove the existence of a compact f-invariant set Q_I (i. e. $f(Q_I) = Q_I$), such that $f\restriction_{Q_I}$ is semiconjugate to the one-sided shift on m symbols. Furthermore, in [71, Lemma 4] (Chaos Lemma), the existence of a smaller compact invariant set $Q_* \subseteq Q_I$ is obtained, on which the map f is sensitive and such that each forward itinerary on m symbols is realized by the f-itinerary generated by some point of Q_*.

Before showing in Theorem 7.2 what we are able to prove in this direction, let us see which is the link between the method of stretching along the paths from Section 4 and the notion of chaotic dynamics in the sense of Definition 7.2:

Theorem 7.1. *Let $\widetilde{\mathcal{R}} := (\mathcal{R}, \mathcal{R}^-)$ be an oriented (N-dimensional) rectangle of a metric space X and let $\mathcal{D} \subseteq \mathcal{R} \cap D_\psi$, with D_ψ as the domain of a map $\psi\colon X \supseteq D_\psi \to X$. If $\mathcal{K}_0, \dots, \mathcal{K}_{m-1}$ are $m \geq 2$ pairwise disjoint compact sets contained in $\mathcal{D}$ and*

$$(\mathcal{K}_i, \psi)\colon \widetilde{\mathcal{R}} \Leftrightarrow\!\!\!\!\rightarrow \widetilde{\mathcal{R}}, \quad \text{for } i = 0, \dots, m-1,$$

then ψ induces chaotic dynamics on m symbols on the set $\mathcal{D}$ relative to $\mathcal{K}_0, \dots, \mathcal{K}_{m-1}$.

Proof. Recalling Definition 7.2, the thesis is just a reformulation of the second and the forth conclusions in Theorem 4.3 (or in Theorem 5.7 in the case of an N-dimensional setting).[4] □

4 Also, recalling the third conclusion in Theorem 4.3 (or in Theorem 5.7), we could complement Theorem 7.1 with the further information about the existence, for any forward sequence on m symbols, of a corresponding continuum joining the lower and the upper sides of $\mathcal{R}$ and consisting of the points whose forward ψ-itinerary realizes the given sequence. However, for the sake of simplicity and since in the applications we don't use this fact, we have decided to omit it. The same observation also applies to the results obtained in Section 10 in relation to the framework of the linked twist maps.

The same conclusions on chaotic dynamics could be obtained in the framework of Theorem 4.2 when assuming condition (4.7). As discussed in [124], such a remark looks useful in view of possible applications of our method to the detection of chaos via computer-assisted proofs.

When we enter the setting of Definition 7.2, many interesting properties for the map ψ can be proved. They are gathered in Theorem 7.2 below. The precise explanation of some concepts (like topological entropy, sensitivity, transitivity, etc.) will be given in Sections 8–9, where the reader can find a more detailed discussion about various classical notions of chaos considered in the literature.

Theorem 7.2. *Let ψ be a map inducing chaotic dynamics on $m \geq 2$ symbols on a set $\mathcal{D} \subseteq X$ and which is continuous on*

$$\mathcal{K} := \bigcup_{i=0}^{m-1} \mathcal{K}_i \subseteq \mathcal{D},$$

where $\mathcal{K}_0, \ldots, \mathcal{K}_{m-1}, \mathcal{D}$ and X are as in Definition 7.2. Setting

$$\mathcal{I}_\infty := \bigcap_{n=0}^{\infty} \psi^{-n}(\mathcal{K}), \tag{7.3}$$

there exists a nonempty compact set

$$\mathcal{I} \subseteq \mathcal{I}_\infty \subseteq \mathcal{K},$$

on which the following are fulfilled:
(i) *$\mathcal{I}$ is invariant for ψ (that is, $\psi(\mathcal{I}) = \mathcal{I}$);*
(ii) *$\psi \restriction_{\mathcal{I}}$ is semiconjugate to the one-sided Bernoulli shift on m symbols, i. e. there exists a continuous map π of $\mathcal{I}$ onto $\Sigma_m^+ := \{0, \ldots, m-1\}^{\mathbb{N}}$, endowed with the distance*

$$\hat{d}(\mathbf{s}', \mathbf{s}'') := \sum_{i \in \mathbb{N}} \frac{|s_i' - s_i''|}{m^{i+1}}, \tag{7.4}$$

for $\mathbf{s}' = (s_i')_{i\in\mathbb{N}}$ and $\mathbf{s}'' = (s_i'')_{i\in\mathbb{N}} \in \Sigma_m^+$ such that the diagram

$$\begin{array}{ccc} \mathcal{I} & \xrightarrow{\ \psi\ } & \mathcal{I} \\ {\scriptstyle \pi}\downarrow & & \downarrow{\scriptstyle \pi} \\ \Sigma_m^+ & \xrightarrow[\ \sigma\ ]{} & \Sigma_m^+ \end{array} \tag{7.5}$$

commutes, that is, $\pi \circ \psi = \sigma \circ \pi$, where $\sigma : \Sigma_m^+ \to \Sigma_m^+$ is the Bernoulli shift defined by $\sigma((s_i)_i) := (s_{i+1})_i, \ \forall i \in \mathbb{N}$;
(iii) *The set $\mathcal{P}$ of the periodic points of $\psi \restriction_{\mathcal{I}_\infty}$ is dense in $\mathcal{I}$ and the preimage $\pi^{-1}(\mathbf{s}) \subseteq \mathcal{I}$ of every k-periodic sequence $\mathbf{s} = (s_i)_{i\in\mathbb{N}} \in \Sigma_m^+$ contains at least one k-periodic point.*

Furthermore, from conclusion (ii), *it follows that*

(iv) $h_{\rm top}(\psi) \geq h_{\rm top}(\psi \restriction_{\mathcal{I}}) \geq h_{\rm top}(\sigma) = \log(m)$, *where* $h_{\rm top}$ *is the topological entropy;*

(v) *There exists a compact invariant set* $\Lambda \subseteq \mathcal{I}$ *such that* $\psi|_{\Lambda}$ *is semiconjugate to the one-sided Bernoulli shift on* m *symbols, topologically transitive and has sensitive dependence on initial conditions.*

Proof. Let us begin by checking that the set $\mathcal{I}_\infty$ in (7.3) is compact and nonempty. By the continuity of the map ψ on $\mathcal{K}$, it follows that $\mathcal{I}_\infty$ is closed and, being contained in the compact set $\mathcal{K}$, it is compact as well. The fact that $\mathcal{I}_\infty$ is nonempty follows from Definition 7.2 on chaotic dynamics by observing that $z \in \mathcal{I}_\infty \Leftrightarrow \psi^n(z) \in \mathcal{K}, \forall n \geq 0$. This remark also implies that $\psi(\mathcal{I}_\infty) \subseteq \mathcal{I}_\infty$: indeed, it is straightforward to see that if $z \in \mathcal{I}_\infty$, then also $\psi(z) \in \mathcal{I}_\infty$.

Calling $\mathcal{P}$ the subset of $\mathcal{I}_\infty$ consisting of the periodic points of $\psi \restriction_{\mathcal{I}_\infty}$, that is,

$$\mathcal{P} := \{w \in \mathcal{I}_\infty : \exists\, k \in \mathbb{N}_0\,, \quad \psi^k(w) = w\}\,, \tag{7.6}$$

we claim that $\psi(\mathcal{P}) = \mathcal{P}$. Indeed, if $z \in \mathcal{P}$, then there exists $l \in \mathbb{N}_0$ such that $\psi^l(z) = z$. Hence, on the one hand, $\psi(z) = \psi(\psi^l(z)) = \psi^{l+1}(z) = \psi^l(\psi(z))$ and thus $\psi(z) \in \mathcal{P}$ as well. This shows that $\psi(\mathcal{P}) \subseteq \mathcal{P}$. Notice that, repeating the same argument, it is possible to prove that if $z \in \mathcal{P}$, then $\psi^h(z) \in \mathcal{P}$ for any $h \geq 1$. On the other hand, if $\psi^l(z) = z$, for some $l \in \mathbb{N}_0$, then two possibilities can occur for l, that is, $l = 1$ or $l \geq 2$. In the former case, we get $\psi(z) = z$ and so $z \in \psi(\mathcal{P})$, while in the latter, we obtain $z = \psi^l(z) = \psi(\psi^{l-1}(z))$. Hence, since $\psi^{l-1}(z) \in \mathcal{P}$ whenever $z \in \mathcal{P}$, we find again $z \in \psi(\mathcal{P})$. In any case, we have proved that if $z \in \mathcal{P}$, then $z \in \psi(\mathcal{P})$, i. e. $\mathcal{P} \subseteq \psi(\mathcal{P})$. The claim is thus checked.

At this point, we observe that since $\mathcal{P}$ is contained in the compact set $\mathcal{I}_\infty$, then

$$\mathcal{I} := \overline{\mathcal{P}} \subseteq \mathcal{I}_\infty\,, \tag{7.7}$$

and moreover $\mathcal{I}$ is compact as it is closed in a compact set. From $\psi(\mathcal{P}) = \mathcal{P}$, it follows that

$$\psi(\mathcal{I}) = \psi(\overline{\mathcal{P}}) \supseteq \psi(\mathcal{P}) = \mathcal{P}\,.$$

However, again, by the compactness of $\psi(\mathcal{I})$, it holds that

$$\psi(\mathcal{I}) \supseteq \overline{\mathcal{P}} = \mathcal{I}\,.$$

Let us show that the reverse inclusion is also fulfilled for $\mathcal{I}$, that is, $\psi(\mathcal{I}) \subseteq \mathcal{I}$. Indeed, since ψ is continuous, we have

$$\psi(\mathcal{I}) = \psi(\overline{\mathcal{P}}) \subseteq \overline{\psi(\mathcal{P})} = \overline{\mathcal{P}} = \mathcal{I}\,.$$

Hence, the invariance of $\mathcal{I}$ is verified, in agreement with conclusion (i).

Let us now consider the diagram

$$\begin{array}{ccc} \mathcal{I}_\infty & \xrightarrow{\psi} & \mathcal{I}_\infty \\ \Big\downarrow{\scriptstyle \pi} & & \Big\downarrow{\scriptstyle \pi} \\ \Sigma_m^+ & \xrightarrow[\sigma]{} & \Sigma_m^+ \end{array}$$

with (Σ_m^+, σ) the Bernoulli system, and define the map $\pi\colon \mathcal{I}_\infty \to \Sigma_m^+$ by associating to any $w \in \mathcal{I}_\infty$, the sequence $(s_n)_{n\in\mathbb{N}} \in \Sigma_m^+$ such that $s_n = j$ if $\psi^n(w) \in \mathcal{K}_j$ for $j = 0, \dots, m-1$. More formally, we notice that for any $w \in \mathcal{I}_\infty$, there exists a forward itinerary $(w_i)_{i\in\mathbb{N}}$ such that $w_0 = w$ and $\psi(w_i) = w_{i+1} \in \mathcal{K}$ for every $i \in \mathbb{N}$. Hence, the function $g_1\colon \mathcal{I}_\infty \to \mathcal{I}_\infty^{\mathbb{N}}$, which maps any $w \in \mathcal{I}_\infty$ into the one-sided sequence of points from the set $\mathcal{I}_\infty$

$$\mathbf{s}_w := (w_i)_{i\in\mathbb{N}}\,, \quad \text{where } w_i := \psi^i(w)\,, \quad \forall\, i \in \mathbb{N}\,,$$

with the usual convention $\psi^0 = \mathrm{Id}_{\mathcal{I}_\infty}$ and $\psi^1 = \psi$ being well-defined. Since the sets $\mathcal{K}_0, \dots, \mathcal{K}_{m-1}$ are pairwise disjoint, for every term w_i of $\mathbf{s}_w$, there exists a unique index

$$s_i = s_i(w_i)\,, \quad \text{with } s_i \in \{0, \dots, m-1\}$$

such that $w_i \in \mathcal{K}_{s_i}$. Therefore, the map $g_2\colon \mathcal{I}_\infty^{\mathbb{N}} \to \Sigma_m^+$,

$$g_2\colon \mathbf{s}_w \mapsto (s_i)_{i\in\mathbb{N}} \in \Sigma_m^+$$

is also well-defined. Thus, by Definition 7.2, the map

$$\pi := g_2 \circ g_1\colon \mathcal{I}_\infty \to \Sigma_m^+$$

is a surjection that makes the diagram (7.5) commute, and the preimage through π of any k-periodic sequence in Σ_m^+ contains at least one k-periodic point of $\mathcal{I}_\infty$. To check that π is continuous, we prove the continuity in a generic $\bar{z} \in \mathcal{I}_\infty$ by showing that for any $\varepsilon > 0$, there exists $\delta > 0$ such that $\forall z \in \mathcal{I}_\infty$ with $d_X(z, \bar{z}) < \delta$, and then $\hat{d}(\pi(z), \pi(\bar{z})) < \varepsilon$, where d_X is the distance on X and $\hat{d}$ is the metric defined in (7.4). Let us fix $\varepsilon > 0$ and let $n \in \mathbb{N}$ such that $0 < 1/m^n < \varepsilon$. We notice that it is sufficient to prove that $(\pi(z))_i = (\pi(\bar{z}))_i$ for any $i = 0, \dots, n$. Indeed, if this is the case, by the definition of $\hat{d}$, it follows that $\hat{d}(\pi(z), \pi(\bar{z})) \le 1/m^n < \varepsilon$.

Since $\bar{z} \in \mathcal{I}_\infty$, there exists a sequence $(s_0, \dots, s_n) \in \{0, \dots, m-1\}^{n+1}$ such that

$$\bar{z} \in \mathcal{K}_{s_0},\ \psi(\bar{z}) \in \mathcal{K}_{s_1}, \dots, \psi^n(\bar{z}) \in \mathcal{K}_{s_n}\,.$$

By the pairwise disjointness of the sets $\mathcal{K}_i$'s, it holds that

$$\eta := \min\{d_X(\mathcal{K}_i, \mathcal{K}_j)\colon i, j = 0, \dots, m-1\} > 0\,.$$

Hence, for any $z \in \mathcal{I}_\infty$ with $d_X(z, \bar{z}) < \eta/2$, it follows that $z \in \mathcal{K}_{s_0}$ as well. By the continuity of ψ in $\bar{z}$, there exists $\delta_1 > 0$ such that $\forall z \in \mathcal{I}_\infty$ with $d_X(z, \bar{z}) < \delta_1$, and then $d_X(\psi(z), \psi(\bar{z})) < \eta/2$. However, this means that $\psi(z) \in \mathcal{K}_{s_1}$. Analogously, by the continuity of ψ^2 in $\bar{z}$, there exists $\delta_2 > 0$ such that $\forall z \in \mathcal{I}_\infty$ with $d_X(z, \bar{z}) < \delta_2$, and then $d_X(\psi^2(z), \psi^2(\bar{z})) < \eta/2$ and thus $\psi^2(z) \in \mathcal{K}_{s_2}$ for any such z. Proceeding in such a way until the n-th iterate of ψ and setting

$$\delta := \min\left\{\frac{\eta}{2}, \delta_1, \ldots, \delta_n\right\},$$

we find that for any $z \in \mathcal{I}_\infty$ with $d_X(z, \bar{z}) < \delta$, it holds that

$$z \in \mathcal{K}_{s_0}, \ \psi(z) \in \mathcal{K}_{s_1}, \ldots, \psi^n(z) \in \mathcal{K}_{s_n},$$

exactly as for $\bar{z}$. However, this means that $(\pi(z))_i = (\pi(\bar{z}))_i$ for any $i = 0, \ldots, n$, and hence $\hat{d}(\pi(z), \pi(\bar{z})) \leq 1/m^n < \varepsilon$. The continuity of π is thus proved.

Considering in diagram (7.5) the restriction of ψ to $\mathcal{P} \subseteq \mathcal{I}_\infty$, we find the commutative diagram

$$\begin{array}{ccc} \mathcal{P} & \xrightarrow{\ \psi\ } & \mathcal{P} \\ {\scriptstyle \pi}\big\downarrow & & \big\downarrow{\scriptstyle \pi} \\ \mathcal{P}_m^+ & \xrightarrow[\ \sigma\ ]{} & \mathcal{P}_m^+ \end{array}$$

where $\mathcal{P}_m^+ \subseteq \Sigma_m^+$ is the set of the periodic sequences of m symbols. Notice that $\pi(\mathcal{P}) = \mathcal{P}_m^+$ lends itself to Definition 7.2. Recalling the well-known fact that $\mathcal{P}_m^+$ is dense in Σ_m^+ [70, Proposition 1.9.1], by the continuity of π, it follows that

$$\pi(\mathcal{I}) = \pi(\overline{\mathcal{P}}) \subseteq \overline{\mathcal{P}_m^+} = \Sigma_m^+ .$$

On the other hand, $\pi(\mathcal{I})$ is a compact set containing $\pi(\mathcal{P}) = \mathcal{P}_m^+$, and hence

$$\pi(\mathcal{I}) \supseteq \overline{\mathcal{P}_m^+} = \Sigma_m^+ .$$

Therefore, we can conclude that $\pi(\mathcal{I}) = \Sigma_m^+$, and the diagram

$$\begin{array}{ccc} \mathcal{I} & \xrightarrow{\ \psi\ } & \mathcal{I} \\ {\scriptstyle \pi}\big\downarrow & & \big\downarrow{\scriptstyle \pi} \\ \Sigma_m^+ & \xrightarrow[\ \sigma\ ]{} & \Sigma_m^+ \end{array}$$

still commutes. Moreover, the preimage through π of any k-periodic sequence in Σ_m^+ contains at least one k-periodic point of $\mathcal{I}$, as $\mathcal{P} \subseteq \mathcal{I}$. Conclusions (ii) and (iii) are thus proved.

Assertion (iv) about the positive topological entropy comes from property (ii) about the semiconjugacy to the Bernoulli shift. For a proof, see [167, Theorem 7.12].

Finally, conclusion (v) regarding the existence of the compact invariant set $\Lambda \subseteq \mathcal{I}$ follows by applying Theorem 9.1 with the positions $(X, f) = (\mathcal{I}, \psi\restriction_{\mathcal{I}})$ and $(Y, g) = (\Sigma_m^+, \sigma)$. □

We point out that results similar to Theorem 7.2 are almost known in the literature (see, e. g. [156, Theorem 3] for a related statement), maybe with exception of conclusion (v). However, for the sake of completeness, we have decided to prove it in full detail. Notice that in frameworks more general than ours, the set $\mathcal{I}_\infty$ in (7.3), and a fortiori $\mathcal{P}$ in (7.6), can be empty and thus it is not possible to define $\mathcal{I}$ as in (7.7) in order to have an invariant set. In our case, we can like that since by Definition 7.2, the sequences of symbols are realized by points in $\mathcal{I}_\infty$ and the periodic sequences of symbols are reproduced by points in $\mathcal{P}$. We also remark that the density of the periodic points of ψ in $\mathcal{I}$ does not imply that the periodic points are dense in the smaller set Λ, on which ψ is transitive and sensitive. Therefore, we cannot conclude the presence of Devaney chaos (cf. Definition 9.2) for ψ on Λ.

Remark 7.1. If, in addition to the hypotheses of Theorem 7.2, the map ψ is also injective on $\mathcal{K}$, then one can deal with bi-infinite sequences of m symbols instead of forward ones and prove the stronger property for ψ of semiconjugacy to the two-sided Bernoulli shift $\sigma\colon \Sigma_m := \{0, \ldots, m-1\}^{\mathbb{Z}} \to \Sigma_m$, defined as $\sigma((s_i)_i) := (s_{i+1})_i$, $\forall i \in \mathbb{Z}$. Indeed, it is sufficient to replace the set $\mathcal{I}_\infty$ in (7.3) with

$$\mathcal{I}_\infty := \bigcap_{n=-\infty}^{\infty} \psi^{-n}(\mathcal{K})$$

and consider $\mathcal{I}$ as in (7.7) in order to obtain, via similar steps, the analogue of the conclusions of Theorem 7.2 with respect to two-sided sequences, rather than one-sided sequences. The precise statement reads as follows:

Theorem 7.3. *Let ψ be a map inducing chaotic dynamics on $m \geq 2$ symbols on a set $\mathcal{D} \subseteq X$ and which is continuous and injective on*

$$\mathcal{K} := \bigcup_{i=0}^{m-1} \mathcal{K}_i \subseteq \mathcal{D},$$

where $\mathcal{K}_0, \ldots, \mathcal{K}_{m-1}, \mathcal{D}$ and X are as in Definition 7.2. Setting

$$\mathcal{I}_\infty := \bigcap_{n=-\infty}^{\infty} \psi^{-n}(\mathcal{K}),$$

there exists a nonempty compact set

$$\mathcal{I} \subseteq \mathcal{I}_\infty \subseteq \mathcal{K},$$

on which the following are fulfilled:

(i) *$\mathcal{I}$ is invariant for ψ (that is, $\psi(\mathcal{I}) = \mathcal{I}$);*

(ii) $\psi \restriction_{\mathcal{I}}$ *is semiconjugate to the two-sided Bernoulli shift on* m *symbols, i. e. there exists a continuous map* π *of* $\mathcal{I}$ *onto* $\Sigma_m := \{0,\dots,m-1\}^{\mathbb{Z}}$*, endowed with the distance*

$$\hat{d}(\mathbf{s}',\mathbf{s}'') := \sum_{i\in\mathbb{Z}} \frac{|s_i' - s_i''|}{m^{|i|+1}},$$

for $\mathbf{s}' = (s_i')_{i\in\mathbb{Z}}$ *and* $\mathbf{s}'' = (s_i'')_{i\in\mathbb{Z}} \in \Sigma_m$*, such that the diagram*

$$\begin{array}{ccc} \mathcal{I} & \xrightarrow{\ \psi\ } & \mathcal{I} \\ \downarrow{\scriptstyle \pi} & & \downarrow{\scriptstyle \pi} \\ \Sigma_m & \xrightarrow[\ \sigma\ ]{} & \Sigma_m \end{array}$$

commutes, that is, $\pi \circ \psi = \sigma \circ \pi$*, where* $\sigma : \Sigma_m \to \Sigma_m$ *is the Bernoulli shift defined by* $\sigma((s_i)_i) := (s_{i+1})_i$*,* $\forall\, i \in \mathbb{Z}$*;*

(iii) *The set* $\mathcal{P}$ *of the periodic points of* $\psi \restriction_{\mathcal{I}_\infty}$ *is dense in* $\mathcal{I}$ *and the preimage* $\pi^{-1}(\mathbf{s}) \subseteq \mathcal{I}$ *of every* k*-periodic sequence* $\mathbf{s} = (s_i)_{i\in\mathbb{N}} \in \Sigma_m$ *contains at least one* k*-periodic point.*

Furthermore, from conclusion (ii)*, it follows that*

(iv) $h_{\text{top}}(\psi) \geq h_{\text{top}}(\psi \restriction_{\mathcal{I}}) \geq h_{\text{top}}(\sigma) = \log(m)$*, where* h_{top} *is the topological entropy;*

(v) *There exists a compact invariant set* $\Lambda \subseteq \mathcal{I}$ *such that* $\psi|_\Lambda$ *is semiconjugate to the two-sided Bernoulli shift on* m *symbols, topologically transitive and has sensitive dependence on initial conditions.*

For further details, see [127], where some alternative approaches are expounded, for example, that in [87].

The present observation applies, for instance, when dealing with the original Smale horseshoe map introduced at the beginning of Section 6. Another field of applicability is that of the ODEs with periodic coefficients. Namely, in such a case, if one looks for periodic solutions, the corresponding map ψ is the homeomorphism called the Poincaré map associated to the system. Some examples will be presented in Section 12.

To conclude the discussion about the results in [71], we notice that thanks to the fact that the kind of chaos in Definition 7.2 is stricter than the one detected in [71, 73], we are able to obtain a result (Theorem 7.2) that covers the Chaos Lemma ([71, Lemma 4]), adding further information. In fact, the set $\mathcal{I}$ in Theorem 7.2 plays the role of Q_I in [71] and the subsets $\Lambda \subseteq \mathcal{I}$ and $Q_* \subseteq Q_I$ share some common features, like the invariance and the sensitivity of the maps defined on them, but we have, in addition, the density of the periodic points on $\mathcal{I}$, while, as already observed, in [71, 73], the existence of periodic points is not ensured at all. Also, the transitivity is missing in the statement of the Chaos Lemma ([71, Lemma 4]), even if after a look at its proof, it is evident that such property holds in that framework as well. Indeed, the invariant set Q_* is the ω-limit set of a certain point x^* of Q_I and thus the orbit of x^* is dense in Q_*. On the other hand, by the invariance of Q_*, this is enough to infer the transitivity on Q_*. As we shall see in Section 9, the same argument is employed in the proof of Theorem 9.1, where the reader can find the missing definitions and details.

Finally, a natural question that can arise is whether our stretching condition in Theorem 7.1 is strong enough to imply a conjugacy to the Bernoulli shift rather than just a semiconjugacy as stated in Theorem 7.2. The answer is generally negative. However, it is possible to add further assumptions in order to get it. A first step in this direction is represented by some recent works by Zgliczyński and collaborators [38, 185], where they combine the theory of covering relations from [186] with certain cone conditions to establish the properties implied by hyperbolicity, for example, the existence of stable and unstable manifolds. Moreover, in their framework, every periodic sequence of symbols is realized by the itinerary generated by a unique point: in symbols using the notation introduced in the statement of Theorem 7.2, this means that the preimage $\pi^{-1}(\mathbf{s}) \subseteq \mathcal{I}$ of every k-periodic sequence $\mathbf{s} = (s_i)_{i\in\mathbb{N}} \in \Sigma_m^+$ contains exactly one k-periodic point.

8 Symbolic dynamics

In this section, we present some tools from symbolic dynamics that will be useful to more deeply investigate the relationship between the notion of chaos in the sense of Definition 7.2 and other ones widely considered in the literature. In order to accomplish such a task, we need to recall a few basic definitions: some of them have already been introduced in Section 7, but we prefer to gather them here for the sake of consistency.

Given an integer $m \geq 2$, we denote by $\Sigma_m := \{0,\dots,m-1\}^{\mathbb{Z}}$ the set of the two-sided sequences of m symbols and by $\Sigma_m^+ := \{0,\dots,m-1\}^{\mathbb{N}}$, the set of one-sided sequences of m symbols. These compact spaces are usually endowed with the distance

$$\hat{d}(\mathbf{s}',\mathbf{s}'') := \sum_{i\in\mathbb{I}} \frac{|s_i' - s_i''|}{m^{|i|+1}}, \quad \text{for } \mathbf{s}' = (s_i')_{i\in\mathbb{I}}, \quad \mathbf{s}'' = (s_i'')_{i\in\mathbb{I}}, \tag{8.1}$$

where $\mathbb{I} = \mathbb{Z}$ or $\mathbb{I} = \mathbb{N}$, respectively. The metric in (8.1) could be replaced with

$$\tilde{d}(\mathbf{s}',\mathbf{s}'') := \sum_{i\in\mathbb{I}} \frac{d(s_i', s_i'')}{m^{|i|+1}}, \quad \text{for } \mathbf{s}' = (s_i')_{i\in\mathbb{I}}, \quad \mathbf{s}'' = (s_i'')_{i\in\mathbb{I}},$$

where $d(\cdot\,,\cdot)$ is the discrete distance on $\{0,\dots,m-1\}$, that is, $d(s_i', s_i'') = 0$ for $s_i' = s_i''$ and $d(s_i', s_i'') = 1$ for $s_i' \neq s_i''$. The significance of this second choice reveals when one needs to look at the elements from $\{0,\dots,m-1\}$ as symbols instead of numbers.

On such spaces, we define the *one-sided Bernoulli shift* $\sigma\colon \Sigma_m^+ \to \Sigma_m^+$ and the *two-sided Bernoulli shift* $\sigma\colon \Sigma_m \to \Sigma_m$ on m symbols as $\sigma((s_i)_i) := (s_{i+1})_i$, $\forall i \in \mathbb{I}$, for $\mathbb{I} = \mathbb{N}$ or $\mathbb{I} = \mathbb{Z}$, respectively. Both maps are continuous and the two-sided shift is a homeomorphism.

A useful tool for the detection of complex dynamics is the *topological entropy* and, indeed, its positivity is generally considered as one of the trademarks of chaos. Such an object can be introduced for any continuous self-map f of a compact topological space X and we indicate it with the symbol $h_{\text{top}}(f)$. Its original definition due to Adler, Konheim and McAndrew [1] is based on the open coverings. More precisely, for an open cover α of X, we define the *entropy of* α as $H(\alpha) := \log N(\alpha)$, where $N(\alpha)$ is the minimal number of elements in a finite subcover of α. Given two open covers α and β of X, we define their *join* $\alpha \vee \beta$ as the open cover of X made by all sets of the form $A \cap B$, with $A \in \alpha$ and $B \in \beta$. Similarly, one can define the join $\vee_{i=1}^{n} \alpha_i$ of any finite collection of open covers of X. If α is an open cover of X and $f\colon X \to X$ is a continuous map, we denote by $f^{-1}\alpha$ the open cover consisting of all sets $f^{-1}(A)$ with $A \in \alpha$. By $\vee_{i=0}^{n-1} f^{-i}\alpha$, we mean $\alpha \vee f^{-1}\alpha \vee \cdots \vee f^{-n+1}\alpha$. Finally, we have

$$h_{\text{top}}(f) := \sup_{\alpha} \left(\lim_{n\to\infty} \frac{1}{n} \left(H\left(\vee_{i=0}^{n-1} f^{-i}\alpha \right) \right) \right),$$

where α ranges over all open covers of X.

Among the various properties of topological entropy, we recall just the ones that are useful in view of the subsequent discussion.

Regarding the (one-sided or two-sided) Bernoulli shift σ on m symbols, it holds that

$$h_{\text{top}}(\sigma) = \log(m).$$

Given a continuous self-map f of a compact topological space X and a *invariant* (resp. *positively invariant*) subset $\mathcal{I} \subseteq X$, that is, such that $f(\mathcal{I}) = \mathcal{I}$ (resp. $f(\mathcal{I}) \subseteq \mathcal{I}$), then

$$h_{\text{top}}(f) \geq h_{\text{top}}(f \restriction_{\mathcal{I}}). \tag{8.2}$$

Denoting by f^n the n-th iterate of the continuous self-map f of a compact topological space X, we have

$$h_{\text{top}}(f^n) = n h_{\text{top}}(f),\ \forall n \geq 1. \tag{8.3}$$

Given two continuous self-maps $f\colon X \to X$ and $g\colon Y \to Y$ of the compact topological spaces X and Y and a continuous onto map $\phi\colon X \to Y$ that makes the diagram

$$\begin{array}{ccc} X & \xrightarrow{\ f\ } & X \\ {\scriptstyle\phi}\downarrow & & \downarrow{\scriptstyle\phi} \\ Y & \xrightarrow[\ g\ ]{} & Y \end{array} \tag{8.4}$$

commute, i. e. such that $\phi \circ f = g \circ \phi$, it holds that

$$h_{\text{top}}(f) \geq h_{\text{top}}(g).$$

If ϕ is also injective, the above inequality is indeed an equality.

When the diagram in (8.4) commutes, we say that f and g are *topologically semiconjugate* and that ϕ is a *semiconjugacy* between them. If ϕ is also one-to-one, then f and g are called *topologically conjugate* and ϕ is referred to as a *conjugacy*.

Thus, when for a continuous self-map f of a compact topological space X and a (positively) invariant subset $\mathcal{I} \subseteq X$, it holds that $f \restriction_{\mathcal{I}}$ is semiconjugate to the (one-sided or two-sided) Bernoulli shift σ on m symbols, and then

$$h_{\text{top}}(f) \geq h_{\text{top}}(f \restriction_{\mathcal{I}}) \geq h_{\text{top}}(\sigma) = \log(m)\,. \tag{8.5}$$

If $f \restriction_{\mathcal{I}}$ is conjugate to σ, then the second inequality is indeed an equality.

We notice that although the topological entropy can be defined for continuous self-maps of topological spaces, we confine ourselves to the case of metric spaces. More precisely, when dealing with chaotic dynamics, we will consider *dynamical systems*, i. e. couples (X, f), where X is a compact metric space and $f\colon X \to X$ is continuous and surjective.

For further features of h_{top} and additional details, see [1, 70, 167]. With reference to the case of compact metric spaces, alternative definitions of entropy can be found in [28, 48].

Let us begin our excursus on complex dynamics by returning to the concepts of "covering" and "crossing" introduced at the beginning of Section 6 and fundamental in the theory of symbolic dynamics.

For the reader's convenience, we recall that given a continuous mapping $f\colon \mathbb{R} \to \mathbb{R}$ and two intervals $I, J \subseteq \mathbb{R}$, we say that I *f-covers* J if $f(I) \supseteq J$, or equivalently, if there exists a subinterval $I_0 \subseteq I$ such that $f(I_0) = J$. We also say that I *f-covers J m times* if there are $m \geq 2$ subintervals $I_0, \dots, I_{m-1} \subseteq I$, with pairwise disjoint interiors such that $f(I_k) = J$ for $k = 0, \dots, m-1$ ([25]). In particular, we will focus on the case in which $I = J$ and there exist m compact intervals $I_0, \dots, I_{m-1} \subseteq I$, with pairwise disjoint interiors such that I_i f-covers I_j for some (maybe all) $i, j \in \{0, \dots, m-1\}$.

The natural extension of the concept of f-covering to the setting of a continuous self-map f of a generic metric space X can be obtained by replacing intervals with compact subsets of X in the previous definitions. Indeed, one can assume the existence of m pairwise disjoint compact sets $C_0, \dots, C_{m-1} \subseteq X$ such that $f(C_i) \supseteq C_j$ for some $i, j \in \{0, \dots, m-1\}$. On the other hand, as suggested in [30], instead of the above covering relation, one could deal with a weaker condition of the form $f(C_i)$ "goes across" C_j, which does not require the map f to be surjective on the C_j's. For example, in our approach, we interpret the expression "goes across" in terms of the stretching along the paths effect from Definition 4.1 or Definition 5.5, that is, $f(C_i)$ "goes across" C_j means for us that $f\colon \widetilde{C}_i \Leftrightarrow\!\!\!\!\to \widetilde{C}_j$, where $\widetilde{C}_i$ and $\widetilde{C}_j$ are (N-dimensional) oriented rectangles of X. In this respect, an interesting framework is the one described in Definition 4.2 for $\widetilde{A} = \widetilde{B} = \widetilde{\mathcal{R}}$, or in the statement of Theorem 7.1 with

$X = \mathbb{R}^2$.[5] In such situations, there exist $m \geq 2$ pairwise disjoint compact subsets $\mathcal{K}_0, \ldots, \mathcal{K}_{m-1}$ of a generalized planar rectangle $\mathcal{R}$ for which $(\mathcal{K}_i, \psi)\colon \widetilde{\mathcal{R}} \Leftrightarrow \widetilde{\mathcal{R}}$ holds, where $\widetilde{\mathcal{R}} = (\mathcal{R}, \mathcal{R}^-)$. Then, in view of Remark 4.2, it is possible to find m pairwise disjoint vertical slabs $\widetilde{\mathcal{R}}_i$ of $\widetilde{\mathcal{R}}$ (cf. Definition 4.3), with $\mathcal{R}_i \supseteq \mathcal{K}_i$, for $i = 0, \ldots, m-1$, which satisfy the relation $\psi\colon \widetilde{\mathcal{R}}_i \Leftrightarrow \widetilde{\mathcal{R}}_j$, $\forall i, j \in \{0, \ldots, m-1\}$.

The pretext for considering the above covering and crossing relations comes from the previously observed fact that whenever a map ψ induces chaotic dynamics in the sense of Definition 7.2, then its entropy is positive (cf. conclusion (iv) in Theorem 7.2) and this happens, for instance, when the stretching condition in Theorem 7.1 is fulfilled. Actually, we are going to show that a positive entropy can be obtained even under weaker assumptions.

For clarity's sake, we first explain the idea in the one-dimensional setting in order to simplify the comprehension of the generic N-dimensional case (for $N \geq 2$). Hence, let us suppose that $f\colon I \to I$ is a continuous self-mapping of a compact interval $I \subset \mathbb{R}$ and assume there exists $a \in I$ such that, calling $b = f(a)$, $c = f(b) = f^2(a)$ and $d = f(c) = f^2(b) = f^3(a)$, it holds that

$$d \leq a < b < c \quad \text{or} \quad d \geq a > b > c\,. \tag{8.6}$$

In particular, if $d = a$, then the map f has in I a point of period three. This is in fact the case mentioned in the title of the well-known paper [93]. Notice that, setting

$$I_0 := [\min\{a, b\}, \max\{a, b\}] \quad \text{and} \quad I_1 := [\min\{b, c\}, \max\{b, c\}]\,, \tag{8.7}$$

it follows that I_0 f-covers I_1 and I_1 f-covers I_0, as well as I_1 f-covers I_1. Under condition (8.6), the authors in [93] can prove the presence in I of periodic points of any period and also the existence of an uncountable scrambled set. We will return to this concept in Definition 9.1. Before that, we want to show the positivity of the topological entropy of f in the recently described framework. To such an end, we need some tools from the theory of symbolic dynamics [77, 94].

Given a continuous map $g\colon J \to J$ defined on a compact interval $J \subset \mathbb{R}$ and $n \geq 2$ closed subintervals $J_0, \ldots, J_{n-1} \subseteq J$, with pairwise disjoint interiors, we associate to the dynamical system (J, g) the $n \times n$ *transition matrix* $T = T(i, k)$, for $i, k = 0, \ldots, n-1$, defined as

$$T(i,k) = \begin{cases} 1 & \text{if } J_i \ g\text{-covers } J_k, \\ 0 & \text{else}\,. \end{cases} \tag{8.8}$$

---

5 The case of sets homeomorphic to $[0, 1]^2$, but contained in a generic metric space X could be considered as well. However, since in the applications the space X is the Euclidean plane and the generalized rectangles are compact regions bounded by graphs of continuous functions, for simplicity, we confine ourselves to the planar setting.

Moreover, let

$$\Sigma_T^+ := \{(s_i)_{i\in\mathbb{N}} \in \Sigma_n^+ : T(s_k, s_{k+1}) = 1,\ \forall k \in \mathbb{N}\} \tag{8.9}$$

be the *space of the admissible sequences for T* and

$$\sigma_T : \Sigma_T^+ \to \Sigma_T^+ , \quad \sigma_T := \sigma \restriction_{\Sigma_T^+} \tag{8.10}$$

be the *subshift of finite type for the matrix T*. In particular, according to [30], we say that such subshift of finite type is *nontrivial* if T is an *irreducible* matrix (that is, for every couple of integers $i,\ k \in \{0,\dots,n-1\}$, there exists a positive integer l such that $T^l(i,k) > 0$), which is not a permutation on n symbols.[6] The space Σ_T^+ inherits the metric from Σ_n^+ and, with this choice, σ_T is continuous. Furthermore, it is possible to prove (cf. [77, Observation 1.4.2]) that when T is irreducible, $h_{\text{top}}(\sigma_T) = \log(\overline{\lambda})$, where $\overline{\lambda}$ is the largest real eigenvalue of T in absolute value, also called the *Perron eigenvalue of T*. Therefore, if there exists a (positively) invariant set $\mathcal{I}^* \subseteq J$ such that $g \restriction_{\mathcal{I}^*}$ is semiconjugate to σ_T, it holds that

$$h_{\text{top}}(g) \geq h_{\text{top}}(g \restriction_{\mathcal{I}^*}) \geq h_{\text{top}}(\sigma_T) = \log(\overline{\lambda}) . \tag{8.11}$$

In the case under investigation, we can take

$$\mathcal{I}^* := \bigcap_{i=0}^{+\infty} g^{-i}(J) \tag{8.12}$$

and define the semiconjugacy $\pi : \mathcal{I}^* \to \Sigma_T^+$ between $g \restriction_{\mathcal{I}^*}$ and σ_T as

$$(\pi(x))_i = k \quad \text{iff } g^i(x) \in J_k , \quad \text{for } i \in \mathbb{N} . \tag{8.13}$$

We recall that, according to [30, Lemma 1.2], for the Perron eigenvalue, it holds that $\overline{\lambda}$ is strictly greater than 1 unless T is a permutation matrix (otherwise, $\overline{\lambda} = 1$). Thus, the topological entropy of any nontrivial subshift of finite type is positive.

We now have all of the ingredients to deal with a map f with the properties expressed in (8.6) available. The transition matrix for f associated to the intervals I_0 and I_1 in (8.7) is

$$T_f = \begin{pmatrix} 0 & 1 \\ 1 & 1 \end{pmatrix} . \tag{8.14}$$

Defining the (positively) invariant set $\mathcal{I}^*$ for f as in (8.12), it is possible to obtain a semiconjugacy as in (8.13) between $f \restriction_{\mathcal{I}^*}$ and the subshift of finite type σ_{T_f} for the matrix T_f. Thus, we can conclude that

$$h_{\text{top}}(f) \geq h_{\text{top}}(\sigma_{T_f}) = \log\left(\frac{1+\sqrt{5}}{2}\right) \tag{8.15}$$

6 Notice that, when T is the matrix whose entries are all 1, then $\Sigma_T^+ = \Sigma_n^+$ and σ_T coincides with the classical Bernoulli shift σ, also known as *full shift* [77].

since the Perron eigenvalue of T_f is $\overline{\lambda} = \frac{1+\sqrt{5}}{2}$. This quantity coincides with the *golden mean ratio* and the map σ_{T_f} is also known under the name of the *golden mean shift*.

We stress that more general situations can be handled in a similar manner. For instance, a further case that can be considered is when for a continuous map $g\colon J \to J$, some multiple coverings among the subintervals $J_0, \dots, J_{n-1}$ of J occur. In such a framework, the transition matrix T is replaced by the $n \times n$ *adjacency matrix* $A = A(i,k)$, for $i, k = 0, \dots, n-1$, defined as

$$A(i,k) = \begin{cases} l & \text{if } J_i \ g\text{-covers } J_k \ l \text{ times}, \\ 0 & \text{if } J_i \text{ does not } g\text{-cover } J_k \,, \end{cases} \tag{8.16}$$

where l is a positive integer.[7] The entries of A are now nonnegative integers, possibly different from $0, 1$. Also in this case, one can associate a suitable subshift of finite type to A. To understand how to proceed, it is necessary to recall some notions from graph theory [94, 138].

A *graph* G consists of a finite set $\mathcal{V} = \mathcal{V}(G)$ of *vertices* and of a finite set of *edges* $\mathcal{E} = \mathcal{E}(G)$ among them. Every edge $e \in \mathcal{E}(G)$ starts at a vertex denoted by $i(e)$ (*initial state*) and ends at a vertex denoted by $t(e)$ (*terminal state*), which can possibly coincide with $i(e)$. Of course, there may be several edges between the same initial and terminal states. In the special case that $A = A(i,k)$ is an adjacency matrix, we call *graph of* A the graph $G = G(A)$, with vertex set $\mathcal{V}(G) = \{0, \dots, n-1\}$ and with $A(i,k)$ edges with initial state in i and terminal state in k, for $i, k = 0, \dots, n-1$. The cardinality of the set of the edges is $|\mathcal{E}(G)| = \sum_{i,k \in \{0,\dots,n-1\}} A(i,k)$. We name this quantity N so that $\mathcal{E}(G) = \{e_1, \dots, e_N\}$ is the set of the edges.

The shift σ_T in (8.10) for the transition matrix T in (8.8) is called a *vertex subshift*. We now want to define an *edge subshift* for the adjacency matrix A in (8.16). In order to realize it, we construct the transition matrix $T' = T'(i,k)$ on $\mathcal{E}(G)$, that is, the $N \times N$ matrix with entries

$$T'(i,k) = \begin{cases} 1 & \text{if } t(e_i) = i(e_k), \\ 0 & \text{else}. \end{cases}$$

The desired edge subshift for A corresponds to the vertex subshift for T', $\sigma_{T'}\colon \Sigma_{T'}^+ \to \Sigma_{T'}^+$, $\sigma_{T'} := \sigma \restriction_{\Sigma_{T'}^+}$, where the definition of $\Sigma_{T'}^+$ is analogous to the one of Σ_T^+ in (8.9).[8]

---

7 If $l = 1$, condition "J_i g-covers J_k l times" simply reads as "J_i g-covers J_k."

8 Observe that when the entries of A are already from $\{0, 1\}$, it is still possible to construct an associated transition matrix T' as described above that will be possibly different from A (also, the dimensions n and N won't coincide in general). On the other hand, it is not difficult to show that the vertex subshift σ_A and the corresponding edge subshift $\sigma_{T'}$ are conjugate. Indeed, the conjugacy $h\colon \Sigma_{T'}^+ \to \Sigma_A^+$ can be defined as $h((e_j)_{j\in\mathbb{N}}) = (v_j)_{j\in\mathbb{N}}$, where $v_j = i(e_j)$, $\forall j \in \mathbb{N}$, with $v_j \in \mathcal{V}$ and $e_j \in \mathcal{E}$.

The edge subshifts can thus be seen as a counterpart of the vertex subshifts in relation to matrices with nonnegative integer entries, possibly different from $0, 1$. An example of matrices of such kind is represented by the powers of a given transition matrix and it is particularly important to handle them when dealing with the iterates of a given function.

Notice that both in the case of edge subshifts and in the case of vertex subshifts, we have confined ourselves to the one-sided sequences of symbols since we do not assume g to be one-to-one on its domain. Analogous results hold true for bi-infinite sequences under the additional hypothesis of injectivity for g (cf. Remark 7.1).

If, in place of self-maps on intervals, one deals with functions $f : X \to X$ defined on a generic metric space X, and considers, instead of the one-dimensional covering relation for intervals, some covering or crossing relation among pairwise disjoint compact subsets (or at least compact subsets with pairwise disjoint interiors, in the case of Markov partitions [140]) of X, then, to such framework, it is possible to associate a transition matrix or an adjacency matrix exactly as in (8.8) and (8.16), respectively, with intervals replaced by compact sets.

In particular, when the hypotheses of Theorem 7.1 are fulfilled with $X = \mathbb{R}^2$ and $m = 2$, we have already observed that there exist two vertical slabs $\widetilde{\mathcal{R}}_0$ and $\widetilde{\mathcal{R}}_1$ of $\widetilde{\mathcal{R}}$, with $\mathcal{R}_0 \supseteq \mathcal{K}_0$ and $\mathcal{R}_1 \supseteq \mathcal{K}_1$, which satisfy $\psi : \widetilde{\mathcal{R}}_i \Leftrightarrow\!\!\!\to \widetilde{\mathcal{R}}_j$, for $i, j = 0, 1$. Thus, the transition matrix for ψ associated with $\widetilde{\mathcal{R}}_0$ and $\widetilde{\mathcal{R}}_1$ is

$$T_\psi = \begin{pmatrix} 1 & 1 \\ 1 & 1 \end{pmatrix} . \tag{8.17}$$

Analogous conclusions can be drawn when restricting our stretching along the paths relation to the one-dimensional setting. In such a case, a pair $\widetilde{I} = (I, I^-)$, where $I = [a, b]$ is a compact interval and $I^- = \{a, b\}$ is the set of its extreme points, may be seen as a degenerate kind of oriented rectangle. Accordingly, the stretching property $(\mathcal{K}, \psi) : \widetilde{I} \Leftrightarrow\!\!\!\to \widetilde{I}$ is equivalent to the fact that $\mathcal{K}$ contains a compact interval I_0 such that $\psi(I_0) = I$. This shows that, in the one-dimensional case, our stretching property reduces to the classical covering relation discussed above. Moreover, noticing that in the just described framework, we have $(I_0, \psi) : \widetilde{I} \Leftrightarrow\!\!\!\to \widetilde{I}$, then, if I contains two disjoint compact subintervals I_0 and I_1 such that $\psi(I_i) = I$, for $i = 0, 1$, we enter the setting of Theorem 7.1, with $\widetilde{I}$ playing the role of $\widetilde{\mathcal{R}}$ and with $\mathcal{K}_i := I_i$.

A look at the matrices T_f in (8.14) and T_ψ in (8.17) shows that the assumptions of Theorem 7.1, when restricted to the one-dimensional setting, are stronger than the conditions in (8.6).[9] Recalling (8.15), this verifies our claim that a positive topological

---

9 At first sight, this conclusion may seem rather obvious since a map ψ as in Theorem 7.1 has periodic points of all periods and thus, in particular, also a point of period three. However, things are not so simple. Indeed, as a consequence of the Sharkowskiĭ Theorem [147, 160], any continuous self-map f

entropy can be obtained under hypotheses weaker than the ones in Theorem 7.1. On the other hand, with respect to [93], the assumptions in Theorem 7.1 allow one to obtain sharper consequences from a dynamical point of view (cf. Section 9).

Before abandoning the one-dimensional setting, let us employ the previously introduced tools from symbolic dynamics to study the behavior of the logistic map

$$F\colon [0,1] \to \mathbb{R}\,, \quad F(x) := \mu x(1-x)\,, \tag{8.18}$$

when the parameter μ ranges in $[0, +\infty]$.

It is easy to see that for $\mu > 4$, the interval $[0,1]$ F-covers $[0,1]$ twice. Indeed, the maximum of F is attained when $x = 1/2$ and $F(1/2) = \mu/4$, and thus $F(1/2) \geq 1$ for $\mu \geq 4$. Since $F(0) = F(1) = 0$, setting $\alpha := F^{-1}(\{1\}) \cap ([0,1/2])$ and $\beta := F^{-1}(\{1\}) \cap ([1/2,1])$, we have $(I_i, F)\colon \widetilde{I} \Leftrightarrow \widetilde{I}$, $i = 0,1$, where $\widetilde{I} := ([0,1], \{0,1\})$, $I_0 := [0,\alpha]$ and $I_1 := [\beta, 1]$.[10] Unfortunately, in such a case, F does not map $[0,1]$ into itself and, for almost every initial point, the iterates F^n of the map limit to $-\infty$ as $n \to \infty$. This generates problems with numerical simulations: indeed, although Theorems 7.1 and 7.2 ensure the existence of a chaotic invariant set in $[0,1]$, such a region is not visible on the computer screen because almost all points eventually leave the domain through the iterates of F.

When $\mu \leq 4$, the existence of multiple coverings can be established for some iterate of F through geometric arguments analogous to the ones employed for the case $\mu > 4$ that allow one to conclude that the conditions of Theorem 7.1 are verified for sufficiently large values of the parameter $\mu \leq 4$. A corresponding example is illustrated in Figure 11, which depicts the second iterate of F for $\mu = 3.88$. The interval $I \subseteq [0,1]$, drawn with a thicker line on the y-axis, is oriented by choosing as I^- its extreme points. For $\widetilde{I} = (I, I^-)$ and calling I_0 and I_1 the two subintervals of I highlighted on the x-axis, it is immediate to verify that $F^2(I_i) = I$, for $i = 0,1$, and hence $(I_i, F^2)\colon \widetilde{I} \Leftrightarrow \widetilde{I}$, $i = 0,1$. Therefore, the map F^2 induces chaotic dynamics on two symbols relative to I_0, I_1 and possesses all the chaotic properties listed in Theorem 7.2. In particular, from this, something can still be deduced for F. Indeed, by the semiconjugacy between F^2 restricted to a suitable invariant set and the one-sided shift σ on two symbols (cf. Theorem 7.2, conclusions (ii) and (v)), it holds that

$$h_{\text{top}}\left(F^2\right) \geq \log(2)$$

defined on a compact interval and possessing a point of period three has periodic points of each period. Moreover, the conditions in (8.6) do not necessarily imply the existence of a point of period three, being in fact more general. This is the reason that led us to introduce some elements from symbolic dynamics in order to compare our framework to the one in [93].

10 Actually, $[0,1]$ F-covers $[0,1]$ twice also for the "watershed" value $\mu = 4$. However, in this case I_0 and I_1 are not disjoint and thus Theorem 7.1 does not apply directly.

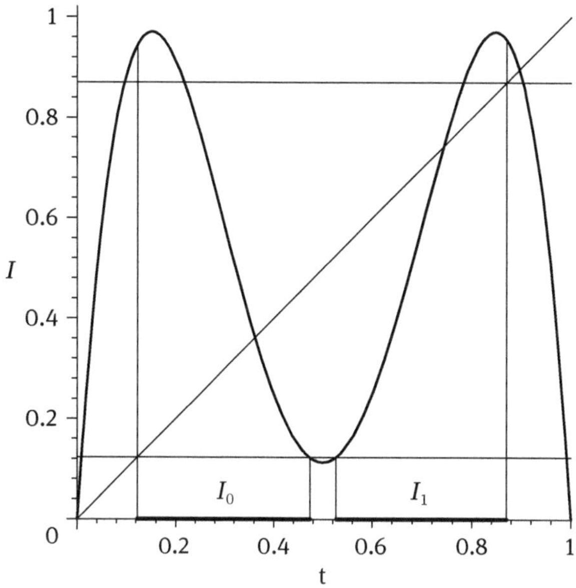

Fig. 11: The graph of the second iterate of the logistic map F in (8.18) for $\mu = 3.88$.

and, since by (8.3), we have $h_{\text{top}}(F^2) = 2h_{\text{top}}(F)$, it follows

$$h_{\text{top}}(F) \geq \log(\sqrt{2}) > 0\,.$$

We conclude this section by reconsidering Definition 7.2 and Theorem 7.1 in light of the above results from symbolic dynamics.

Indeed, when a map satisfies the hypotheses of Theorem 7.1 with $X = \mathbb{R}^2$ and for some $m \geq 2$, then the associated transition matrix has all elements equal to one (see (8.17) for the case $m = 2$). On the other hand, one could face the more general situation in which the transition matrix T associated with the system is irreducible, but possibly contains also some zeros [87]. The topological entropy can then be estimated as in (8.11) and one could say that the map ψ induces chaotic dynamics when the topological entropy of the corresponding subshift of finite type σ_T is positive. As already mentioned, this happens, for example, when σ_T is nontrivial [30].

According to [70, 77], if T is irreducible, the map σ_T is Devaney chaotic (Definition 9.2) on Σ_T^+ and thus it displays transitivity and sensitivity with respect to initial conditions and the set of periodic points of σ_T is dense in Σ_T^+. However, in order to proceed as in the proof of Theorem 7.1 and transfer the chaotic features of σ_T to ψ (via Theorem 9.1), we need the periodic sequences of σ_T to be realized by periodic points of ψ. In view of Theorem 7.1, this happens, for instance, when our stretching relation is fulfilled with respect to some oriented rectangles, but maybe not all the possible coverings are realized, so that the conclusions of Theorem 7.2 hold with the full shift σ on Σ_m^+ replaced by σ_T on Σ_T^+.

At last, we notice that a concrete framework in which it is possible to obtain chaotic dynamics in the sense of Definition 7.2 is when there exist $m \geq 2$ pairwise disjoint subsets C_i, $i = 0, \ldots, m-1$ of a metric space X, each with the Fixed Point Property (FPP)[11] and such that there exists a homeomorphism φ defined on $\bigcup_{i=0}^{m-1} C_i$, with $\varphi(C_i) \supseteq C_j, \forall i, j = 0, \ldots, m-1$. Then, since $\varphi(C_i) \supseteq C_i$ is equivalent to $\varphi^{-1}(C_i) \subseteq C_i$, by the continuity of $\varphi^{-1}: C_i \to C_i$, it follows that there exists $x^* \in C_i$ with $\varphi^{-1}(x^*) = x^*$, i. e. $\varphi(x^*) = x^*$. Thus, the presence of at least a fixed point is ensured in each C_i, $i = 0, \ldots, m-1$. Working with the iterates of the map φ, the existence of periodic points in the C_i's can be shown in an analogous manner. To check that the periodic sequences on m symbols are realized by periodic itineraries, it is then sufficient to follow the same steps as in the proof of the second and the forth conclusions in Theorem 4.3. Due to the invertibility of φ, we observe that it is possible to deal with two-sided sequences (cf. Remark 7.1). In view of the previous discussion, one could also consider the more general case in which $\varphi(C_i) \supseteq C_j$, for *some* $i, j = 0, \ldots, m-1$, but the associated transition matrix T is still irreducible. Obviously, in this setting, only the admissible (one-sided or two-sided) sequences can be realized.

9 On various notions of chaos

Let us further pursue the discussion on chaotic dynamics started in Sections 7–8. In particular, we will try to show the mutual relationships among some of the most classical definitions of chaos (such as the ones by Li–Yorke, Devaney, etc.), considering also the notion of chaotic dynamics in the sense of Definition 7.2.

Firstly, we would like to conclude the comparison between our results in Section 7 and the achievements in [93]. As we have seen, the presence of a point of period three for a continuous self-map f defined on a compact interval, or more generally the conditions (8.6) from [93], ensure the positivity of the topological entropy for f. Thanks to some tools from symbolic dynamics, we have also realized that our assumptions in Theorem 7.1 are stronger than the ones in (8.6). Now, we are going to show that the consequences of Theorem 7.1, listed in Theorem 7.2, are strictly sharper than the conclusions in [93].

The main result obtained by Li and Yorke in [93] is often recalled as a particular case of the Sharkowskiĭ Theorem [147, 160] but, actually, the authors in [93] proved much more than the presence of periodic points of any period for an interval map possessing a point of period three. Indeed, the existence of an uncountable *scrambled* set (cf. Definition 9.1) was obtained for such a map, which was called "chaotic" in [93] for the first time in the literature, even if the precise corresponding definition of chaos (now known as *Li–Yorke chaos*) was not given there.

11 We say that a topological space W has FPP if every continuous self-map of W has a fixed point in W.

Definition 9.1. Let (X, d_X) be a metric space and $f: X \to X$ be a continuous map. We say that $S \subseteq X$ is a *scrambled set for* f if for any $x, y \in S$, with $x \neq y$, it holds that

$$\liminf_{n\to\infty} d_X(f^n(x), f^n(y)) = 0 \quad \text{and} \quad \limsup_{n\to\infty} d_X(f^n(x), f^n(y)) > 0 .$$

If the set S is uncountable, we say that f is *chaotic in the sense of Li–Yorke*.

We remark that according to [93], the scrambled set S should satisfy an extra assumption, i. e.

$$\limsup_{n\to\infty} d_X(f^n(x), f^n(p)) > 0 ,$$

for any $x \in S$ and for any periodic point $p \in X$. However, in [10], this condition has been proved to be redundant in any compact metric space and therefore it is usually omitted.

We also point out that the original framework in [93] was one-dimensional. The subsequent extension to generic metric spaces is due to different authors (e. g. [10, 22, 67, 79]) that have collated the concept of chaos from [93] to other ones available in the literature. We will try to present some of these connections in the next pages.

We warn the reader that in what follows, the term chaotic will be referred to without distinction to a dynamical system, meant as a couple (X, f), where X is a compact metric space and $f: X \to X$ is continuous and surjective, as well as only to the map f defining it. In this respect, with a slight abuse of terminology, some properties of the map f (such as transitivity, sensitivity, etc.) will be transferred to the system (X, f). When we need to specify the distance d_X on X, we will also write (X, f, d_X) in place of (X, f). Since the map f has to be onto, if we are in the framework described in Theorem 7.2, the dynamical system we usually consider is given by $(\mathcal{I}, \psi \restriction_{\mathcal{I}})$, where $\mathcal{I}$ is the invariant set in (7.7).

In order to understand the relationship between the kind of chaos expressed in Definition 7.2 and the Li–Yorke chaos, a key role is played by the topological entropy. Indeed, as we have seen in Theorem 7.2, conclusion (iv), thanks to the semiconjugacy with the Bernoulli shift, the topological entropy of ψ is positive in the setting described in Definition 7.2. On the other hand, in [22, Theorem 2.3], it is established that any dynamical system with positive topological entropy admits an uncountable scrambled set and therefore it is chaotic in the sense of Li–Yorke. Hence, we can conclude that our notion of chaos is stronger than the one in Definition 9.1 since any chaotic system, according to Definition 7.2, is also Li–Yorke chaotic, while the vice versa does not hold in general. In fact, there exist maps that are Li–Yorke chaotic, but with zero topological entropy: for an example on the unit interval, see [152].

After the discussion on the concepts introduced in [93] and on the results obtained therein, the next feature related to the presence of chaos we take into consideration is the *sensitivity with respect to initial data*. This is one of the three requirements in Devaney's definition of chaos, together with the *topological transitivity* and the *density of periodic points*, and it is perhaps the most intuitive among them. Indeed, it is pretty natural to associate the idea of chaos to a certain unpredictability of the forward behavior of the system under consideration. The sensitivity with respect to initial data expresses exactly such a concept: no matter how close two points start, there exists an instant in the future in which they are at a given positive distance. However, although its intuitiveness, the sensitivity on initial conditions have been proved to be redundant in Devaney's definition of chaos in any infinite metric space [16, 148].[12]

In view of the subsequent treatment, we present the complete definition of *Devaney chaos* for the reader's convenience, even if at first we will focus only on the third condition.

Definition 9.2. Given a metric space (X, d_X) and a continuous function $f: X \to X$, we say that f is *chaotic in the sense of Devaney* if:
- f is *topologically transitive*, i. e. for any couple of nonempty open subsets $U, V \subseteq X$, there exists an integer $n \geq 1$ such that $U \cap f^n(V) \neq \emptyset$;
- The set of the periodic points for f is dense in X;
- f is *sensitive with respect to initial data* (or f displays *sensitive dependence on initial conditions*) on X, i. e. there exists $\delta > 0$ such that for any $x \in X$, there is a sequence $(x_i)_{i \in \mathbb{N}}$ of points in X such that $x_i \to x$ when $i \to \infty$, and for each $i \in \mathbb{N}$, there exists a positive integer m_i with $d_X(f^{m_i}(x_i), f^{m_i}(x)) \geq \delta$.

We stress that some authors call Devaney chaotic maps for which the three conditions above hold true only with respect to a compact positively invariant subset of the domain [10, 92]. We also remark that sometimes a map f on X is named topologically transitive if there exists a dense orbit for f in X [80]. In the case of compact metric spaces without isolated points, the two definitions turn out to be equivalent, but in general, they are independent. The precise relationship between such notions can be found in [148, Proposition 1.1]: in any metric space without isolated points, the existence of a dense orbit implies the topological transitivity; and vice versa, in separable and second category metric spaces, the topological transitivity implies the existence of a dense orbit. According to [11, Lemma 3], another case in which the two definitions of transitivity coincide is when f is onto and this happens, for instance, in any dynamical system. Such facts will find an application in Theorem 9.1.

Regarding the sensitivity, we notice that it can be equivalently defined using neighborhoods instead of sequences: given a metric space (X, d_X), the continuous map $f: X \to X$ is said to display sensitive dependence on initial conditions if there exists $\delta > 0$ such that for any $x \in X$ and for every open set $X \supseteq \mathcal{O}_x \ni x$, there

12 We recall that in the special case of intervals, the density of periodic points is also superfluous according to [165] and therefore Devaney chaos coincides with transitivity in the one-dimensional framework.

exist $y \in O_X$ and a positive integer m with $d_X(f^m(x), f^m(y)) \geq \delta$ [148]. We also recall that it is possible to give a pointwise version of the definition of sensitivity using neighborhoods or sequences. Regarding this latter characterization, any $x \in X$ admitting a sequence $(x_i)_{i\in\mathbb{N}}$ of points of X with $x_i \to x$ for $i \to \infty$ and a sequence $(m_i)_{i\in\mathbb{N}}$ of positive integers with $d_X(f^{m_i}(x_i), f^{m_i}(x)) \geq \delta$, $\forall i \in \mathbb{N}$ and for some $\delta > 0$ is sometimes called δ*-unstable* (or simply *unstable*) [11]. For an equivalent definition of unstable points based on neighborhoods, see [22], where it is also observed that, in general, the instability of f at every point of X does not imply f to be sensitive on X. Indeed, there could be no positive δ such that all points of X are δ-unstable. However, the previous inference is true if the map is, e. g. transitive [22], such as in Theorem 9.1, where we use the concept of sensitivity in both its equivalent versions and also the definition of unstable point based on sequences.

The positivity of the topological entropy and the sensitivity on initial conditions are someway related since both are signals of a certain instability of the system. The topological entropy is, however, a "locally detectable" feature in the sense that, according to (8.2), it is sufficient to find a (positively) invariant subset of the domain where it is positive in order to infer its positivity on the whole domain. Therefore, in general, we cannot expect the system to be sensitive at each point if the entropy is positive. Adding a global property, such as transitivity, this implication holds true. Indeed, in [22], it is argued that any transitive map with positive topological entropy displays sensitivity with respect to initial data. On the other hand, if we are content with the presence of sensitivity only on an invariant subset of the domain, then some authors have obtained corresponding results by considering, instead of the positivity of the entropy, the stronger property of chaos in the sense of coin-tossing (cf. Section 7), or at least the semiconjugacy to the Bernoulli shift σ for the map f defining the dynamical system or one of its iterates. For example, the already cited [71, Lemma 4] (Chaos Lemma), under hypotheses for a map f similar to the ones for ψ in Theorem 7.1, establishes the existence of a compact f-invariant set Q_*, on which f is sensitive and such that each forward itinerary on m symbols is realized by the itinerary generated by some point of Q_* (here, m is the crossing number introduced in Section 4.1). A similar result is mentioned, without proof, by Aulbach and Kieninger in [10] and it asserts that if a continuous self-map f of a compact metric space X is *chaotic in the sense of Block and Coppel*, that is, if there exist an iterate f^k (with $k \geq 1$) of f and a compact subset $Y \subseteq X$ positively f^k-invariant, such that $f^k \restriction_Y$ is semiconjugate to the one-sided Bernoulli shift on two symbols,[13] then $f \restriction_Z$ is Block–Coppel

13 We observe that, although the definition of Block–Coppel chaos is presented in [10] with respect to the shift on two symbols, in all the related results stated below, one could deal with the more general case in which $m \geq 2$ symbols are involved. Indeed, the situation is exactly the same as for Definitions 7.1 and 7.2, where the difference is only a formal matter.

chaotic, transitive and sensitive on Z, where Z is a suitable compact positively f-invariant subset of X. As suggested in [10], such conclusions rely on [11, Theorem 3]. Since the proof of this latter result is only sketched in [11], we provide the details in Theorem 9.1 below for the sake of completeness. Notice that the original requirement on the existence of a dense orbit in the statement of [11, Theorem 3] has been here replaced with the one of transitivity from Definition 9.2 since such notions coincide in every dynamical system (cf. [11, Lemma 3]). Moreover, we have replaced the pointwise instability in [11, Theorem 3] with the (generally) stronger property of sensitivity: indeed, as mentioned before, such concepts turn out to be equivalent in any transitive system [22].[14] Finally, we stress that the set X_0 in the statement of [11, Theorem 3] is claimed to be only positively invariant. However, after a look at its proof, since by construction X_0 is the ω-limit set of a certain point of X, it turns out to be invariant by the compactness of X (cf. [167, Theorem 5.5]).

Theorem 9.1 (Auslander–Yorke). *Let (X, f, d_X) and (Y, g, d_Y) be dynamical systems and let $\pi: X \to Y$ be a continuous and surjective map such that $\pi \circ f = g \circ \pi$. Assume that g is sensitive and transitive on Y. Then, there exists a closed f-invariant subset $X_0 \subseteq X$ such that $\pi(X_0) = Y$ and such that f is sensitive and transitive on X_0.*

Proof. First of all, we use Zorn Lemma to show the existence of a closed positively f-invariant subset $X_0 \subseteq X$ such that $\pi(X_0) = Y$ and X_0 is minimal with respect to these properties. Indeed, let us consider a chain $C = \{X_i\}_{i\in\mathcal{F}}$, whose elements are closed positively f-invariant subsets of X and $\pi(X_i) = Y$, $\forall i \in \mathcal{F}$. Then, we have to check that $X_\infty := \bigcap_{i\in\mathcal{F}} X_i$ is still closed, positively f-invariant and $\pi(X_\infty) = Y$. Notice that X_∞ is nonempty by the strong Cantor property, and valid for compact spaces (in fact, equivalent to compactness in any topological space).[15] Since X_∞ is the intersection of closed sets, it is closed as well. The positive f-invariance of X_∞ follows easily by observing that

$$f(X_\infty) \subseteq \bigcap_{i\in\mathcal{F}} f(X_i) \subseteq \bigcap_{i\in\mathcal{F}} X_i = X_\infty \,.$$

In order to verify that $\pi(X_\infty) = Y$, we have to check two inclusions. Since $\pi(X_i) = Y$, $\forall i \in \mathcal{F}$, then $\pi(X_\infty) \subseteq \bigcap_{i\in\mathcal{F}} \pi(X_i) = \bigcap_{i\in\mathcal{F}} Y = Y$. For the reverse inclusion, let us fix $y \in Y$ and consider the compact sets $C_i := \pi^{-1}(y) \cap X_i$, $i \in \mathcal{F}$. Notice that they are nonempty as $\pi(X_i) = Y$, $\forall i \in \mathcal{F}$. Moreover, the family $\{C_i\}_{i\in\mathcal{F}}$ has the finite intersection property: indeed, recalling that C is a chain, given $C_{i_1}, \dots, C_{i_k}$,

14 Actually, the sensitivity on initial conditions is equivalent to the instability of a point with dense orbit [11], as it will emerge from the proof of Theorem 9.1.

15 We recall that a topological space X has the strong Cantor property if for any family $\{C_\alpha\}_{\alpha\in\mathcal{I}}$ of closed subsets of X with the finite intersection property (i. e. such that for every finite subset $\mathcal{J} \subseteq \mathcal{I}$, it holds that $\bigcap_{\alpha\in\mathcal{J}} C_\alpha \neq \emptyset$), we have $\bigcap_{\alpha\in\mathcal{I}} C_\alpha \neq \emptyset$.

with $i_1, \dots, i_k \in \mathcal{F}$, it holds that

$$\bigcap_{j=1}^{k} C_{i_j} = C_{i^*}, \quad \text{for some } i^* = i_1, \dots, i_k$$

and thus such an intersection is nonempty. By the strong Cantor property, there exists

$$x \in \bigcap_{i\in\mathcal{F}} C_i = \bigcap_{i\in\mathcal{F}} \left(\pi^{-1}(y) \cap X_i\right) = \pi^{-1}(y) \cap \left(\bigcap_{i\in\mathcal{F}} X_i\right) = \pi^{-1}(y) \cap X_\infty ,$$

that is, there exists $x \in X_\infty$ such that $\pi(x) = y$. Hence, $Y \subseteq \pi(X_\infty)$. The existence of the minimal set X_0 is then ensured by Zorn Lemma.

Notice that since $\pi(X_0) = Y$ and $f(X_0) \subseteq X_0$, the relation $\pi\restriction_{X_0} \circ f\restriction_{X_0} = g \circ \pi\restriction_{X_0}$ holds. Indeed, for any $x_0 \in X_0 \subseteq X$, we have $\pi(f\restriction_{X_0}(x_0)) = \pi(f(x_0)) = g(\pi(x_0)) = g(\pi\restriction_{X_0}(x_0))$. On the other hand, since $f(x_0) \in X_0$, then $\pi(f\restriction_{X_0}(x_0)) = \pi\restriction_{X_0}(f\restriction_{X_0}(x_0))$, from which $\pi\restriction_{X_0}(f\restriction_{X_0}(x_0)) = g(\pi\restriction_{X_0}(x_0))$. The validity of $\pi\restriction_{X_0} \circ f\restriction_{X_0} = g \circ \pi\restriction_{X_0}$ is thus checked.

By the transitivity of the onto map g on Y, there exists $y^* \in Y$ with dense g-orbit, that is, $\overline{\gamma(y^*)} = Y$, for $\gamma(y^*) := \{g^n(y^*) : n \in \mathbb{N}\}$. Let $x_0^* \in X_0$ be such that $\pi(x_0^*) = y^*$. We will show that

$$\pi(\omega(x_0^*)) = \omega(y^*) = Y ,$$

where, for a dynamical system (Z, l) with $z \in Z$, we have

$$\omega(z) := \{x \in Z : \exists\, n_j \nearrow \infty \text{ with } l^{n_j}(z) \to x\} .$$

According to [167, Theorem 5.5], it holds that $\omega(z)$, called *ω-limit set of z*, is closed nonempty and invariant by the compactness of Z. Let us start with the verification of $\omega(y^*) = Y$. By the surjectivity of g on Y, there exists $\tilde{y} \in Y$ such that $g(\tilde{y}) = y^*$. Since the g-orbit of y^* is dense in Y and recalling that $\overline{\gamma(y^*)} = \gamma(y^*) \cup \omega(y^*)$, there are two possibilities for $\tilde{y}$, that is, $\tilde{y} \in \gamma(y^*)$ or $\tilde{y} \in \omega(y^*)$. In the former case, we find $\tilde{y} = g^k(y^*)$, for some $k \geq 0$. Since $g(\tilde{y}) = y^*$, this means that $g^{k+1}(y^*) = y^*$ and thus y^* is a periodic point, from which $Y = \overline{\gamma(y^*)} = \gamma(y^*) = \omega(y^*)$ is a finite set. On the other hand, if $\tilde{y} \in \omega(y^*)$, it follows that $g(\tilde{y}) = y^* \in \omega(y^*)$, by the invariance of $\omega(y^*)$. Recalling that it is also closed, then $Y = \overline{\gamma(y^*)} \subseteq \omega(y^*)$. Therefore, we find $\omega(y^*) = Y$ also in the latter case.

Regarding the equality $\pi(\omega(x_0^*)) = \omega(y^*)$, at first we observe that since $\pi\restriction_{X_0} \circ f\restriction_{X_0} = g \circ \pi\restriction_{X_0}$, then it follows that $\pi\restriction_{X_0} \circ (f\restriction_{X_0})^m = g^m \circ \pi\restriction_{X_0}, \forall m \geq 2$. Let us start with the inclusion $\pi(\omega(x_0^*)) \subseteq \omega(y^*)$. If $x^* \in \omega(x_0^*)$, then there exists a sequence $n_j \nearrow \infty$ such that $f^{n_j}(x_0^*) \to x^*$. By the continuity of the map π, it holds that $\pi(f^{n_j}(x_0^*)) \to \pi(x^*)$ for $j \to \infty$, but $\pi(f^{n_j}(x_0^*)) = g^{n_j}(\pi(x_0^*)) = g^{n_j}(y^*)$, i. e. $\lim_{j\to\infty} g^{n_j}(y^*) = \pi(x^*)$ and thus $\pi(x^*) \in \omega(y^*)$. For the reverse inclusion, we have to check that any $\bar{y} \in \omega(y^*)$ can be written as $\bar{y} = \pi(\bar{x})$, with

$\bar{x} \in \omega(x_0^*)$. Indeed, if $\bar{y} \in \omega(y^*)$, then there exists a sequence $m_k \nearrow \infty$ such that $g^{m_k}(y^*) \to \bar{y}$. Therefore, $\pi(f^{m_k}(x_0^*)) = g^{m_k}(\pi(x_0^*)) = g^{m_k}(y^*) \to \bar{y}$ as $k \to \infty$. Since $(f^{m_k}(x_0^*))_{k\in\mathbb{N}}$ is a sequence contained in the compact set X_0, there exist a subsequence $m_{k_j} \nearrow \infty$ of $(m_k)_{k\in\mathbb{N}}$ and $\bar{x} \in X_0$ such that $f^{m_{k_j}}(x_0^*) \to \bar{x}$ for $j \to \infty$. This means that $\bar{x} \in \omega(x_0^*)$. Notice that also for the subsequence $(m_{k_j})_{j\in\mathbb{N}}$, it holds that $\pi(f^{m_{k_j}}(x_0^*)) = g^{m_{k_j}}(\pi(x_0^*)) = g^{m_{k_j}}(y^*) \to \bar{y}$ as $j \to \infty$. Hence, passing to the limit, we find $\pi(\bar{x}) = \lim_{j\to\infty} g^{m_{k_j}}(\pi(x_0^*)) = \bar{y}$, from which one obtains $\bar{y} = \pi(\bar{x})$, with $\bar{x} \in \omega(x_0^*)$. The equality $\pi(\omega(x_0^*)) = \omega(y^*)$ is thus established.

Then, since $\pi(\omega(x_0^*)) = Y$ and recalling that $\omega(x_0^*)$ is a closed f-invariant subset of X_0, by the minimality of X_0, it follows that $\omega(x_0^*) = X_0$. By the invariance of $\omega(x_0^*)$, then $f(\omega(x_0^*)) = \omega(x_0^*)$ and so X_0 is invariant as well.

Moreover, from $\overline{\gamma(x_0^*)} = \gamma(x_0^*) \cup \omega(x_0^*)$, we find that the f-orbit of x_0^* is dense in X_0. In fact, by the minimality of X_0, the f-orbit of any point of X_0 is dense in X_0.[16]

Let us now check that f is sensitive on X_0. By the sensitivity of g on Y, there exists $\varepsilon > 0$ such that for every $y \in Y$ with dense g-orbit (actually, for any point of Y), there exist a sequence $(y_j)_{j\in\mathbb{N}}$ of points of Y and a sequence $(n_j)_{j\in\mathbb{N}}$ of positive integers such that $y_j \to y$, but $d_Y(g^{n_j}(y), g^{n_j}(y_j)) > \varepsilon$, $\forall\, j \in \mathbb{N}$. Let $(x_j)_{j\in\mathbb{N}}$ be a sequence in X_0 such that $\pi(x_j) = y_j$, $\forall\, j \in \mathbb{N}$, and let $(x_{j_k})_{k\in\mathbb{N}}$ be a converging subsequence of $(x_j)_{j\in\mathbb{N}}$, with $x_{j_k} \to \overline{x_0}$, for some $\overline{x_0} \in X_0$. By the continuity of π, it follows that $\pi(\overline{x_0}) = y$. We want to show that $\overline{x_0}$ is unstable, that is, there exists $\bar{\delta} > 0$ such that $d_X(f^{n_{j_k}}(\overline{x_0}), f^{n_{j_k}}(x_{j_k})) > \bar{\delta}$, $\forall k \in \mathbb{N}$, where $(n_{j_k})_{k\in\mathbb{N}}$ is the subsequence of $(n_j)_{j\in\mathbb{N}}$ corresponding to $(y_{j_k})_{k\in\mathbb{N}}$. Indeed, if by contradiction there exists a subsequence of (n_{j_k}), that by notational convenience we still denote by (n_{j_k}), such that $d_X(f^{n_{j_k}}(\overline{x_0}), f^{n_{j_k}}(x_{j_k})) \to 0$ for $k \to \infty$, by the continuity of π, we would find $0 < \varepsilon < d_Y(g^{n_{j_k}}(y), g^{n_{j_k}}(y_{j_k})) = d_Y(g^{n_{j_k}}(\pi(\overline{x_0})), g^{n_{j_k}}(\pi(x_{j_k}))) = d_Y(\pi(f^{n_{j_k}}(\overline{x_0})), \pi(f^{n_{j_k}}(x_{j_k}))) \to 0$. Therefore, $d_X(f^{n_{j_k}}(\overline{x_0}), f^{n_{j_k}}(x_{j_k})) > \bar{\delta}$, $\forall k \in \mathbb{N}$ for some $\bar{\delta} > 0$, and $\overline{x_0} \in X_0$ is an unstable point. Moreover, since X_0 is minimal, the f-orbit of $\overline{x_0}$ is dense in X_0. Let us prove that these two facts about $\overline{x_0}$ are sufficient to conclude that there exists a $\tilde{\delta} > 0$ such that any point of X_0 is $\tilde{\delta}$-unstable.

By the continuity of f, since $\overline{x_0}$ is $\bar{\delta}$-unstable, then also $f(\overline{x_0})$ is, and thus the whole f-orbit $\gamma(\overline{x_0})$ is $\bar{\delta}$-unstable. Calling $\mathcal{U}_{\bar{\delta}} \subseteq X_0$ the set of $\bar{\delta}$-unstable points of X_0, we show that $\overline{\mathcal{U}_{\bar{\delta}}} \subseteq \mathcal{U}_{\bar{\delta}/2}$. If this is true, since $X_0 = \overline{\mathcal{U}_{\bar{\delta}}}$, the thesis is achieved for $\tilde{\delta} := \bar{\delta}/2$. Therefore, let $x \in \overline{\mathcal{U}_{\bar{\delta}}}$. Then, there exists a sequence $(x_i)_{i\in\mathbb{N}}$ in $\mathcal{U}_{\bar{\delta}}$ with $x_i \to x$. For any such x_i and for every open set $X_0 \supseteq O_i \ni x_i$, there exist $\widetilde{x_i} \in O_i$ and a positive integer m_i with $d_X(f^{m_i}(x_i), f^{m_i}(\widetilde{x_i})) \geq \bar{\delta}$. In particular, we can choose $O_i := B(x_i, 1/i) \cap X_0$ so that $d_X(x_i, \widetilde{x_i}) < 1/i \to 0$, as $i \to \infty$. Now, there

16 A subset M of the dynamical system (Z, l) is called *minimal* if it is closed nonempty positively l-invariant and contains no proper closed nonempty positively l-invariant subsets. An equivalent definition is that M is minimal if each of its points has dense l-orbit in M.

are two alternatives: if $d_X(f^{m_i}(x_i), f^{m_i}(x)) \geq \bar{\delta}/2$, then x is $\bar{\delta}/2$-unstable and we are done. Otherwise, it holds that $d_X(f^{m_i}(x), f^{m_i}(\widetilde{x_i})) \geq \bar{\delta}/2$. Indeed, if this would fail, then $d_X(f^{m_i}(x_i), f^{m_i}(\widetilde{x_i})) \leq d_X(f^{m_i}(x_i), f^{m_i}(x)) + d_X(f^{m_i}(x), f^{m_i}(\widetilde{x_i})) < \bar{\delta}$, a contradiction. Since $\widetilde{x_i} \to x$, it follows also in this latter case that $x \in \mathcal{U}_{\bar{\delta}/2}$. The proof is complete. □

By the similarity between our point of view and the frameworks in [10, 71], it is not difficult to show that both the above quoted results on chaotic invariant sets obtained therein still hold in the setting of Definition 7.2. Therefore, the sensitivity is for us ensured at least on some subset of the domain. More precisely, as discussed in Section 7, since our notion of chaos in Definition 7.2 is stricter than the one considered in [71, 73], we can prove the existence of the chaotic invariant set Q_* mentioned above, exactly as in [71, Lemma 4]. On the other hand, with reference to [10] and [11], we notice that any map chaotic according to Definition 7.1 is also chaotic in the sense of Block–Coppel, thanks to Theorem 7.2, conclusion (ii). Thus, following the suggestion in [10], we have used Theorem 9.1 to get the existence of the set Λ in Theorem 7.2, conclusion (v). More precisely, we point out that in Theorem 7.2, conclusion (ii), a property stronger than the notion of chaos in the sense of Block–Coppel is obtained. Indeed, the semiconjugacy with the Bernoulli shift is established for the map ψ itself and not for one of its iterates. The same remark applies to Theorem 7.2, conclusion (v), where again, a sharper feature than the Block–Coppel chaos is deduced.

Pursuing further this discussion on the Block–Coppel chaos, we notice that every system (X, f) Block–Coppel chaotic has positive topological entropy. Indeed, by the postulated semiconjugacy between an iterate f^k (with $k \geq 1$) of the map f, restricted to a suitable positively invariant subset of the domain, and the one-sided Bernoulli shift on two symbols, by (8.3) and (8.5), it follows that $k h_{\text{top}}(f) = h_{\text{top}}(f^k) \geq \log(2)$, from which $h_{\text{top}}(f) \geq \log(2)/k > 0$. Thus, by the previously quoted [22, Theorem 2.3], any such system is also Li–Yorke chaotic. However, we observe that the Block–Coppel chaos is strictly weaker than chaos in the sense of coin-tossing and, a fortiori, also than the concept in Definition 7.1: indeed, according to [10, Remark 3.2], there exist systems (X, f) such that f^2 restricted to some f-invariant subset of X is semiconjugate to the one-sided Bernoulli shift on two symbols, while such a property does not hold for f.

Returning back to the analysis on the relationship between the topological entropy and the sensitivity on initial conditions, we recall that, except for the special case of continuous self-mappings of compact intervals, in general, the sensitivity does not imply a positive entropy [79]. Actually, even more can be said. Indeed, if instead of the sensitivity alone, we take into account the definition of Devaney chaos in its completeness, then it is possible to prove that the Devaney chaoticity and the positivity of topological entropy are independent, as none of the two implies the other [12, 86]. In particular, this means that, in generic metric spaces, Devaney chaos does not imply chaos in the sense of Definition 7.2 (because, otherwise, the topological entropy would be positive in any Devaney chaotic system). Vice versa, it is not clear if our notion of chaos implies the one by Devaney. Indeed, on the one hand, in [10], it is presented as a dynamical system Block–Coppel chaotic, but not chaotic according to Devaney since it has no periodic points. On the other hand, we have already noticed that our notion of chaos is strictly stronger than the one in the Block–Coppel sense.

When confining ourselves to the one-dimensional case, most of the definitions of chaos are known to be equivalent, while it is in the higher dimensional setting that the relationship among them becomes more involved. Indeed, in [92], it was proved that for a continuous self-mapping f of a compact interval to have positive topological entropy is equivalent to be chaotic in the sense of Devaney on some closed (positively) invariant subset of the domain. The positivity of the topological entropy is also equivalent to the fact that some iterate of f is turbulent (or even strictly turbulent, cf. Section 8 for the corresponding definitions) or to the chaoticity in the Block–Coppel sense [10]. Such equivalence, however, does not extend to chaos in the sense of Li–Yorke: indeed, as already pointed out, there exist Li–Yorke chaotic interval maps, but with zero topological entropy [152].

When moving to more general frameworks, the previous discussion should suggest that several links among the various notions of chaos get lost. Nonetheless, something can still be said. In addition to the facts already expounded (e. g. chaos according to Definition 7.1 ⇒ chaos in the sense of coin-tossing ⇒ Block–Coppel chaos ⇒ $h_{\mathrm{top}} > 0$ ⇒ Li–Yorke chaos), we mention that Devaney chaos implies Li–Yorke chaos in any compact metric space: actually, in order to prove this implication, the hypothesis on the density of periodic points in Definition 9.2 could be replaced with the weaker condition that at least one periodic point does exist [67]. On the other hand, this weaker requirement is necessary because transitivity and sensitivity alone are not sufficient to imply Li–Yorke chaos and vice versa, as shown in [22]. We recall that a dynamical system that is both sensitive and transitive is sometimes named the *Auslander–Yorke chaotic* [11, 22]. Therefore, we can rephrase the previous sentence by saying that the concepts of Li–Yorke chaos and Auslander–Yorke chaos are independent.

Of course, the above treatment is just meant to give an idea of the intricate network of connections among some of the most well-known definitions of chaos. A useful and almost complete survey on the existing relationships can be found in [110].

10 Linked twist maps

In this last section, we present a different geometrical context where it is possible to apply our method of "stretching along the paths" from Section 4 and the corresponding results on chaotic dynamics from Section 7. More precisely, we will be concerned with the study of the so-called *linked twist maps* (for short, LTMs). In such framework, instead of considering a single map that expands the arcs along a domain homeomorphic to a rectangle, one deals with a geometric configuration characterized by the alternation of two planar homeomorphisms (or diffeomorphisms) which twist two circular annuli (or two families of them [133]) intersecting in two disjoint generalized rectangles $\mathcal{A}$ and $\mathcal{B}$. Each annulus is turned onto itself by a homeomorphism which leaves the boundaries of the annulus invariant. Both the maps act in their domain so that a twist effect is produced. This happens, for instance, when the angular speed is monotone with respect to the radius. Considering the composition of the two movements in the common regions, we obtain a resulting function which is what we call a "linked twist map" (refer to [170, 171] for a detailed description of the geometry of the domain of a LTM). Such maps furnish a geometrical setting for the existence of Smale horseshoes: in fact, under certain conditions, it is possible to prove the presence of a Smale horseshoe inside $\mathcal{A}$ and $\mathcal{B}$ [46]. Usual assumptions on the twist mappings require, among others, their smoothness, monotonicity of the angular speed with respect to the radial coordinate and preservation of the Lebesgue measure. On the other hand, since our approach is purely topological, we just need a twist condition on the boundary (cf. Example 10.1).

In the past decades, a growing interest has concerned LTMs. In the 1980s, they were studied from a theoretical point of view by Devaney [46], Burton and Easton [33], Przytycki [132, 133] and Wojtkowski [175] (just to cite a few contributions in this direction), proving some mathematical properties like ergodicity, hyperbolicity and conjugacy to the Bernoulli shift. However, as observed in [46], such maps naturally appear in various applicative contexts, for instance, in mathematical models for particle motions in a magnetic field as well as in differential geometry in the study of diffeomorphisms of surfaces. Special configurations related to LTMs can also be found in the restricted three-body problem [9, pp. 231–237] and [111, pp. 90–94]. In more recent years, thanks to the work of Ottino, Sturman and Wiggins, significant applications of LTMs have been performed in the area of fluid mixing [161, 162, 170, 171]. See also [153, 154] for recent contributions on this subject.

Our purpose is to adapt the general results from Sections 4–7 to a geometrical framework which is connected to and generalizes (in a direction explained in Example 10.2) the case of the LTMs. The corresponding Theorems 10.1 and 10.2 below will be applied to some nonlinear ODEs with periodic coefficients in Section 12.

Theorem 10.1. *Let X be a metric space and assume that $\varphi\colon X \supseteq D_\varphi \to X$ and $\psi\colon X \supseteq D_\psi \to X$ are continuous maps defined on the sets D_φ and D_ψ, respectively. Let also $\widetilde{\mathcal{A}} := (\mathcal{A}, \mathcal{A}^-)$ and $\widetilde{\mathcal{B}} := (\mathcal{B}, \mathcal{B}^-)$ be oriented rectangles of X. Suppose that the following conditions are satisfied:*

(H_φ) *There are $m \geq 2$ pairwise disjoint compact sets $\mathcal{H}_0, \dots, \mathcal{H}_{m-1} \subseteq \mathcal{A} \cap D_\varphi$ such that $(\mathcal{H}_i, \varphi)\colon \widetilde{\mathcal{A}} \Leftrightarrow\!\!\!\!\!\rightarrow \widetilde{\mathcal{B}}$, for $i = 0, \dots, m-1$;*

(H_ψ) *$\mathcal{B} \subseteq D_\psi$ and $\psi : \widetilde{\mathcal{B}} \Leftrightarrow\!\!\!\!\!\rightarrow \widetilde{\mathcal{A}}$.*

Then, the map $\phi := \psi \circ \varphi$ induces chaotic dynamics on m symbols in the set $\mathcal{H} \cap \varphi^{-1}(\mathcal{B})$, where $\mathcal{H} := \bigcup_{i=0}^{m-1} \mathcal{H}_i$, and thus satisfies properties (i)–(v) *from Theorem 7.2* *(or the stronger properties* (i)–(v) *from Theorem 7.3, if* ϕ *is also injective on* $\mathcal{H} \cap \varphi^{-1}(\mathcal{B})$[17]*).*

Proof. We show that

$$(\mathcal{H}_i \cap \varphi^{-1}(\mathcal{B}), \phi) \colon \widetilde{\mathcal{A}} \Leftrightarrow\!\!\!\to \widetilde{\mathcal{A}}, \quad \forall i = 0, \ldots, m-1, \tag{10.1}$$

from which the thesis about the chaotic dynamics is an immediate consequence of Theorem 7.1.

To check condition (10.1), let us consider a path $\gamma \colon [0,1] \to \mathcal{A}$ such that $\gamma(0) \in \mathcal{A}_\ell^-$ and $\gamma(1) \in \mathcal{A}_r^-$ (or with $\gamma(0) \in \mathcal{A}_r^-$ and $\gamma(1) \in \mathcal{A}_\ell^-$) and let us fix $i \in \{0, \ldots, m-1\}$. By (H_φ), there exists a compact interval $[t', t''] \subseteq [0,1]$ such that $\gamma(t) \in \mathcal{H}_i$ and $\varphi(\gamma(t)) \in \mathcal{B}$, for every $t \in [t', t'']$, with $\varphi(\gamma(t'))$ and $\varphi(\gamma(t''))$ belonging to different components of $\mathcal{B}^-$. Define now

$$\omega \colon [t', t''] \to \mathcal{B}, \quad \omega(t) := \varphi(\gamma(t)).$$

By (H_ψ), there is a compact interval $[s', s''] \subseteq [t', t'']$ such that $\psi(\omega(t)) \in \mathcal{A}$, for every $t \in [s', s'']$, with $\psi(\omega(s'))$ and $\psi(\omega(s''))$ belonging to different components of $\mathcal{A}^-$.

Rewriting all in terms of γ, we have thus proved that

$$\gamma(t) \in \mathcal{H}_i \cap \varphi^{-1}(\mathcal{B}) \quad \text{and} \quad \phi(\gamma(t)) \in \mathcal{A}, \quad \forall t \in [s', s''],$$

with $\phi(\gamma(s'))$ and $\phi(\gamma(s''))$ belonging to different components of $\mathcal{A}^-$. The continuity of the composite mapping $\phi = \psi \circ \varphi$ on $\mathcal{H}_i \cap \varphi^{-1}(\mathcal{B})$ follows from the continuity of φ on $D_\varphi \supseteq \mathcal{H}_i$ and from the continuity of ψ on $D_\psi \supseteq \mathcal{B}$. By the arbitrariness of the path γ and of $i \in \{0, \ldots, m-1\}$, the verification of (10.1) is complete. □

We remark that in condition (H_φ), we could consider the case in which there exists a set $\mathcal{D} \subseteq \mathcal{A} \cap D_\varphi$ with $\mathcal{H} = \bigcup_{i=0}^{m-1} \mathcal{H}_i \subseteq \mathcal{D}$, in analogy with Theorem 7.1. From the proof of Theorem 10.1, it is also clear that the continuity of φ on D_φ and of ψ on D_ψ could be weakened, requiring that φ is continuous only on $\mathcal{H}$ and ψ on $\mathcal{B}$. For simplicity's sake, we have preferred to present the easier framework.

Moreover, regarding condition (H_ψ), we could assume that the stretching condition therein holds with respect to a compact set $\mathcal{K} \subseteq \mathcal{B} \subseteq D_\psi$, that is, $(\mathcal{K}, \psi) \colon \widetilde{\mathcal{B}} \Leftrightarrow\!\!\!\to \widetilde{\mathcal{A}}$. Then, the chaotic dynamics would be localized in the set

$$\mathcal{H}^* := \bigcup_{i=0}^{m-1} \left(\mathcal{H}_i \cap \varphi^{-1}(\mathcal{K})\right).$$

Such a case is analyzed in the more general Theorem 10.2.

17 This is, for instance, the case in Section 12 since there we deal with the Poincaré map, that is, a homeomorphism.

We now present two simple examples for the application of Theorem 10.1, which are meant to show how this result is well-fit for studying LTMs. Example 10.1 is fairly classical as it concerns two overlapping annuli subject to twist rotations, while Example 10.2 describes the composition of a map which twists an annulus with a longitudinal motion along a strip. A similar geometric setting was already considered by Kennedy and Yorke in [72] in the framework of the theory of fluid mixing, studying planar functions obtained as composition of a squeezing map and a stirring rotation.

We stress that such examples are only of "pedagogical" nature and, in fact, the chaotic-like dynamics that we obtain could be also proved using different approaches already developed in various papers (for instance, [107, 156, 182, 186]).

Example 10.1. A classical kind of LTM is represented by the composition of two planar maps φ and ψ which act as twist rotations around two given points. For instance, a possible choice is that of considering two continuous functions expressed by means of complex variables as

$$\varphi(z) := -r + (z + r)\, e^{\imath(c_1 + d_1|z+r|)} \tag{10.2}$$

and

$$\psi(z) := r + (z - r)\, e^{\imath(c_2 + d_2|z-r|)}\,, \tag{10.3}$$

where $\imath$ is the imaginary unit, while $r > 0$, c_j $(j = 1, 2)$ and $d_j \neq 0$ are real coefficients. Such maps twist around the centers $(-r, 0)$ and $(r, 0)$, respectively. We denote by p_1 and p_2 the inner and the outer radii for the annulus around $(-r, 0)$ and by q_1 and q_2 the inner and the outer radii for the annulus around $(r, 0)$.

For a suitable choice of r of the radii $p_1 < p_2$ and $q_1 < q_2$, as well as of the parameters determining φ and ψ, it is possible to apply Theorem 10.1 in order to obtain chaotic dynamics (Figures 12–14).

Example 10.2. A nonstandard LTM is represented by the composition of two planar maps φ and ψ, where φ is a twist rotation around a given point and ψ produces a longitudinal motion along a strip with different velocities on the upper and lower components of the boundary of the strip.

First, we introduce, for any pair of real numbers $a < b$, the real valued function

$$Pr_{[a,b]}(t) := \frac{1}{b-a}\min\{b - a, \max\{0, t - a\}\}\,,$$

and then we set

$$f(t) := c_1 + d_1 Pr_{[p_1,p_2]}(t)\,, \quad g(t) := c_2 + d_2 Pr_{[q_1,q_2]}(t)\,,$$

where c_j $(j = 1, 2)$ and $d_j \neq 0$ are real coefficients, while

$$0 < p_1 < p_2 \quad \text{and} \quad -p_1 < q_1 < q_2 < p_1$$

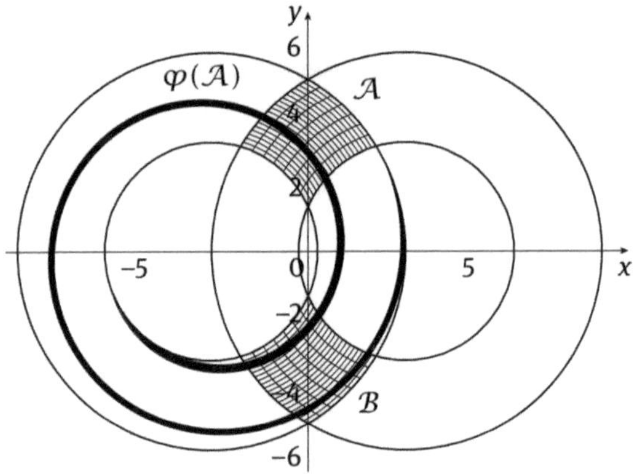

Fig. 12: A pictorial comment to condition (H_φ) of Theorem 10.1, with reference to Example 10.1. In order to determine the two annuli, we have set $r = 3$, $p_1 = q_1 = 3.3$ and $p_2 = q_2 = 6$ (this choice is just to simplify the explanation, since no special symmetry is needed). For the map φ in (10.2), we have taken $c_1 = -1.5$ and $d_1 = 3.3$. We choose, as a generalized rectangle $\mathcal{A}$ in Theorem 10.1, the upper intersection of the two annuli and select as components of $\mathcal{A}^-$ the intersections of $\mathcal{A}$ with the inner and outer boundaries of the annulus at the left-hand side. The set $\mathcal{B}$ is defined as the lower intersection of the two annuli and the two components of $\mathcal{B}^-$ are the intersections of $\mathcal{B}$ with the inner and outer boundaries of the annulus at the right-hand side. The narrow strip spiraling inside the left annulus is the image of $\mathcal{A}$ under φ. Clearly, any path in $\mathcal{A}$ joining the two components of $\mathcal{A}^-$ is transformed by φ onto a path crossing $\mathcal{B}$ twice in the correct manner (that is, from a side of $\mathcal{B}^-$ to the other side).

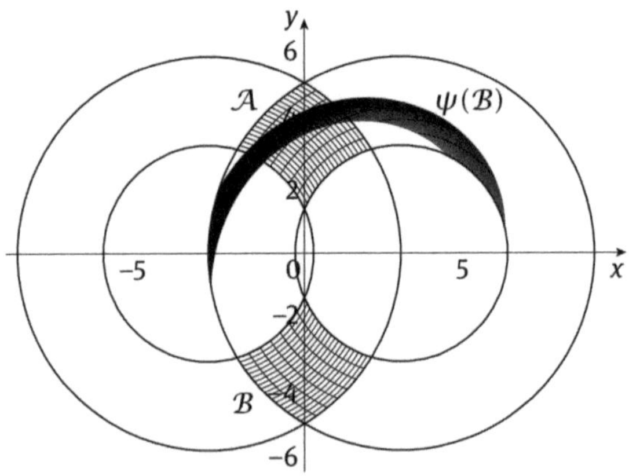

Fig. 13: A pictorial comment to condition (H_ψ) of Theorem 10.1, with reference to Example 10.1. For r, p_1, p_2, q_1, q_2 as in Figure 12, we have also fixed the parameters of ψ in (10.3) by taking $c_2 = 0$ and $d_2 = 0.9$. It is evident that the image of $\mathcal{B}$ under ψ crosses $\mathcal{A}$ once in the correct way.

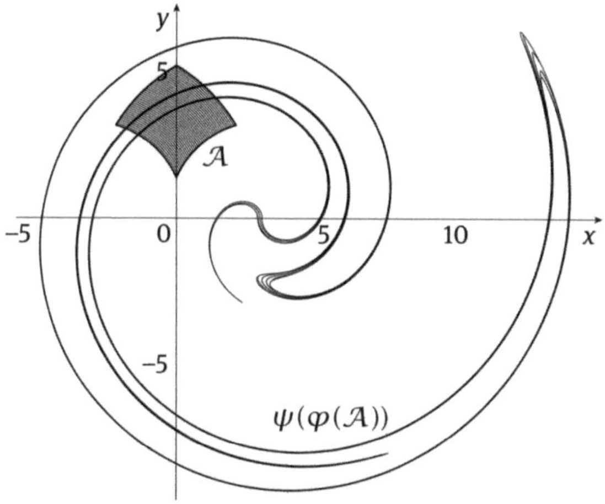

Fig. 14: For $r, p_1, p_2, q_1, q_2, \varphi$ as in Figure 12 and ψ as in Figure 13, we show the image of $\mathcal{A}$ under the composite map $\psi \circ \varphi$. It is clear that $\psi(\varphi(\mathcal{A}))$ crosses $\mathcal{A}$ twice in the correct way.

are real parameters which determine the inner and the outer radii of the circular annulus

$$C[p_1, p_2] := \{(x, y) \colon\ p_1^2 \le x^2 + y^2 \le p_2^2\}$$

and the position of the strip

$$S[q_1, q_2] := \{(x, y) \colon\ q_1 \le y \le q_2\}\,.$$

Finally, we define two continuous maps expressed by means of complex variables as

$$\varphi(z) := z\, e^{\imath f(|z|)} \tag{10.4}$$

(which twists around the origin) and, for $\Im(z)$ the imaginary part of z,

$$\psi(z) := z + g(\Im(z)) \tag{10.5}$$

(which shifts the points along the horizontal lines).

For a suitable choice of the parameters determining φ and ψ, it is possible to apply Theorem 10.1 in order to obtain chaotic dynamics (Figures 15–18).

With a straightforward modification in the proof of Theorem 10.1, one could check that (H_φ) and (H_ψ) imply the existence of chaotic dynamics also for $\varphi \circ \psi$. Actually, both Theorem 10.1 and such variant of it can be obtained as corollaries of the following more general result, whose proof is only sketched by the similarity with the verification of Theorem 10.1.

Theorem 10.2. *Let X be a metric space and assume that $\varphi\colon X \supseteq D_\varphi \to X$ and $\psi\colon X \supseteq D_\psi \to X$ are continuous maps defined on the sets D_φ and D_ψ, respectively. Let also $\widetilde{\mathcal{A}} := (\mathcal{A}, \mathcal{A}^-)$ and $\widetilde{\mathcal{B}} := (\mathcal{B}, \mathcal{B}^-)$ be oriented rectangles of X. Suppose that the following conditions are satisfied:*

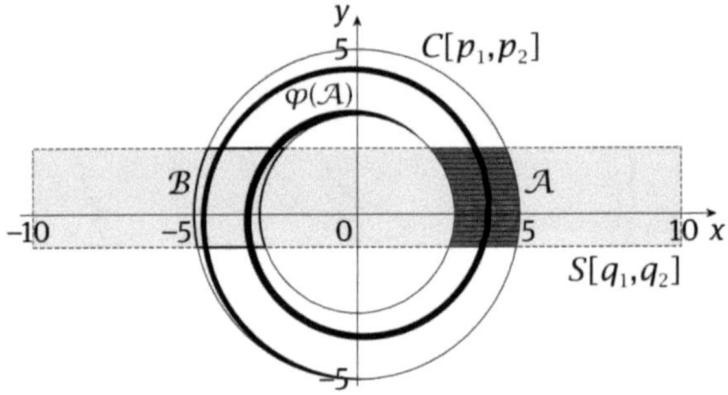

Fig. 15: A pictorial comment to condition (H_φ) of Theorem 10.1, with reference to Example 10.2. We have set $p_1 = 3$, $p_2 = 5$, $q_1 = -1$ and $q_2 = 2$ in order to determine the circular annulus $C[p_1, p_2]$ centered at the origin and the strip $S[q_1, q_2]$ which goes across it. For the map φ in (10.4), we have taken $c_1 = 0.4\,\pi$ and $d_1 = 3\,\pi$. We choose as set $\mathcal{A}$ in Theorem 10.1 the right-hand side intersection of the annulus with the strip and select as components of $\mathcal{A}^-$, the intersections of $\mathcal{A}$ with the inner and outer boundaries of the annulus. The set $\mathcal{B}$ is defined as the left-hand side intersection of the annulus with the strip and the two components of $\mathcal{B}^-$ are the intersections of $\mathcal{B}$ with the lower and upper sides of the strip. The narrow band spiraling inside the annulus is the image of $\mathcal{A}$ under φ. Clearly, any path in $\mathcal{A}$ joining the two components of $\mathcal{A}^-$ is transformed by φ onto a path crossing $\mathcal{B}$ twice in the correct manner.

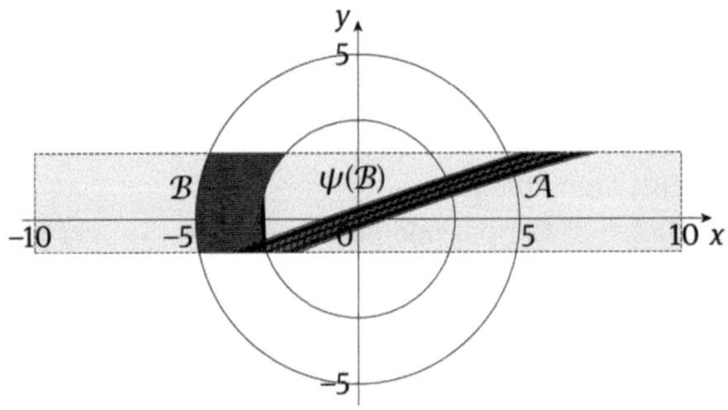

Fig. 16: A pictorial comment to condition (H_ψ) of Theorem 10.1 with reference to Example 10.2. The sets $C[p_1, p_2]$ and $S[q_1, q_2]$ and the oriented rectangles $\widetilde{\mathcal{A}}$ and $\widetilde{\mathcal{B}}$ are as in Figure 15. For the map ψ in (10.5), we have chosen $c_2 = 1$ and $d_2 = 8.6$. The parallelogram-shaped narrow figure inside the strip $S[q_1, q_2]$ is the image of $\mathcal{B}$ under ψ. Any path in $\mathcal{B}$ joining the two components of $\mathcal{B}^-$ is transformed by ψ onto a path crossing $\mathcal{A}$ once in the right way.

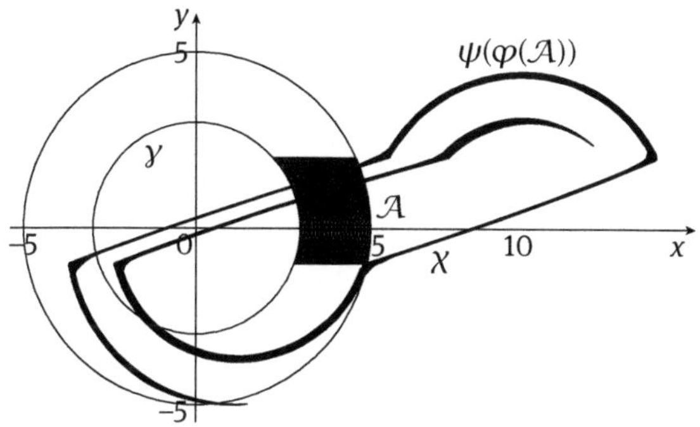

Fig. 17: For p_1, p_2, q_1, q_2 and φ as in Figure 15 and ψ as in Figure 16, we show the image of $\mathcal{A}$ under the composite map $\psi \circ \varphi$. It is evident that $\psi(\varphi(\mathcal{A}))$ crosses $\mathcal{A}$ twice in the correct manner.

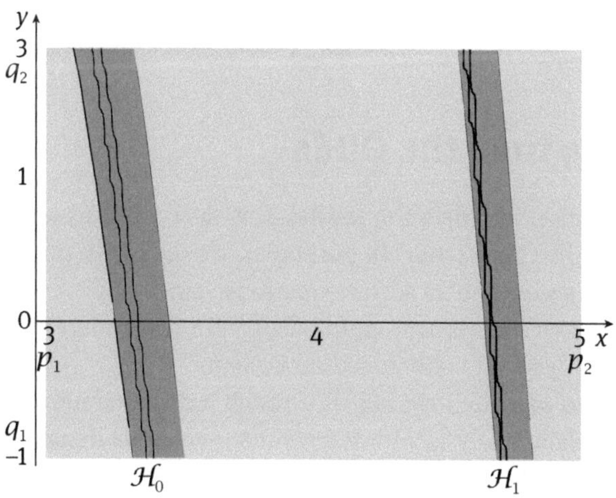

Fig. 18: In the simple case of Example 10.2, for φ and ψ with the coefficients chosen for drawing Figure 17, we can determine the sets $\mathcal{H}_0$ and $\mathcal{H}_1$ from Theorem 10.1. Indeed, inside the set $\mathcal{A}$, transformed in the rectangle $[p_1, p_2] \times [q_1, q_2]$ by a suitable change of coordinates, we have found two subsets (drawn with a darker color) whose image under φ is contained in $\mathcal{B}$. Inside these subsets, we have localized two smaller narrow domains such that their images under $\psi \circ \varphi$ are contained again in $\mathcal{A}$. Clearly, any path contained in the rectangle $[p_1, p_2] \times [q_1, q_2]$ and joining the left and the right sides contains two subpaths (inside the smaller narrow subsets) which are stretched by the composite mapping onto paths again connecting the left and the right sides of the rectangle. Such smaller subsets can thus be taken as $\mathcal{H}_0$ and $\mathcal{H}_1$. Up to a homeomorphism between $\mathcal{A}$ and $[p_1, p_2] \times [q_1, q_2]$, this is precisely what happens in $(\mathcal{A}, \mathcal{A}^-)$ for $\psi \circ \varphi$.

(H_φ) *There are $m \geq 1$ pairwise disjoint compact sets $\mathcal{H}_0, \dots, \mathcal{H}_{m-1} \subseteq \mathcal{A} \cap D_\varphi$ such that $(\mathcal{H}_i, \varphi)\colon \widetilde{\mathcal{A}} \Leftrightarrow \widetilde{\mathcal{B}}$, for $i = 0, \dots, m-1$;*

(H_ψ) *There exist $l \geq 1$ pairwise disjoint compact sets $\mathcal{K}_0, \dots, \mathcal{K}_{l-1} \subseteq \mathcal{B} \cap D_\psi$ such that $(\mathcal{K}_i, \psi)\colon \widetilde{\mathcal{B}} \Leftrightarrow \widetilde{\mathcal{A}}$, for $i = 0, \dots, l-1$.*

If at least one between m and l is greater or equal than 2, then the composite map $\phi := \psi \circ \varphi$ induces chaotic dynamics on $m \times l$ symbols in the set

$$\mathcal{H}^* := \bigcup_{i,j} \mathcal{H}'_{i,j}, \quad \textit{with } \mathcal{H}'_{i,j} := \mathcal{H}_i \cap \varphi^{-1}(\mathcal{K}_j)\,, i = 0, \dots, m-1,\ j = 0, \dots, l-1\,,$$

and thus satisfies properties (i)–(v) *from Theorem 7.2.*

Proof. In order to get the thesis, it suffices to show that

$$(\mathcal{H}'_{i,j}, \phi)\colon \widetilde{\mathcal{A}} \Leftrightarrow \widetilde{\mathcal{A}}\,, \quad \forall i = 0, \dots, m-1 \text{ and } j = 0, \dots, l-1\,.$$

Such a condition can be checked by steps analogous to the ones in Theorem 10.1. The details are omitted since they are straightforward. □

As we shall see, this latter result will find an application in Section 12 when dealing with a periodic version of the Volterra predator–prey model.

11 Examples from the ODEs

In the present section, we apply the results from Section 10 to some nonlinear ODE models with periodic coefficients. In particular, we deal with planar systems of the first order or with second order scalar equations: since any second order equation can be written as a system of two equations of the first order, in this introductory discussion, we will generally refer to planar systems.

The systems we consider are studied through a combination of a careful, but elementary, phase-plane analysis with the results on chaotic dynamics for LTMs from Section 10. We recall that a similar strategy has already been employed in [122, 178] with respect to second order equations. More precisely, we prove in a rigorous way (i. e. without the need of computer assistance, differently from several works on related topics [15, 17, 57, 58, 97, 108, 130, 172, 176, 177, 183]) the presence of infinitely many periodic solutions for our systems as well as of a chaotic behavior in the sense of Definition 7.2 for the associated Poincaré map Ψ. In fact, a classical approach ([81]) to show the existence of periodic solutions (harmonics or subharmonics) of nonautonomous differential systems like

$$\dot{\zeta} = f(t, \zeta)\,, \tag{11.1}$$

where $f\colon \mathbb{R} \times \mathbb{R}^N \to \mathbb{R}^N$ is a continuous vector field which is T-periodic in the time-variable, that is, $f(t+T, z) = f(t, z)$, $\forall (t, z) \in \mathbb{R} \times \mathbb{R}^N$, under the assumption of uniqueness of the solutions for the Cauchy problems, is based on the search of the fixed points for the Poincaré map $\Psi = \Psi_T$ or for its iterates, where

$$\Psi\colon z \mapsto \zeta(t_0 + T; t_0, z)$$

and $\zeta(\cdot\,; t_0, z)$ is the solution of (11.1), satisfying the initial condition $\zeta(t_0) = z \in \mathbb{R}^N$. As a consequence of the fundamental theory of ODEs, it turns out that Ψ is a homeomorphism of its domain (which is an open subset of $\mathbb{R}^N$) onto its image. Applying our "Stretching Along the Paths" method to Ψ, we are led back to work with discrete dynamical systems. The difference with the examples for discrete dynamical systems coming from the iteration of a continuous map (not necessarily invertible) is that since the Poincaré map is a homeomorphism, it is possible to prove a semi-conjugacy to the two-sided Bernoulli shift (cf. Remark 7.1), while we recall that the controlling functions of the models like those dealing with the discrete logistic map are not injective.

We notice that the kind of chaos in Definition 7.2, when considered in relation to the case of the Poincaré map, looks similar to other ones detected in the literature on complex dynamics for ODEs with periodic coefficients (for instance, [37, 158]).

In more detail, in Section 12 we present an application of the results on LTMs from Section 10 to a modified version of a Volterra predator–prey model in which a periodic harvesting is included [127]. Indeed, when the seasons with fishing alternate with the ones without harvesting in a periodic fashion, it is possible to prove the presence of chaotic features for the system, provided that we have the freedom to tune the switching times between the two regimes. Analogous conclusions could be drawn for those time-periodic planar Kolmogorov systems [78]

$$x' = X(t, x, y)\,, \quad y' = Y(t, x, y)$$

which possess dynamical features similar to the ones of Volterra model.

12 Predator–prey model

12.1 The effects of a periodic harvesting

The classical Volterra predator–prey model concerns the first order planar differential system

$$\begin{cases} x' = x(a - by) \\ y' = y(-c + dx)\,, \end{cases} \tag{E_0}$$

where $a, b, c, d > 0$ are constant coefficients. The study of system (E_0) is confined to the open first quadrant $(\mathbb{R}_0^+)^2$ of the plane since $x(t) > 0$ and $y(t) > 0$ represent the size (number of individuals or density) of the prey and the predator populations, respectively. Such a model was proposed by Vito Volterra in 1926, answering D'Ancona's question about the percentage of selachians and fish caught in the northern Adriatic Sea during a range of years covering the period of the World War I (refer to [29, 103] for a more detailed historical account).

System (E_0) is conservative and its phase-portrait is that of a global center at the point

$$P_0 := \left(\frac{c}{d}, \frac{a}{b}\right),$$

surrounded by periodic orbits (run in the counterclockwise sense) which are the level lines of the first integral

$$\mathcal{E}_0(x, y) := dx - c \log x + by - a \log y$$

that we call "energy" in analogy to mechanical systems. The choice of the sign in the definition of the first integral implies that $\mathcal{E}_0(x, y)$ achieves a strict absolute minimum at the point P_0.

According to Volterra's analysis of (E_0), the average of a periodic solution $(x(t), y(t))$, evaluated over a time-interval corresponding to its natural period, coincides with the coordinates of the point P_0.

In order to include the effects of fishing in the model, one can suppose that during the harvesting time, both the prey and the predator populations are reduced at a rate proportional to the size of the population itself. This assumption leads to the new system

$$\begin{cases} x' = x(a_\mu - by) \\ y' = y(-c_\mu + dx), \end{cases} \tag{E_μ}$$

where

$$a_\mu := a - \mu \quad \text{and} \quad c_\mu := c + \mu$$

are the modified growth coefficients which take into account the fishing rates $-\mu x(t)$ and $-\mu y(t)$, respectively. The parameter μ is assumed to be positive, but small enough $(\mu < a)$ in order to prevent the extinction of the populations. System (E_μ) has the same form like (E_0) and hence its phase-portrait is that of a global center at

$$P_\mu := \left(\frac{c + \mu}{d}, \frac{a - \mu}{b}\right).$$

The periodic orbits surrounding P_μ are the level lines of the first integral

$$\mathcal{E}_\mu(x, y) := dx - c_\mu \log x + by - a_\mu \log y.$$

The coordinates of P_μ coincide with the average values of the prey and the predator populations under the effect of fishing (Figure 19). A comparison between the coordinates of P_0 and P_μ motivates the conclusion of () Volterra's principle that a moderate harvesting has a favorable effect for the prey population [29].

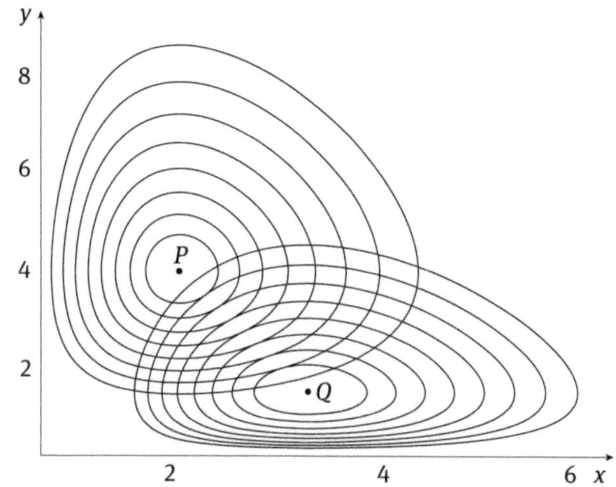

Fig. 19: In this picture, we show some periodic orbits of the Volterra system (E_0) with a center at $P = P_0$, as well as of the perturbed system (E_μ) with a center at $Q = P_\mu$ (for a certain $\mu \in]0, a[$).

If we are interested in incorporating the consequences of a cyclic environment in the Volterra original model (E_0), we can assume a seasonal effect on the coefficients, which leads us to consider a system of the form

$$\begin{cases} x' = x(a(t) - b(t)y) \\ y' = y(-c(t) + d(t)x), \end{cases} \tag{E}$$

where $a(\cdot), b(\cdot), c(\cdot), d(\cdot)\colon \mathbb{R} \to \mathbb{R}$ are periodic functions with a common period $T > 0$. In such a framework, it is natural to look for harmonic (i. e. T-periodic) or m-th order subharmonic (i. e. mT-periodic, for some integer $m \geq 2$, with mT the minimal period in the set $\{jT\colon j = 1, 2, \dots\}$) solutions with range in the open first quadrant (*positive solutions*).

The past forty years have witnessed a growing attention towards such models and several results have been obtained regarding the existence, multiplicity and stability of for Lotka–Volterra type predator–prey systems with periodic coefficients [6, 34, 36, 43, 49, 50, 63, 64, 96, 164]. Further references are available in [127], where the reader can also find historical details about the fortune of this model as well as some related results from [49, 50].

Let us come back for a moment to the original Volterra system with constant coefficients and suppose that the interaction between the two populations is governed by system (E_0) for a certain period of the season (corresponding to a time-interval of length r_0) and by system (E_μ) for the rest of the season (corresponding to a time-interval of length r_μ). Assume also that such alternation between (E_0) and (E_μ) occurs in a periodic fashion, and thus,

$$T := r_0 + r_\mu$$

is the period of the season. In other terms, first we consider system (E_0) for $t \in [0, r_0[$. Next, we switch to system (E_μ) at time r_0 and assume that (E_μ) rules the dynamics for $t \in [r_0, T[$. Finally, we suppose that we switch back to system (E_0) at time $t = T$ and repeat the cycle with T-periodicity.

Such two-state alternating behavior can be equivalently described in terms of system (E) by assuming

$$a(t) = \hat{a}_\mu(t) := \begin{cases} a & \text{for } 0 \le t < r_0\,, \\ a - \mu & \text{for } r_0 \le t < T\,, \end{cases}$$

$$c(t) = \hat{c}_\mu(t) := \begin{cases} c & \text{for } 0 \le t < r_0\,, \\ c + \mu & \text{for } r_0 \le t < T\,, \end{cases}$$

as well as

$$b(t) \equiv b\,, \quad d(t) \equiv d\,,$$

with a, b, c, d positive constants and μ a parameter with $0 < \mu < a$. Hence, we can consider the system

$$\begin{cases} x' = x(\hat{a}_\mu(t) - by) \\ y' = y(-\hat{c}_\mu(t) + dx)\,, \end{cases} \tag{E^*}$$

where the piecewise constant functions $\hat{a}_\mu$ and $\hat{c}_\mu$ are supposed to be extended to the whole real line by T-periodicity.

It is our aim now to prove that (E^*) generates chaotic dynamics in the sense of Definition 7.2. To this end, as explained at the beginning of Section 11, we apply the results on LTMs from Section 10 to the Poincaré map

$$\Psi\colon (\mathbb{R}_0^+)^2 \to (\mathbb{R}_0^+)^2\,, \quad \Psi(z) := \zeta(T, z)\,,$$

where $\zeta(\cdot, z) = (x(\cdot, z), y(\cdot, z))$ is the solution of system (E^*) starting from $z = (x_0, y_0) \in (\mathbb{R}_0^+)^2$ at the time $t = 0$. As a consequence, we will have ensured all the chaotic features listed in Theorem 7.3 (like, for instance, a semiconjugacy to the two-sided Bernoulli shift and thus positive topological entropy, sensitivity with respect to initial conditions, topological transitivity, a compact invariant set containing a dense subset of periodic points).

With the aid of Figure 20, we try to explain how to enter the setting of Theorem 10.2 for the switching system (E^*).

As a first step, we take two closed overlapping annuli consisting of level lines of the first integrals associated to system (E_0) and (E_μ), respectively. In particular, the inner and outer boundaries of each annulus are closed trajectories surrounding the

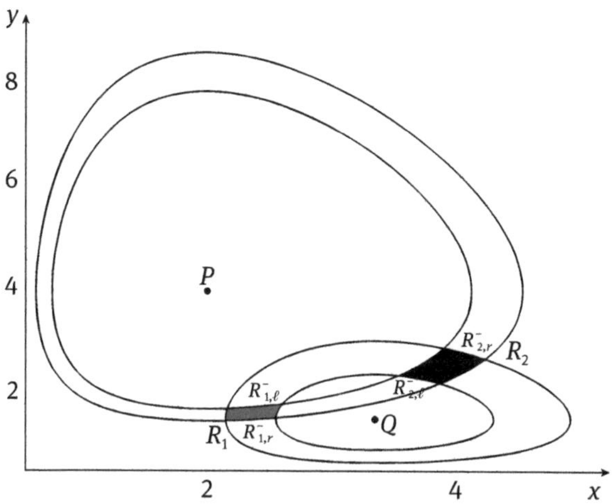

Fig. 20: The two annular regions $\mathcal{A}_P$ and $\mathcal{A}_Q$ (centered at P and Q, respectively) are linked together. We have drawn with a darker color the two rectangular sets $\mathcal{R}_1$ and $\mathcal{R}_2$ where they meet.

equilibrium point ($P = P_0$ for system (E_0) and $Q = P_\mu$ for system (E_μ)). Such annuli, that we will from now on call $\mathcal{A}_P$ and $\mathcal{A}_Q$, intersect in two compact disjoint sets $\mathcal{R}_1$ and $\mathcal{R}_2$, which are generalized rectangles. The way in which we label the two regions (as $\mathcal{R}_1/\mathcal{R}_2$) is completely arbitrary. However, the choice of an order will effect some details in the argument that we describe below. Whenever we enter a framework like that visualized in Figure 20, we say that the annuli $\mathcal{A}_P$ and $\mathcal{A}_Q$ are *linked together*. Technical conditions on the energy level lines defining $\mathcal{A}_P$ and $\mathcal{A}_Q$, sufficient to guarantee the linking between them, are presented in Section 12.2.

As a second step, we give an "orientation" to $\mathcal{R}_i$ (for $i = 1, 2$) by selecting the boundary sets $\mathcal{R}_i^- = \mathcal{R}_{i,\ell}^- \cup \mathcal{R}_{i,r}^-$. In the specific example of Figure 20, we take, as $\mathcal{R}_1^-$, the intersection of $\mathcal{R}_1$ with the inner and outer boundaries of $\mathcal{A}_P$ and, as $\mathcal{R}_2^-$, the intersection of $\mathcal{R}_2$ with the inner and outer boundaries of $\mathcal{A}_Q$. The way in which we name (as left/right) the two components of $\mathcal{R}_i^-$ is inessential for the rest of the discussion. Just to fix ideas, let us say that we choose, as $\mathcal{R}_{1,\ell}^-$, the component of $\mathcal{R}_1^-$ which is closer to P and, as $\mathcal{R}_{2,\ell}^-$, the component of $\mathcal{R}_2^-$ which is closer to Q (of course, the "right" components $\mathcal{R}_{1,r}^-$ and $\mathcal{R}_{2,r}^-$ are the remaining ones).

As a third step, we observe that the Poincaré map associated to (E^*) can be decomposed as

$$\Psi = \Psi_\mu \circ \Psi_0 ,$$

where Ψ_0 is the Poincaré map of system (E_0) on the time-interval $[0, r_0]$ and Ψ_μ is the Poincaré map for (E_μ) on the time-interval $[0, r_\mu] = [0, T - r_0]$. Consider a path $\gamma\colon [0,1] \to \mathcal{R}_1$ with $\gamma(0) \in \mathcal{R}_{1,\ell}^-$ and $\gamma(1) \in \mathcal{R}_{1,r}^-$. As we will see in Section 12.2, the points of $\mathcal{R}^-_{1,\ell}$ move faster than those belonging to $\mathcal{R}^-_{1,r}$ under the action of system (E_0). Hence, for a choice of the first switching time r_0 that is large enough, it is possible to make the path

$$[0,1] \ni s \mapsto \Psi_0(\gamma(s))$$

turn in a spiral-like fashion inside the annulus $\mathcal{A}_P$ and cross at least twice the rectangular region $\mathcal{R}_2$ from $\mathcal{R}^-_{2,\ell}$ to $\mathcal{R}^-_{2,r}$. Thus, we can select two subintervals of $[0,1]$ such that $\Psi_0 \circ \gamma$ restricted to each of them is a path contained in $\mathcal{R}_2$ and connecting the two components of $\mathcal{R}^-_2$. We observe that the points of $\mathcal{R}^-_{2,\ell}$ move faster than those belonging to $\mathcal{R}^-_{2,r}$ under the action of system (E_μ). Therefore, we can repeat the same argument as above and conclude that, for a suitable choice of $r_\mu = T - r_0$ sufficiently large, we can transform, via Ψ_μ, any path in $\mathcal{R}_2$ joining the two components of $\mathcal{R}^-_2$ onto a path which crosses at least once $\mathcal{R}_1$ from $\mathcal{R}^-_{1,\ell}$ to $\mathcal{R}^-_{1,r}$.

As a final step, we complete the proof about the existence of chaotic-like dynamics by applying Theorem 10.1. Actually, in order to obtain a more complex behavior, in place of Theorem 10.1, it is possible to employ the more general Theorem 10.2.

In fact, our main result can be stated as follows.

Theorem 12.1. *For any choice of positive constants a, b, c, d, μ with $\mu < a$ and for every pair $(\mathcal{A}_P, \mathcal{A}_Q)$ of linked together annuli, the following conclusion holds:*

For every integer $m \geq 2$, there exist two positive constants α and β such that for each

$$r_0 > \alpha \quad \textit{and} \quad r_\mu > \beta\,,$$

the Poincaré map associated to system (E^*) induces chaotic dynamics on m symbols in $\mathcal{R}_1$ and $\mathcal{R}_2$.

The proof will be performed only for the first rectangular region. The proof for the second region is analogous.

As remarked in [120], if we consider Definition 7.2 and its consequences in the context of concrete examples of ODEs (for instance, when ψ turns out to be the Poincaré map), condition (7.2) may sometimes be interpreted in terms of the oscillatory behavior of the solutions. Such a situation occurred in [117, 121] and takes place also for system (E^*). Indeed, as it will be clear from the proof of Theorem 12.1, it is possible to draw more precise conclusions in the statement of our main result. Namely, the following additional properties can be obtained:

For every decomposition of the integer $m \geq 2$ as

$$m = m_1 m_2\,, \quad \textit{with } m_1, m_2 \in \mathbb{N}\,,$$

there exist integers $\kappa_1, \kappa_2 \geq 1$ (with $\kappa_1 = \kappa_1(r_0, m_1)$ and $\kappa_2 = \kappa_2(r_\mu, m_2)$) such that, for each two-sided sequence of symbols

$$\mathbf{s} = (s_i)_{i\in\mathbb{Z}} = (p_i, q_i)_{i\in\mathbb{Z}} \in \{0, \ldots, m_1 - 1\}^{\mathbb{Z}} \times \{0, \ldots, m_2 - 1\}^{\mathbb{Z}}\,,$$

there exists a solution

$$\zeta_{\mathbf{s}}(\cdot) = (x_{\mathbf{s}}(\cdot), y_{\mathbf{s}}(\cdot))$$

of (E^*) *with* $\zeta_{\mathbf{s}}(0) \in \mathcal{R}_1$ *such that* $\zeta_{\mathbf{s}}(t)$ *crosses* $\mathcal{R}_2$ *exactly* $\kappa_1 + p_i + 1$ *times for* $t \in]iT, r_0 + iT[$ *and crosses* $\mathcal{R}_1$ *exactly* $\kappa_2 + q_i + 1$ *times for* $t \in]r_0 + iT, (i+1)T[$. *Moreover, if* $(s_i)_{i\in\mathbb{Z}} = (p_i, q_i)_{i\in\mathbb{Z}}$ *is a periodic sequence, that is,* $s_{i+k} = s_i$, *for some* $k \geq 1$, *then* $\zeta_{\mathbf{s}}(t + kT) = \zeta_{\mathbf{s}}(t)$, $\forall t \in \mathbb{R}$.

In particular, taking $m = m_1 \geq 2$ *and* $m_2 = 1$, *we obtain the dynamics on the set of* m *symbols* $\{0, \dots, m-1\} \equiv \{0, \dots, m-1\} \times \{0\}$ *as in the statement of Theorem* 12.1.

The solutions $(x(t), y(t))$ are meant in the Carathéodory sense, i. e. $(x(t), y(t))$ is absolutely continuous and satisfies system (E^*) for almost every $t \in \mathbb{R}$. Of course, such solutions are of class C^1 if the coefficients are continuous.

The constants α and β in Theorem 12.1, representing the lower bounds for r_0 and r_μ, can be estimated in terms of m_1 and m_2 and other geometric parameters such as the fundamental periods of the orbits bounding the linked annuli ((12.1) and (12.4)).

We end this introductory discussion with a few observations about our main result.

First of all, we notice that, according to Theorem 12.1, there is an abundance of chaotic regimes for system (E^*), provided that the time-interval lengths r_0 and r_μ (and, consequently, the period T) are sufficiently large. More precisely, we are able to prove the existence of chaotic invariant sets inside each intersection of two annular regions linked together. One could conjecture the presence of Smale horseshoes contained in such intersections, like in the classical case of linked twist maps with circular domains [46], even if we recall that our purely topological approach just requires one to check a twist hypothesis on the boundary, without the need of verifying any hyperbolicity condition. This does not prevent the possibility of a further deeper analysis using more complex computations.

We also stress that our result is stable with respect to small perturbations of the coefficients. In fact, as it will emerge from the proof, whenever $r_0 > \alpha$ and $r_\mu > \beta$ are chosen so as to achieve the conclusion of Theorem 12.1, it follows that there exists a constant $\varepsilon > 0$ such that Theorem 12.1 applies to equation (E) too, provided that

$$\int_0^T |a(t) - \hat{a}_\mu(t)|\, dt < \varepsilon\,, \quad \int_0^T |c(t) - \hat{c}_\mu(t)|\, dt < \varepsilon\,,$$

$$\int_0^T |b(t) - b|\, dt < \varepsilon\,, \quad \int_0^T |d(t) - d|\, dt < \varepsilon\,.$$

Here, the T-periodic coefficients may be in $L^1([0, T])$ or even continuous or smooth functions, possibly of class C^∞.

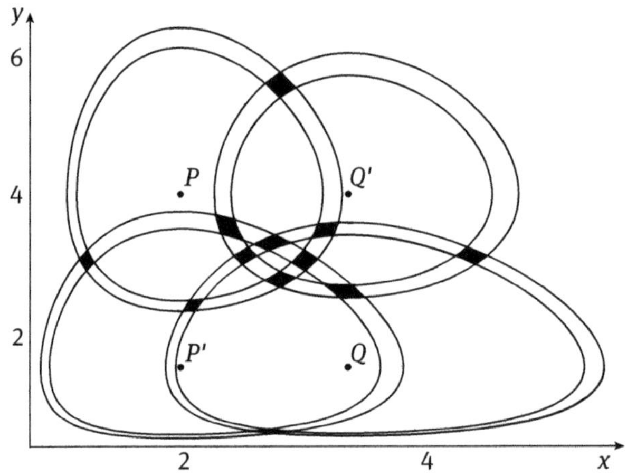

Fig. 21: We have depicted four linked annular regions bounded by energy level lines corresponding to Volterra systems with centers at $P = P_0$, $P' = (c/d, a_\mu/b)$, $Q = P_\mu$ and $Q' = (c_\mu/d, a/b)$ by putting the regions of mutual intersection in evidence, where it is possible to locate the chaotic invariant sets.

A final remark concerns the fact that, in our model, we have assumed that the harvesting period starts and ends for both the species at the same moment. With this respect, one could face a more general situation in which some phase-shift between the two harvesting intervals occurs. Such cases have been already explored in some biological models, mostly from a numerical point of view: see [112] for an example on competing species and [137] for a predator–prey system. If we assume a phase-shift in the periodic coefficients, that is, if we consider

$$a(t) := \hat{a}_\mu(t - \theta_1) \quad \text{and} \quad c(t) := \hat{c}_\mu(t - \theta_2)\,,$$

for some $0 < \theta_1, \theta_2 < T$, and we also suppose that the length r_0 of the time-intervals without harvesting may differ for the two species (say $r_0 = r_a \in\,]0, T[$ in the definition of $\hat{a}_\mu$ and $r_0 = r_c \in\,]0, T[$ in the definition of $\hat{c}_\mu$), then the geometry of our problem turns out to be a combination of linked twist maps on two, three or four annuli, which are mutually linked together. In this manner, we increase the possibility of chaotic configurations, provided that the system is subject to the different regimes for a sufficiently long time. For a pictorial comment, see Figure 21, where all the possible links among four annuli are realized.

12.2 Technical details and proofs

Let us consider system (E_0) and let

$$\ell > \chi_0 := \mathcal{E}_0(P_0) = \min\{\mathcal{E}_0(x,y) \colon x > 0, y > 0\}\,.$$

The level line

$$\Gamma_0(\ell) := \left\{(x,y) \in (\mathbb{R}_0^+)^2 : \mathcal{E}_0(x,y) = \ell\right\}$$

is a closed orbit (surrounding P_0) which is run counterclockwise, completing one turn in a fundamental period that we denote by $\tau_0(\ell)$. According to classical results on the period of the Lotka–Volterra system [141, 166], the map

$$\tau_0 : \,]\chi_0, +\infty[\, \to \mathbb{R}$$

is strictly increasing with $\tau_0(+\infty) = +\infty$ and satisfies

$$\lim_{\ell \to \chi_0^+} \tau_0(\ell) = T_0 := \frac{2\pi}{\sqrt{ac}}\,.$$

Similarly, considering system (E_μ) with $0 < \mu < a$, we denote by $\tau_\mu(h)$ the minimal period associated to the orbit

$$\Gamma_\mu(h) := \left\{(x,y) \in (\mathbb{R}_0^+)^2 : \mathcal{E}_\mu(x,y) = h\right\}\,,$$

for

$$h > \chi_\mu := \mathcal{E}_\mu(P_\mu) = \min\{\mathcal{E}_\mu(x,y) \colon x > 0, y > 0\}\,.$$

Also, in this case, the map $h \mapsto \tau_\mu(h)$ is strictly increasing with $\tau_\mu(+\infty) = +\infty$ and

$$\lim_{h \to \chi_\mu^+} \tau_\mu(h) = T_\mu := \frac{2\pi}{\sqrt{a_\mu c_\mu}}\,.$$

Before giving the details for the proof of our main result, we describe conditions on the energy level lines of two annuli, $\mathcal{A}_P$ and $\mathcal{A}_Q$, centered at $P = P_0 = (\frac{c}{d}, \frac{a}{b})$ and $Q = P_\mu = (\frac{c+\mu}{d}, \frac{a-\mu}{b})$, respectively, sufficient to ensure that they are linked together. With this respect, we have to consider the intersections among the closed orbits around the two equilibria and the straight line r passing through the points P and Q, whose equation is $by + dx - a - c = 0$. We introduce an orientation on such a line by defining an order "$\preceq$" among its points. More precisely, we set $A \preceq B$ (resp. $A \prec B$) if and only if $x_A \le x_B$ (resp. $x_A < x_B$), where $A = (x_A, y_A)$, $B = (x_B, y_B)$. In this manner, the order on r is that inherited from the oriented x-axis by projecting the points of r onto the abscissa. Assume now that we have two closed orbits $\Gamma_0(\ell_1)$ and $\Gamma_0(\ell_2)$ for system (E_0), with $\chi_0 < \ell_1 < \ell_2$. Let us call the intersection points among r and such level lines $P_{1,-}, P_{1,+}$, with reference to ℓ_1, and $P_{2,-}, P_{2,+}$, with reference to ℓ_2, where

$$P_{2,-} \prec P_{1,-} \prec P \prec P_{1,+} \prec P_{2,+}\,.$$

Analogously, when we consider two orbits $\Gamma_\mu(h_1)$ and $\Gamma_\mu(h_2)$ for system (E_μ), with $\chi_\mu < h_1 < h_2$, we name the intersection points among r and these level lines $Q_{1,-}, Q_{1,+}$, with reference to h_1, and $Q_{2,-}, Q_{2,+}$, with reference to h_2, where

$$Q_{2,-} \prec Q_{1,-} \prec Q \prec Q_{1,+} \prec Q_{2,+} .$$

Then, the two annuli $\mathcal{A}_P$ and $\mathcal{A}_Q$ turn out to be linked together if

$$P_{2,-} \prec P_{1,-} \preceq Q_{2,-} \prec Q_{1,-} \preceq P_{1,+} \prec P_{2,+} \preceq Q_{1,+} \prec Q_{2,+} .$$

Proof of Theorem 12.1. Consistently with the notation introduced above, we denote, by $\Gamma_0(\ell)$, with $\ell \in [\ell_1, \ell_2]$ for some $\chi_0 < \ell_1 < \ell_2$ the level lines filling $\mathcal{A}_P$, and thus

$$\mathcal{A}_P = \bigcup_{\ell_1 \le \ell \le \ell_2} \Gamma_0(\ell) .$$

Analogously, we indicate the level lines filling $\mathcal{A}_Q$ by $\Gamma_\mu(h)$, with $h \in [h_1, h_2]$ for some $\chi_\mu < h_1 < h_2$, and thus we can write

$$\mathcal{A}_Q = \bigcup_{h_1 \le h \le h_2} \Gamma_\mu(h) .$$

By construction, such annular regions turn out to be invariant for the dynamical systems generated by (E_0) and (E_μ), respectively. We now consider the two regions in which each annulus is cut by the line r (passing through P and Q) and we call such sets $\mathcal{A}_P^t$, $\mathcal{A}_P^b$, $\mathcal{A}_Q^t$ and $\mathcal{A}_Q^b$ in order to have $\mathcal{A}_P = \mathcal{A}_P^t \cup \mathcal{A}_P^b$ and $\mathcal{A}_Q = \mathcal{A}_Q^t \cup \mathcal{A}_Q^b$, where the sets with superscript t are the "upper" ones and the sets with superscript b are the "lower" ones, with respect to the line r. We name $\mathcal{R}^b$ the rectangular region in which $\mathcal{A}_P^b$ and $\mathcal{A}_Q^b$ meet and analogously we denote, by $\mathcal{R}^t$, the rectangular region belonging to the intersection between $\mathcal{A}_P^t$ and $\mathcal{A}_Q^t$ (Figure 22).

Let

$$m_1 \ge 2 \quad \text{and} \quad m_2 \ge 1$$

be two fixed integers. The case $m_1 = 1$ and $m_2 \ge 2$ can be treated in a similar manner and therefore is omitted.

As a first step, we are interested in the solutions of system (E_0) starting from $\mathcal{A}_P^b$ and crossing $\mathcal{A}_P^t$ at least m_1 times. After having performed the rototranslation of the plane $\mathbb{R}^2$ that brings the origin to the point P and makes the x-axis coincide with the line r, whose equations are

$$\begin{cases} \tilde{x} = (x - \frac{c}{d}) \cos \vartheta + (y - \frac{a}{b}) \sin \vartheta \\ \tilde{y} = (\frac{c}{d} - x) \sin \vartheta + (y - \frac{a}{b}) \cos \vartheta, \end{cases}$$

where $\vartheta := \arctan(\frac{d}{b})$, it is possible to use the Prüfer transformation and introduce generalized polar coordinates. Therefore, we can express the solution $\zeta(\cdot, z) =$

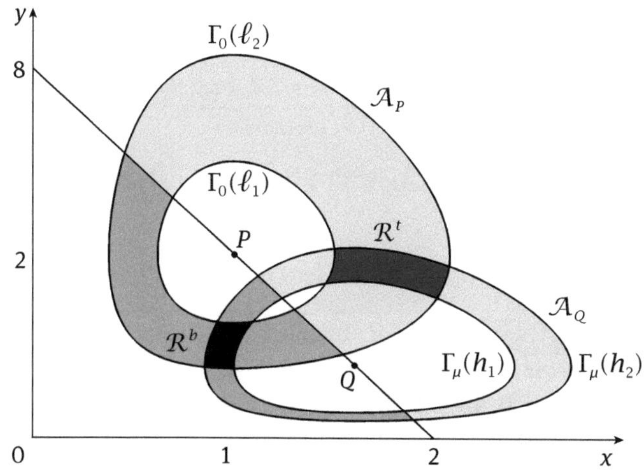

Fig. 22: For the two linked annuli in the picture (one around P and the other around Q), we have drawn in different colors the upper and lower parts (with respect to the line r), as well as the intersection regions $\mathcal{R}^t$ and $\mathcal{R}^b$ between them. As a guideline for the proof, we recall that $\mathcal{A}_P$ is the annulus around P, having as inner and outer boundaries the energy level lines $\Gamma_0(\ell_1)$ and $\Gamma_0(\ell_2)$. Similarly, $\mathcal{A}_Q$ is the annulus around Q, having as inner and outer boundaries the energy level lines $\Gamma_\mu(h_1)$ and $\Gamma_\mu(h_2)$.

$(x(\cdot,z),y(\cdot,z))$ of system (E_0) with initial point in $z=(x_0,y_0)\in\mathcal{A}_P^b$ through the radial coordinate $\rho(t,z)$ and the angular coordinate $\theta(t,z)$. Thus, we can assume that $\theta(0,z)\in[-\pi,0]$. For any $t\in[0,\tau_0]$ and $z\in\mathcal{A}_P^b$, let us also introduce the *rotation number*, that is, the quantity

$$\mathrm{rot}_0(t,z):=\frac{\theta(t,z)-\theta(0,z)}{2\pi},$$

that indicates the normalized angular displacement along the orbit of system (E_0) starting at z during the time-interval $[0,t]$. The continuous dependence of the solutions from the initial data implies that the function $(t,z)\mapsto\theta(t,z)$ and consequently the map $(t,z)\mapsto\mathrm{rot}_0(t,z)$ are continuous. From the definition of the rotation number and the star-shapedness of the level lines of $\mathcal{E}_0$ with respect to the point P, for every $z\in\Gamma_0(\ell)$, the following properties hold:

$$\begin{aligned}
&\forall\, j\in\mathbb{Z}:\ \mathrm{rot}_0(t,z)=j &&\Longleftrightarrow\quad t=j\,\tau_0(\ell)\,,\\
&\forall\, j\in\mathbb{Z}: j<\mathrm{rot}_0(t,z)<j+1 &&\Longleftrightarrow\quad j\,\tau_0(\ell)<t<(j+1)\,\tau_0(\ell)
\end{aligned}$$

(if the annuli were not star-shaped, the inference "$\Longleftarrow$" would still be true). Although we have implicitly assumed that $\ell_1\le\ell\le\ell_2$, such properties hold for every $\ell>\chi_0$.

Observe that, thanks to the fact that the time-map τ_0 is strictly increasing, we know that $\tau_0(\ell_1)<\tau_0(\ell_2)$. We shall use this condition to show that a twist property for the rotation number holds for sufficiently large time-intervals. Indeed, we claim that if we choose a switching time $r_0 \geq \alpha$, where

$$\alpha := \frac{(m_1 + 3 + \frac{1}{2})\,\tau_0(\ell_1)\,\tau_0(\ell_2)}{\tau_0(\ell_2) - \tau_0(\ell_1)}\,, \tag{12.1}$$

then, for any path $\gamma\colon [0,1] \to \mathcal{A}_P$ with $\gamma(0) \in \Gamma_0(\ell_1)$ and $\gamma(1) \in \Gamma_0(\ell_2)$, the interval inclusion

$$[\theta(r_0,\gamma(1)), \theta(r_0,\gamma(0))] \supseteq [2\pi n^*, 2\pi(n^* + m_1) - \pi] \tag{12.2}$$

is fulfilled for some $n^* = n^*(r_0) \in \mathbb{N}$. To check our claim, at first we notice that for a path $\gamma(s)$ as above, it holds that $\mathrm{rot}_0(t,\gamma(0)) \geq \lfloor t/\tau_0(\ell_1)\rfloor$ and $\mathrm{rot}_0(t,\gamma(1)) \leq \lceil t/\tau_0(\ell_2)\rceil$ for every $t > 0$, and so

$$\mathrm{rot}_0(t,\gamma(0)) - \mathrm{rot}_0(t,\gamma(1)) > t\,\frac{\tau_0(\ell_2) - \tau_0(\ell_1)}{\tau_0(\ell_1)\,\tau_0(\ell_2)} - 2\,, \quad \forall\, t > 0\,.$$

Hence, for $t \geq \alpha$, with α defined as in (12.1), we obtain

$$\mathrm{rot}_0(t,\gamma(0)) > m_1 + 1 + \tfrac{1}{2} + \mathrm{rot}_0(t,\gamma(1))\,,$$

which, in turns, implies

$$\theta(t,\gamma(0)) - \theta(t,\gamma(1)) > 2\pi(m_1 + 1)\,, \quad \forall t \geq \alpha\,.$$

Therefore, recalling the bound $2\pi(\lceil t/\tau_0(\ell_2)\rceil - \frac{3}{2}) < \theta(t,\gamma(1)) \leq 2\pi\lceil t/\tau_0(\ell_2)\rceil$, the interval inclusion (12.2) is achieved for

$$n^* = n^*(r_0) := \left\lceil \frac{r_0}{\tau_0(\ell_2)} \right\rceil.$$

This proves our claim.

By the continuity of the composite mapping $[0,1] \ni s \mapsto \mathrm{rot}_0(r_0,\gamma(s))$, it follows that

$$\{\theta(r_0,\gamma(s)),\ s \in [0,1]\} \supseteq [2\pi n^*, 2\pi(n^* + m_1 - 1) + \pi]\,.$$

As a consequence, by the Bolzano Theorem, there exist m_1 pairwise disjoint maximal intervals $[t_i', t_i''] \subseteq [0,1]$, for $i = 0,\ldots,m_1 - 1$ such that

$$\theta(r_0,\gamma(s)) \in [2\pi n^* + 2\pi i, 2\pi n^* + \pi + 2\pi i],\ \forall s \in [t_i', t_i''],\ i = 0,\ldots,m_1 - 1\,,$$

with $\theta(r_0,\gamma(t_i')) = 2\pi n^* + 2\pi i$ and $\theta(r_0,\gamma(t_i'')) = 2\pi n^* + \pi + 2\pi i$. Setting

$$\mathcal{R}_1 := \mathcal{R}^b \quad \text{and} \quad \mathcal{R}_2 := \mathcal{R}^t\,,$$

we orientate such rectangular regions by choosing

$$\mathcal{R}^-_{1,\ell} := \mathcal{R}_1 \cap \Gamma_0(\ell_1) \quad \text{and} \quad \mathcal{R}^-_{1,r} := \mathcal{R}_1 \cap \Gamma_0(\ell_2)$$

as well as

$$\mathcal{R}_{2,\ell}^{-} := \mathcal{R}_2 \cap \Gamma_\mu(h_1) \quad \text{and} \quad \mathcal{R}_{2,r}^{-} := \mathcal{R}_2 \cap \Gamma_\mu(h_2)\,.$$

See the caption of Figure 22 as a reminder for the corresponding sets.

Introducing, at last, the m_1 nonempty and pairwise disjoint compact sets

$$\mathcal{H}_i := \{z \in \mathcal{A}_P^b \colon \theta(r_0, z) \in [2\pi n^* + 2\pi i, 2\pi n^* + \pi + 2\pi i]\}\,,$$

for $i = 0, \dots, m_1 - 1$, we are ready to prove that

$$(\mathcal{H}_i, \Psi_0) \colon \widetilde{\mathcal{R}}_1 \Leftrightarrow\!\!\!\rightarrow \widetilde{\mathcal{R}}_2\,, \quad \forall i = 0, \dots, m_1 - 1\,, \tag{12.3}$$

where we recall that Ψ_0 is the Poincaré map associated to system (E_0). Indeed, let us take a path $\gamma\colon [0,1] \to \mathcal{R}_1$ with $\gamma(0) \in \mathcal{R}_{1,\ell}^{-}$ and $\gamma(1) \in \mathcal{R}_{1,r}^{-}$. For $r_0 \geq \alpha$ and fixing $i \in \{0, \dots, m_1 - 1\}$, there is a subinterval $[t_i', t_i''] \subseteq [0,1]$ such that $\gamma(t) \in \mathcal{H}_i$ and $\Psi_0(\gamma(t)) \in \mathcal{A}_P^t$, $\quad \forall t \in [t_i', t_i'']$. Noting that $\Gamma_\mu(\Psi_0(\gamma(t_i'))) \leq h_1$ and $\Gamma_\mu(\Psi_0(\gamma(t_i''))) \geq h_2$, there is a subinterval $[t_i^*, t_i^{**}] \subseteq [t_i', t_i'']$ such that $\Psi_0(\gamma(t)) \in \mathcal{R}_2$, $\forall t \in [t_i^*, t_i^{**}]$, with $\Psi_0(\gamma(t_i^*)) \in \mathcal{R}_{2,\ell}^{-}$ and $\Psi_0(\gamma(t_i^{**})) \in \mathcal{R}_{2,r}^{-}$. Hence, condition (12.3) is verified.

Let us turn to system (E_μ). This time, we focus our attention on the solutions of such a system starting from $\mathcal{A}_Q^t$ and crossing $\mathcal{A}_Q^b$ at least m_2 times. Similarly as before, we assume to have performed a rototranslation of the plane that makes the x-axis coincide with the line r and that brings the origin to the point Q. Therefore, we can express the solution $\zeta(\cdot, w)$ of system (E_μ) with starting point in $w \in \mathcal{A}_Q^t$ through polar coordinates $(\tilde{\rho}, \tilde{\theta})$. In particular, it holds that $\tilde{\theta}(0, w) \in [0, \pi]$. For any $t \in [0, r_\mu] = [0, T - r_0]$ and $w \in \mathcal{A}_Q^t$, the rotation number is now defined as

$$\operatorname{rot}_\mu(t, w) := \frac{\tilde{\theta}(t, w) - \tilde{\theta}(0, w)}{2\pi}\,.$$

Since the time-map τ_μ is strictly increasing, it follows that $\tau_\mu(h_1) < \tau_\mu(h_2)$. We claim that, choosing a switching time $r_\mu \geq \beta$ with

$$\beta := \frac{(m_2 + 3 + \frac{1}{2})\, \tau_\mu(h_1)\, \tau_\mu(h_2)}{\tau_\mu(h_2) - \tau_\mu(h_1)}\,, \tag{12.4}$$

that for any path $\omega\colon [0,1] \to \mathcal{A}_Q$, with $\omega(0) \in \Gamma_\mu(h_1)$ and $\omega(1) \in \Gamma_\mu(h_2)$, the interval inclusion

$$\left[\tilde{\theta}(r_\mu, \omega(1)), \tilde{\theta}(r_\mu, \omega(0))\right] \supseteq [\pi(2n^{**} + 1), 2\pi(n^{**} + m_2)] \tag{12.5}$$

is satisfied for some $n^{**} = n^{**}(r_\mu) \in \mathbb{N}$. The claim can be proved with arguments analogous to the ones employed above and thus its verification is omitted. The non-negative integer n^{**} has to be chosen as

$$n^{**} := \left\lceil \frac{r_\mu}{\tau_\mu(h_2)} \right\rceil\,.$$

By (12.5) and the continuity of the composite map $[0,1] \ni s \mapsto \mathrm{rot}_\mu(r_\mu, \omega(s))$, it follows that

$$\left\{\tilde{\theta}(r_\mu, \omega(s)),\ s \in [0,1]\right\} \supseteq [2\pi n^{**} + \pi, 2\pi(n^{**} + m_2)] \ .$$

As a consequence, the Bolzano Theorem ensures the existence of m_2 pairwise disjoint maximal intervals $[s_i{}', s_i{}''] \subseteq [0,1]$ for $i = 0, \ldots, m_2 - 1$ such that

$$\tilde{\theta}(r_\mu, \omega(s)) \in [2\pi n^{**} + \pi + 2\pi i, 2\pi n^{**} + 2\pi + 2\pi i]\,,$$
$$\forall s \in [s_i{}', s_i{}'']\,,\ \ i = 0, \ldots, m_2 - 1\,,$$

with $\tilde{\theta}(r_\mu, \omega(s_i{}')) = 2\pi n^{**} + \pi + 2\pi i$ and $\tilde{\theta}(r_\mu, \omega(s_i{}'')) = 2\pi n^{**} + 2\pi + 2\pi i$.

For $\widetilde{\mathcal{R}}_1$ and $\widetilde{\mathcal{R}}_2$ as above and introducing the m_2 nonempty, compact and pairwise disjoint sets

$$\mathcal{K}_i := \{w \in \mathcal{A}_Q^t \colon \tilde{\theta}(r_\mu, w) \in [2\pi n^{**} + \pi + 2\pi i, 2\pi n^{**} + 2\pi + 2\pi i]\}$$

with $i = 0, \ldots, m_2 - 1$, we are in position to check that

$$(\mathcal{K}_i, \Psi_\mu) \colon \widetilde{\mathcal{R}}_2 \Leftrightarrow\!\!\!\!\!\rightarrow \widetilde{\mathcal{R}}_1, \forall i = 0, \ldots, m_2 - 1\,, \tag{12.6}$$

where Ψ_μ is the Poincaré map associated to system (E_μ). Indeed, taking a path $\omega \colon [0,1] \to \mathcal{R}_2$ with $\omega(0) \in \mathcal{R}_{2,\ell}^-$ and $\omega(1) \in \mathcal{R}_{2,r}^-$ for $r_\mu \geq \beta$ and for any $i \in \{0, \ldots, m_2 - 1\}$ fixed, there exists a subinterval $[s_i{}', s_i{}''] \subseteq [0,1]$, such that $\omega(t) \in \mathcal{K}_i$ and $\Psi_\mu(\omega(t)) \in \mathcal{A}_Q^b\,, \quad \forall t \in [s_i{}', s_i{}'']$. Since $\Gamma_0(\Psi_\mu(\omega(s_i{}'))) \leq \ell_1$ and $\Gamma_0(\Psi_\mu(\omega(s_i{}''))) \geq \ell_2$, there exists a subinterval $[s_i^*, s_i^{**}] \subseteq [s_i{}', s_i{}'']$ such that $\Psi_\mu(\omega(t)) \in \mathcal{R}_1$, $\forall t \in [s_i^*, s_i^{**}]$, with $\Psi_\mu(\omega(s_i^*)) \in \mathcal{R}_{1,\ell}^-$ and $\Psi_\mu(\omega(s_i^{**})) \in \mathcal{R}_{1,r}^-$. Hence, condition (12.6) is proved.

The stretching properties in (12.3) and (12.6) allow one to apply Theorem 10.2 and the thesis immediately follows. □

We observe that in the proof we have chosen as $\mathcal{R}_1$, the "lower" set $\mathcal{R}^b$ and as $\mathcal{R}_2$ the "upper" set is $\mathcal{R}^t$. However, since the orbits of both systems (E_0) and (E_μ) are closed, the same argument works (by slightly modifying some constants, if needed) for $\mathcal{R}_1 = \mathcal{R}^t$ and $\mathcal{R}_2 = \mathcal{R}^b$.

The example we have presented suggests the possibility of proving, in an elementary but rigorous manner, the presence of chaotic dynamics for a broad class of second order nonlinear ODEs with periodic coefficients which are piecewise constant or at least not too far from the piecewise constant case, as long as we have the freedom to tune some switching time parameters within a certain range depending on the coefficients governing the equations. With this respect, our approach shows some resemblances with different techniques based on more sophisticated tools, for instance, on modifications of the methods by Shilnikov or Melnikov in which one looks for a transverse homoclinic point [19, 61, 114]. Indeed, also in these cases, it is possible to prove the presence of complex dynamics for differential equations, provided that the period of the time-dependent coefficients tends to infinity. See [4, 5, 21, 45], where alternative techniques are employed. In [65], the differentiability hypothesis on the Melnikov function is replaced with sign-changing conditions and the proofs make use of the Poincaré–Miranda Theorem.

We conclude by observing that the same arguments employed above seem to also work for more general time-dependent coefficients. However, a rigorous proof in such cases would require a more delicate analysis or possibly the aid of computer assistance (as in [15, 57, 108, 130, 172, 176, 177, 183]) and this is beyond the aims of the present work.

The geometrical configuration of linked annuli $\mathcal{A}_P$ and $\mathcal{A}_Q$ in Theorem 12.1 is (up to a planar homeomorphism) the same as that depicted in Figures 12 and 13. It may be interesting to look whether there exists a simple planar ODE model of Volterra type leading to a geometrical structure of the form described in Figures 15 and 16. A possible realization of such a geometry is obtained for system (E) when

$$a(t) \equiv a\,, \quad c(t) \equiv c$$

and

$$b(t) := \begin{cases} b & \text{for } 0 \le t < r_0\,, \\ 0 & \text{for } r_0 \le t < T\,, \end{cases}$$

$$d(t) := \begin{cases} d & \text{for } 0 \le t < r_0\,, \\ 0 & \text{for } r_0 \le t < T\,, \end{cases}$$

with a, b, c, d positive constants. The piecewise constant functions $b(t)$ and $d(t)$ are supposed to be extended to the whole real line by T-periodicity. We also set $r_l := T - r_0$. A possible interpretation of system (E) with such a choice of the periodic coefficients may be described as follows. The interaction between the two populations is governed by the classical Volterra system (E_0) with constant coefficients for a certain period of the season (corresponding to a time-interval of length r_0) and by the linear system

$$x' = a\,x\,, \quad y' = -cy \tag{E_l}$$

for the rest of the season (corresponding to a time-interval of length r_l). Such alternation between (E_0) and (E_l) occurs in a T-periodic fashion, with

$$T = r_0 + r_l\,.$$

From the point of view of the interaction between the predators and the prey populations, this can be explained, for instance, by assuming that for a certain period of the "year," the two populations are separated or when the prey can take shelter from the predators.

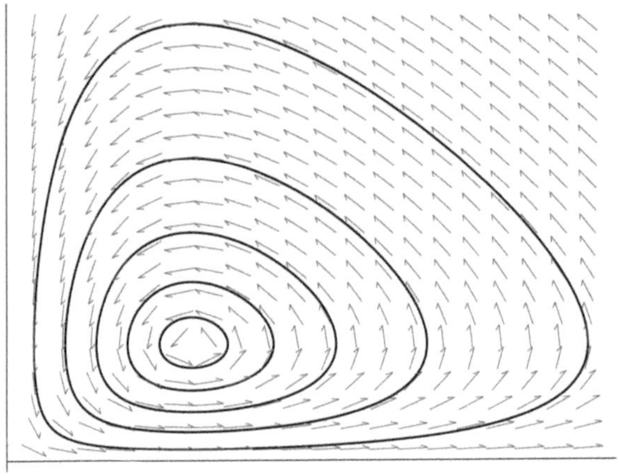

Fig. 23: The classical phase-portrait associated to system (E_0).

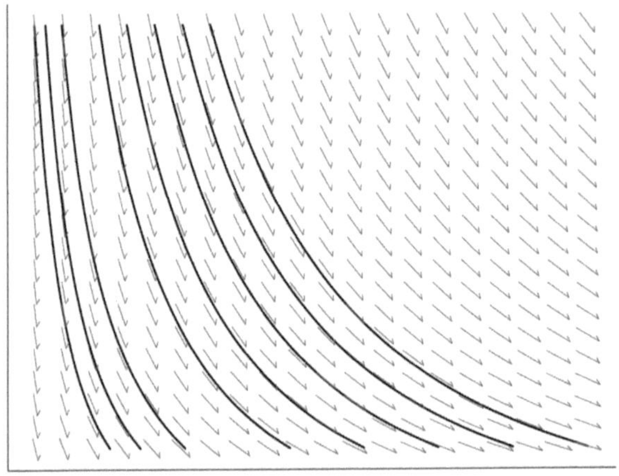

Fig. 24: The phase-portrait associated to system (E_l), where we have put in evidence some trajectories. Along all the motions, $x(t)$ is increasing and $y(t)$ is decreasing.

In this case, the Poincaré map Ψ associated to system (E) can be decomposed as

$$\Psi = \Psi_l \circ \Psi_0 \,,$$

where Ψ_0 is the Poincaré map of system (E_0) on the time-interval $[0, r_0]$ and

$$\Psi_l : (x, y) \mapsto (x \exp(a r_l), y \exp(-c r_l))$$

is the Poincaré map of system (E_l) on the time-interval $[0, r_l]$. From a geometrical point of view, we switch from the phase-portraits of the standard Volterra system of Figure 23 to that of Figure 24 concerning the linear system (E_l).

As an application of Theorem 10.1, we can now prove the following result.

Theorem 12.2. *For any choice of positive constants a, b, c, d and for any time-length $r_l > 0$, the following conclusion holds: For every integer $m \geq 2$, there exists a positive constant α such that for each $r_0 > \alpha$, the Poincaré map associated to system (E) induces chaotic dynamics on m symbols in a suitable rectangular region $\mathcal{A}$ of the positive quadrant.*

In the statement of the theorem, we have been intentionally vague about the localization of the region $\mathcal{A}$ where our result applies. The reason is that once the coefficients a, b, c, d are given and the length of period of noninteraction r_l is fixed, the required regions are easily detected by an elementary argument which is presented below.

Like Theorem 12.1, this last result is also stable with respect to small perturbations of the coefficients in the L^1-norm on $[0, T]$.

Proof. The arguments we are going to employ are similar to the ones used in Theorem 12.1. However, the construction of the oriented rectangles where the SAP technique will be applied requires more attention. Hence, we mainly focus on this step.

As a first step, we take a closed orbit Γ of system (E_0) and on Γ, two points $Q_1 = (x_1, y_1)$ and $Q_2 = (x_2, y_2)$ such that the images of Q_1 and Q_2 by the Poincaré map Ψ_l belong to different components of $\mathbb{R}^2 \setminus \Gamma$. More precisely, we require that $\Psi_l(Q_1)$ is external to Γ (however, we also need for some points of the orbit-path connecting Q_1 to $\Psi_l(Q_1)$ to be internal to Γ) while $\Psi_l(Q_2)$ lies in the part of the plane internal to Γ (Figure 25). By continuity, such a construction is always possible (once r_l is assigned), provided that Γ is a level line of the function $\mathcal{E}_0$ for a suitably chosen (possibly large) "energy" level.

Next, we select a second periodic orbit Γ_1 of system (E_0), internal to Γ and sufficiently close to Γ such that the point $(x_2 \exp(a(t + r_l)), y_2 \exp(-c(t + r_l)))$ is in the open part of the plane internal to Γ_1 as long as $(x_2 \exp(at), y_2 \exp(-ct))$ is in the closed annular region between Γ_1 and Γ. This is a formal way to express the fact that if we denote by Q_3 the first point where the positive semiorbit

$$\gamma^+(Q_2) := \{(x_2 \exp(at), y_2 \exp(-ct)) : t \geq 0\}$$

intersects Γ_1, then we require that $\Psi_l(Q_3)$ belongs to the open part of the plane internal to Γ_1. Clearly, this goal is always accomplished if we take Γ_1 sufficiently close to Γ.

Now that Γ, Q_1, Q_2 and Γ_1 are selected, we can construct two rectangular regions $\mathcal{A}$ and $\mathcal{B}$ as in Figure 26. Such regions are obtained from the intersections of the annular region between Γ_1 and Γ with the positive semiorbits of Q_1 and Q_2. At this step, we can give an orientation to $\mathcal{A}$ and $\mathcal{B}$ by taking $\mathcal{A}_\ell^- := \mathcal{A} \cap \Gamma_1$, $\mathcal{A}_r^- := \mathcal{A} \cap \Gamma$ and, moreover, $\mathcal{B}_\ell^-$ and $\mathcal{B}_r^-$ as the intersections of $\mathcal{B}$ with the orbits of system (E_l) passing through Q_2 and Q_1, respectively.

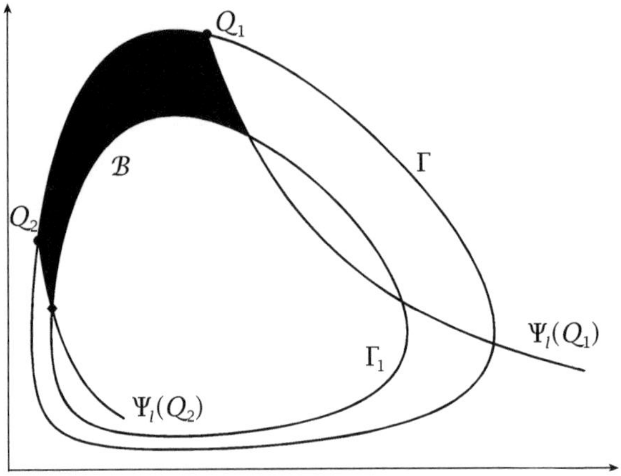

Fig. 25: The present figure illustrates the required geometric configuration with respect to the closed orbit Γ of system (E_0) and the positive semiorbits of (E_l) emanating from Q_1 and Q_2. The point Q_3 is the one marked by a black diamond. The darker region $\mathcal{B}$ is not involved at the present moment, but it will be relevant in the next steps. This figure has been drawn with mathematical software with the choice of the coefficients $a = 2, b = 2, c = 3, d = 1$ for the time length $r_l = 0.5$ and with the points $Q_1 = (3.5, 3.2)$, $Q_2 = (0.8, 1.7)$. For graphical reasons, a slightly different x- and y-scaling has been used.

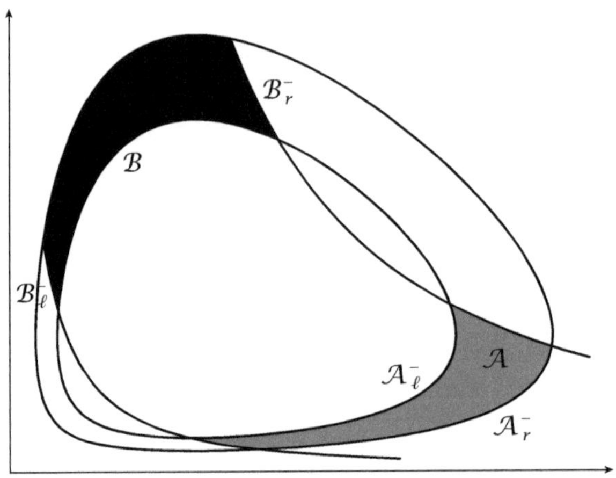

Fig. 26: The present figure illustrates the choice of the oriented rectangles $\widetilde{\mathcal{A}}$ and $\widetilde{\mathcal{B}}$. These two regions are bounded by trajectories of systems (E_0) and (E_l).

Repeating the same argument already exposed in the proof of Theorem 12.1, we can see that for every integer $m \geq 2$, we can find a constant $\alpha = \alpha^*(m)$ such that if $r_0 > \alpha$, then condition (H_ϕ) of Theorem 10.1 is satisfied for $\phi = \Psi_0$ (such a verification is omitted since it is only a repetition, mutatis mutandis, of what has been presented before). A graphical illustration of this property is given in Figure 27.

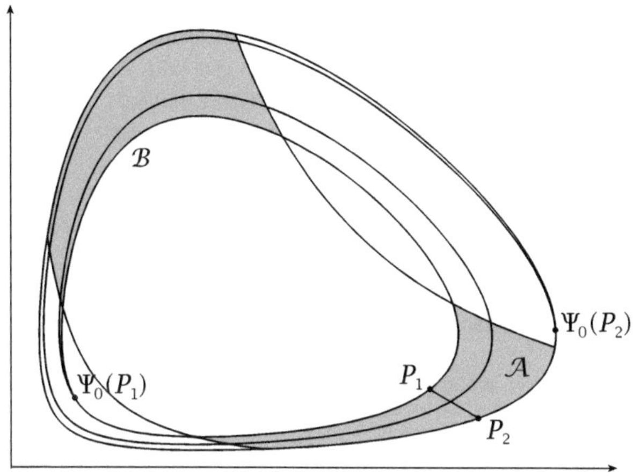

Fig. 27: The present figure illustrates the stretching along the paths effect of Ψ_0 between $\widetilde{\mathcal{A}}$ and $\widetilde{\mathcal{B}}$. The arc connecting in $\mathcal{A}$ the points $P_1 \in \mathcal{A}_\ell^-$ and $P_2 \in \mathcal{A}_2^-$ is twisted and stretched to a spiral-like curve, connecting the points $\Psi_0(P_1) \in \Gamma_1$ and $\Psi_0(P_2) \in \Gamma$. Such a spiral crosses $\mathcal{B}$ twice in the correct direction. This figure has been drawn with a mathematical software with the same choice of the coefficients of Figure 25. In order to produce the desired twist and stretching effect, we have taken $r_0 = 80$. For graphical reasons, a slightly different x- and y-scaling has been used.

Finally, the choice of Γ, Q_1, Q_2 and Γ_1 above guarantees that condition (H_ψ) of Theorem 10.1 is satisfied for $\psi = \Psi_l$ (Figure 25) and the proof is complete. □

Geometrical configurations which are topologically equivalent to the one described for this last example have been exhibited in [121] for a second order ODE related to the Lazer–McKenna suspension bridge model and by Alfonso Ruiz-Herrera in [143] for a prey–predator system with impulsive effects.

The two simple examples presented in this final part of our work show a possible way to obtain infinitely many subharmonic solutions and chaotic-like dynamics in the Volterra system (E) with T-periodic coefficients by using the SAP method. The interested reader can easily find many different other possibilities of varying the coefficients in (E) in order to demonstrate the presence of "chaotic effects" for this classical and yet fascinating model.

Bibliography

[1] R. L. Adler, A. G. Konheim and M. H. McAndrew, *Topological entropy*, Trans. Amer. Math. Soc. **114** (1965), 309–319.

[2] H. N. Agiza and A. A. Elsadany, *Chaotic dynamics in nonlinear duopoly game with heterogeneous players*, Appl. Math. Comput. **149** (2004), 843–860.

[3] J. C. Alexander, *A primer on connectivity*, in: Fixed point theory (Sherbrooke, Que., 1980), pp. 455–483, Lecture Notes in Math., vol. 886, Springer, Berlin-New York, 1981.

[4] A. Ambrosetti and M. Badiale, *Homoclinics: Poincaré–Melnikov type results via a variational approach*, Ann. Inst. H. Poincaré Anal. Non Linéaire **15** (1998), 233–252.

[5] A. Ambrosetti and M. Badiale, *Variational perturbative methods and bifurcation of bound states from the essential spectrum*, Proc. Roy. Soc. Edinburgh Sect. A **128** (1998), 1131–1161.

[6] Z. Amine and R. Ortega, *A periodic prey-predator system*, J. Math. Anal. Appl. **185** (1994), 477–489.

[7] J. Andres, M. Gaudenzi and F. Zanolin, *A transformation theorem for periodic solutions of nondissipative systems*, Rend. Sem. Mat. Univ. Pol. Torino **48** (1990), 171–186 (1992).

[8] J. Andres, L. Górniewicz and M. Lewicka, *Partially dissipative periodic processes*, in: Topology in nonlinear analysis (Warsaw 1994), Banach Center Publ. 35, pp. 109–118, Polish Acad. Sci., Warsaw, 1996.

[9] V. I. Arnold, V. V. Kozlov and A. I. Neishtadt, *Mathematical aspects of classical and celestial mechanics. Dynamical systems III*, Encyclopaedia Math. Sci. 3, Springer, Berlin, 1993.

[10] B. Aulbach and B. Kieninger, *On three definitions of chaos*, Nonlinear Dyn. Syst. Theory **1** (2001), 23–37.

[11] J. Auslander and J. A. Yorke, *Interval maps, factors of maps, and chaos*, Tôhoku Math. J. **32** (1980), 177–188.

[12] F. Balibrea and L. Snoha, *Topological entropy of Devaney chaotic maps*, Topology Appl. **133** (2003), 225–239.

[13] B. Bánhelyi, T. Csendes and B. M. Garay, *Optimization and the Miranda approach in detecting horseshoe-type chaos by computer*, Internat. J. Bifur. Chaos Appl. Sci. Engrg. **17** (2007), 735–747.

[14] B. Bánhelyi, T. Csendes, B. M. Garay and L. Hatvani, *Computer assisted proof of chaotic behavior of the forced damped pendulum*, in: Colloquium on Differential and Difference Equations—CDDE 2006, pp. 9–20, Folia Fac. Sci. Natur. Univ. Masaryk. Brun. Math., vol. 16, Masaryk Univ., Brno, 2007.

[15] B. Bánhelyi, T. Csendes, B. M. Garay and L. Hatvani, *A computer-assisted proof of Σ_3-chaos in the forced damped pendulum equation*, SIAM J. Appl. Dyn. Syst. **7** (2008), 843–867.

[16] J. Banks, J. Brooks, G. Cairns, G. Davis and P. Stacey, *On Devaney's definition of chaos*, Amer. Math. Monthly **99** (1992), 332–334.

[17] M. V. Bartuccelli, G. Gentile and K. V. Georgiou, *On the dynamics of a vertically driven damped planar pendulum*, R. Soc. Lond. Proc. Ser. A Math. Phys. Eng. Sci. **457** (2001), 3007–3022.

[18] B. Batko and M. Mrozek, *The Euler–Poincaré characteristic of index maps*, Topology Appl. **154** (2007), 859–866.

[19] F. Battelli and K. J. Palmer, *Chaos in the Duffing equation*, J. Differential Equations **101** (1993), 276–301.

[20] A. Berarducci, D. Dikranjan and J. Pelant, *Uniform quasi components, thin spaces and compact separation*, in: Proceedings of the International Conference on Topology and its Applications (Yokohama, 1999), Topology Appl. **122** (2002), 51–64.

[21] M. Berti and P. Bolle, *Homoclinics and chaotic behavior for perturbed second order systems*, Ann. Mat. Pura Appl. **176** (1999), 323–378.

[22] F. Blanchard, E. Glasner, S. Kolyada and A. Maass, *On Li–Yorke pairs*, J. Reine Angew. Math. **547** (2002), 51–68.

[23] L. S. Block and W. A. Coppel, *Stratification of continuous maps of an interval*, Trans. Amer. Math. Soc. **297** (1986), 587–604.

[24] L. S. Block and W. A. Coppel, *Dynamics in one dimension*, Lecture Notes in Mathematics, vol. 1513, Springer-Verlag, Berlin, 1992.

[25] L. Block, J. Guckenheimer, M. Misiurewicz and L. S. Young, *Periodic points and topological entropy of one-dimensional maps*, in: Global theory of dynamical systems (Proc. Internat.

Conf., Northwestern Univ., Evanston, Ill., 1979), pp. 18–34, Lecture Notes in Math., vol. 819, Springer, Berlin, 1980.

[26] A. Blokh and E. Teoh, *How little is little enough?*, Discrete Contin. Dyn. Syst. **9** (2003), 969–978.

[27] A. Boscaggin and F. Zanolin, *Positive periodic solutions of second order nonlinear equations with indefinite weight: multiplicity results and complex dynamics*, J. Differential Equations **252** (2012), 2922–2950.

[28] R. Bowen, *Entropy for group endomorphisms and homogeneous spaces*, Trans. Amer. Math. Soc. **153** (1971), 401–414, erratum, **181** (1973), 509–510.

[29] M. Braun, *Differential equations and their applications. An introduction to applied mathematics. Fourth ed.*, Texts in Applied Mathematics, vol. 11, Springer-Verlag, New York, 1993.

[30] K. Burns and H. Weiss, *A geometric criterion for positive topological entropy*, Comm. Math. Phys. **172** (1995), 95–118.

[31] L. Burra and F. Zanolin, *Chaotic dynamics in a simple class of Hamiltonian systems with applications to a pendulum with variable length*, Differential Integral Equations **22** (2009), 927–948.

[32] L. Burra and F. Zanolin, *Chaotic dynamics in a vertically driven planar pendulum*, Nonlinear Anal. **72** (2010), 1462–1476.

[33] R. Burton and R. W. Easton, *Ergodicity of linked twist maps*, in: Global theory of dynamical systems (Proc. Internat. Conf., Northwestern Univ., Evanston, Ill., 1979), pp. 35–49, Lecture Notes in Math., vol. 819, Springer, Berlin, 1980.

[34] T. A. Burton and V. Hutson, *Permanence for nonautonomous predator–prey systems*, Differential Integral Equations **4** (1991), 1269–1280.

[35] G. J. Butler, *Rapid oscillation, nonextendability, and the existence of periodic solutions to second order nonlinear ordinary differential equations*, J. Differential Equations **22** (1976), 467–477.

[36] G. J. Butler and H. I. Freedman, *Periodic solutions of a predator–prey system with periodic coefficients*, Math. Biosci. **55** (1981), 27–38.

[37] A. Capietto, W. Dambrosio and D. Papini, *Superlinear indefinite equations on the real line and chaotic dynamics*, J. Differential Equations **181** (2002), 419–438.

[38] M. Capiński and P. Zgliczyński, *Cone conditions and covering relations for normally hyperbolic invariant manifolds*, Discrete Contin. Dyn. Syst. **30** (2011), 641–670.

[39] M. Carbinatto, J. Kwapisz and K. Mischaikow, *Horseshoes and the Conley index spectrum*, Ergodic Theory Dynam. Systems **20** (2000), 365–377.

[40] W. G. Chinn and N. E. Steenrod, *First concepts of topology*, New Mathematical Library, vol. 18, Random House, New York; The L. W. Singer Co., Syracuse, N. Y., 1966.

[41] C. Conley, *An application of Ważewski's method to a non-linear boundary value problem which arises in population genetics*, J. Math. Biol. **2** (1975), 241–249.

[42] C. Conley, *Isolated invariant sets and the Morse index*, CBMS Regional Conference Series in Mathematics, vol. 38, American Mathematical Society, Providence, R. I., 1978.

[43] J. M. Cushing, *Periodic time-dependent predator–prey systems*, SIAM J. Appl. Math. **32** (1977), 82–95.

[44] W. Dambrosio and D. Papini, *Periodic solutions of asymptotically linear second order equations with indefinite weight*, Ann. Mat. Pura Appl. (4) **183** (2004), 537–554.

[45] M. del Pino, P. Felmer and K. Tanaka, *An elementary construction of complex patterns in nonlinear Schrödinger equations*, Nonlinearity **15** (2002), 1653–1671.

[46] R. L. Devaney, *Subshifts of finite type in linked twist mappings*, Proc. Amer. Math. Soc. **71** (1978), 334–338.

[47] R. L. Devaney, *An introduction to chaotic dynamical systems. Second edition.*, Addison-Wesley Studies in Nonlinearity. Addison-Wesley Publishing Company, Advanced Book Program, Redwood City, CA, 1989.
[48] E. I. Dinaburg, *A correlation between topological entropy and metric entropy*, Dokl. Akad. Nauk SSSR **190** (1970), 19–22; English Translation: Soviet Math. Dokl. **11** (1969), 13–16.
[49] T. R. Ding, H. Huang and F. Zanolin, *A priori bounds and periodic solutions for a class of planar systems with applications to Lotka-Volterra equations*, Discrete Contin. Dynam. Systems **1** (1995), 103–117.
[50] T. R. Ding and F. Zanolin, *Periodic solutions and subharmonic solutions for a class of planar systems of Lotka-Volterra type*, in: World Congress of Nonlinear Analysts '92, vol. I–IV (Tampa, FL, 1992), pp. 395–406, de Gruyter, Berlin, 1996.
[51] J. Dugundji and A. Granas, *Fixed point theory*, Springer-Verlag, New York, 2003.
[52] R. W. Easton, *Isolating blocks and symbolic dynamics*, J. Differential Equations **17** (1975), 96–118.
[53] S. Eilenberg and E. Otto, *Quelques propriétés caractéristiques de la dimension*, Fund. Math. **31** (1938), 149–153.
[54] R. Engelking, *Dimension theory*, North-Holland Publishing Co., Amsterdam, 1978.
[55] V. V. Fedorchuk, M. Levin and E. V. Shchepin, *On the Brouwer definition of dimension*, Uspekhi Mat. Nauk **54** (1999), 193–194.
[56] P. M. Fitzpatrick, I. Massabó and J. Pejsachowicz, *On the covering dimension of the set of solutions of some nonlinear equations*, Trans. Amer. Math. Soc. **296** (1986), 777–798.
[57] Z. Galias and P. Zgliczyński, *Computer assisted proof of chaos in the Lorenz equations*, Phys. D **115** (1998), 165–188.
[58] M. Gameiro, T. Gedeon, W. Kalies, H. Kokubu, K. Mischaikow and H. Oka, *Topological horseshoes of traveling waves for a fast-slow predator–prey system*, J. Dynam. Differential Equations **19** (2007), 623–654.
[59] M. Gidea and C. Robinson, *Topological crossing heteroclinic connections to invariant tori*, J. Differential Equations **193** (2003), 49–74.
[60] P. Glendinning, *Stability, instability and chaos: an introduction to the theory of nonlinear differential equations*, Cambridge Texts in Applied Mathematics. Cambridge University Press, Cambridge, 1994.
[61] J. Hale and X.-B. Lin, *Symbolic dynamics and nonlinear semiflows*, Ann. Mat. Pura Appl. **144** (1986), 229–259.
[62] B. Hasselblatt and A. Katok, *A first course in dynamics. With a panorama of recent developments*, Cambridge University Press, New York, 2003.
[63] A. R. Hausrath, *Periodic integral manifolds for periodically forced Volterra-Lotka equations*, J. Math. Anal. Appl. **87** (1982), 474–488.
[64] A. R. Hausrath and R. F. Manásevich, *Periodic solutions of a periodically perturbed Lotka-Volterra equation using the Poincaré-Birkhoff theorem*, J. Math. Anal. Appl. **157** (1991), 1–9.
[65] M. Henrard and F. Zanolin, *Bifurcation from a periodic orbit in perturbed planar Hamiltonian systems*, J. Math. Anal. Appl. **277** (2003), 79–103.
[66] J. G. Hocking and G. S. Young, *Topology*, Addison-Wesley Publishing Co., Inc., Reading, Mass.-London, 1961.
[67] W. Huang and X. Ye, *Devaney's chaos or 2-scattering implies Li–Yorke's chaos*, Topology Appl. **117** (2002), 259–272.
[68] W. Hurewicz and H. Wallman, *Dimension Theory*, Princeton Mathematical Series, vol. 4, Princeton University Press, Princeton, N. J., 1941.
[69] J. Kampen, *On fixed points of maps and iterated maps and applications*, Nonlinear Anal. **42** (2000), 509–532.

[70] A. Katok and B. Hasselblatt, *Introduction to the modern theory of dynamical systems. With a supplementary chapter by Katok and Leonardo Mendoza*, Encyclopedia of Mathematics and its Applications, 54. Cambridge University Press, Cambridge, 1995.

[71] J. Kennedy, S. Koçak and Y. A. Yorke, *A chaos lemma*, Amer. Math. Monthly **108** (2001), 411–423.

[72] J. Kennedy and J. A. Yorke, *The topology of stirred fluids*, Topology Appl. **80** (1997), 201–238.

[73] J. Kennedy and J. A. Yorke, *Topological horseshoes*, Trans. Amer. Math. Soc. **353** (2001), 2513–2530.

[74] J. Kennedy and J. A. Yorke, *A chaos lemma with applications to Hénon-like difference equations*, in: New trends in difference equations (Temuco, 2000), pp. 173–205, Taylor & Francis, London, 2002.

[75] J. Kennedy and J. A. Yorke, *Generalized Hénon difference equations with delay*, Univ. Iagel. Acta Math. **41** (2003), 9–28.

[76] U. Kirchgraber and D. Stoffer, *On the definition of chaos*, Z. Angew. Math. Mech. **69** (1989), 175–185.

[77] B. P. Kitchens, *Symbolic dynamics. One-sided, two-sided and countable state Markov shifts*, Universitext. Springer-Verlag, Berlin, 1998.

[78] A. N. Kolmogorov, *Sulla teoria di Volterra della lotta per l'esistenza*, Giorn. Ist. Ital. Attuari **7** (1936), 74–80.

[79] S. F. Kolyada, *Li–Yorke sensitivity and other concepts of chaos*, Ukrainian Math. J. **56** (2004), 1242–1257.

[80] S. Kolyada and L. Snoha, *Some aspects of topological transitivity–a survey*, in: Iteration theory (ECIT 94) (Opava), pp. 3–35, Grazer Math. Ber. 334, Karl-Franzens-Univ. Graz, Graz, 1997.

[81] M. A. Krasnosel'skiĭ, *Translation Along Trajectories of Differential Equations*, Transl. Math. Monographs, vol. 19, Amer. Math. Soc., Providence, R. I., 1968.

[82] W. Kulpa, *The Poincaré–Miranda theorem*, Amer. Math. Monthly **104** (1997), 545–550.

[83] W. Kulpa, M. Pordzik, L. Socha and M. Turzański, *L_1 cheapest paths in "Fjord scenery"*, European J. Oper. Res. **161** (2005), 736–753.

[84] W. Kulpa, L. Socha and M. Turzański, *Parametric extension of the Poincaré theorem*, Acta Univ. Carolin. Math. Phys. **41** (2000), 39–46.

[85] K. Kuratowski, *Topology. Volume II*, Academic Press, New York, 1968.

[86] D. Kwietniak and M. Misiurewicz, *Exact Devaney chaos and entropy*, Qual. Theory Dyn. Syst. **6** (2005), 169–179.

[87] B. Lani-Wayda and R. Srzednicki, *A generalized Lefschetz fixed point theorem and symbolic dynamics in delay equations*, Ergodic Theory Dynam. Systems **22** (2002), 1215–1232.

[88] A. Lasota and J. A. Yorke, *On the existence of invariant measures for transformations with strictly turbulent trajectories*, Bull. Acad. Polon. Sci. Sér. Sci. Math. Astronom. Phys. **25** (1977), 233–238.

[89] A. C. Lazer and P. J. McKenna, *Large scale oscillatory behavior in loaded asymmetric systems*, Ann. Inst. Henry Poincaré, Analyse non lineaire **4** (1987), 244–274.

[90] A. C. Lazer and P. J. McKenna, *Large–amplitude periodic oscillations in suspension bridges: some new connections with nonlinear analysis*, SIAM Review **32** (1990), 537–578.

[91] J. Leray and J. Schauder, *Topologie et équations fonctionnelles*, Ann. Sci. Ecol. Norm. Sup. **51** (1934), 45–78.

[92] S. Li, *ω-chaos and topological entropy*, Trans. Amer. Math. Soc. **339** (1993), 243–249.

[93] T. Y. Li and J. A. Yorke, *Period three implies chaos*, Amer. Math. Monthly **82** (1975), 985–992.

[94] D. Lind and B. Marcus, *An introduction to symbolic dynamics and coding*, Cambridge University Press, Cambridge, 1995.

[95] N. G. Lloyd, *Degree theory*, Cambridge Tracts in Mathematics, **No. 73**. Cambridge University Press, Cambridge-New York-Melbourne, 1978.

[96] J. López-Gómez, R. Ortega and A. Tineo, *The periodic predator–prey Lotka-Volterra model*, Adv. Differential Equations **1** (1996), 403–423.

[97] E. N. Lorenz, *A deterministic non periodic flow*, Journal of Atmospheric Science **20** (1963), 130–141.

[98] A. Margheri, C. Rebelo and F. Zanolin, *Chaos in periodically perturbed planar Hamiltonian systems using linked twist maps*, J. Differential Equations **249** (2010), 3233–3257.

[99] A. Margheri, C. Rebelo and F. Zanolin, *Complex dynamics in pendulum-type equations with variable length*, 2011, submitted.

[100] M. M. Marsh, *Fixed point theorems for partially outward mappings*, Topology Appl. **153** (2006), 3546–3554.

[101] J. Mawhin, *Leray–Schauder continuation theorems in the absence of a priori bounds*, Topol. Methods Nonlinear Anal. **9** (1997), 179–200.

[102] J. Mawhin, *Poincaré early use of Analysis situs in nonlinear differential equations: Variations around the theme of Kronecker's integral*, Philosophia Scientiae **4** (2000), 103–143.

[103] J. Mawhin, *The legacy of Pierre-François Verhulst and Vito Volterra in population dynamics*, in: The first 60 years of nonlinear analysis of Jean Mawhin, World Sci. Publ., River Edge, NJ, 2004, pp. 147–160.

[104] A. Medio, *Chaotic dynamics. Theory and applications to economics*, Cambridge University Press, Cambridge, 1992.

[105] A. Medio, M. Pireddu and F. Zanolin, *Chaotic dynamics for maps in one and two dimensions. A geometrical method and applications to economics*, Internat. J. Bifur. Chaos Appl. Sci. Engrg. **19** (2009), 3283–3309.

[106] C. Miranda, *Un'osservazione su un teorema di Brouwer*, Boll. Un. Mat. Ital. **3** (1940), 5–7.

[107] K. Mischaikow and M. Mrozek, *Isolating neighborhoods and chaos*, Japan J. Indust. Appl. Math. **12** (1995), 205–236.

[108] K. Mischaikow and M. Mrozek, *Chaos in the Lorenz equations: a computer-assisted proof*, Bull. Amer. Math. Soc. (N. S.) **32** (1995), 66–72.

[109] K. Mischaikow and M. Mrozek, *Conley index*, Handbook of dynamical systems, vol. 2, pp. 393–460, North-Holland, Amsterdam, 2002.

[110] T. K. Subrahmonian Moothathu, *A ballad on topological dynamics – To be read like a railway guide*, Internet 2004, Communication by Prof. Medio.

[111] J. Moser, *Stable and random motions in dynamical systems. With special emphasis on celestial mechanics*, Hermann Weyl Lectures, the Institute for Advanced Study, Princeton, N. J. Annals of Mathematics Studies, No. 77, Princeton University Press, Princeton, N. J., 1973.

[112] T. Namba, *Bifurcation phenomena appearing in the Lotka-Volterra competition equations: a numerical study*, Math. Biosci. **81** (1986), 191–212.

[113] V. V. Nemytskii and V. V. Stepanov, *Qualitative theory of differential equations*, Princeton Mathematical Series, No. 22, Princeton University Press, Princeton, N. J., 1960.

[114] K. J. Palmer, *Transversal heteroclinic points and Cherry's example of a nonintegrable Hamiltonian system*, J. Differential Equations **65** (1986), 321–360.

[115] D. Papini and F. Zanolin, *A topological approach to superlinear indefinite boundary value problems*, Topol. Methods Nonlinear Anal. **15** (2000), 203–233.

[116] D. Papini and F. Zanolin, *Periodic points and chaotic-like dynamics of planar maps associated to nonlinear Hill's equations with indefinite weight*, Georgian Mathematical Journal **9** (2002), 339–366.

[117] D. Papini and F. Zanolin, *On the periodic boundary value problem and chaotic-like dynamics for nonlinear Hill's equations*, Adv. Nonlinear Stud. **4** (2004), 71–91.

[118] D. Papini and F. Zanolin, *Fixed points, periodic points, and coin-tossing sequences for mappings defined on two-dimensional cells*, Fixed Point Theory Appl. **2004** (2004), 113–134.
[119] D. Papini and F. Zanolin, *Some results on periodic points and chaotic dynamics arising from the study of the nonlinear Hill equations*, Rend. Sem. Mat. Univ. Pol. Torino **65** (2007), 115–157 (Special Issue: Subalpine Rhapsody in Dynamics).
[120] A. Pascoletti, M. Pireddu and F. Zanolin, *Multiple periodic solutions and complex dynamics for second order ODEs via linked twist maps*, Electron. J. Qual. Theory Differ. Equ., Proc. 8th Coll. Qualitative Theory of Diff. Equ. **14** (2008), 1–32.
[121] A. Pascoletti and F. Zanolin, *Example of a suspension bridge ODE model exhibiting chaotic dynamics: a topological approach*, J. Math. Anal. Appl. **339** (2008), 1179–1198.
[122] A. Pascoletti and F. Zanolin, *Chaotic dynamics in periodically forced asymmetric ordinary differential equations*, J. Math. Anal. Appl. **352** (2009), 890–906.
[123] A. Pascoletti and F. Zanolin, *A path crossing lemma and applications to nonlinear second order equations under slowly varying perturbations*, Matematiche (Catania) **65** (2010), 121–168.
[124] M. Pireddu and F. Zanolin, *Fixed points for dissipative-repulsive systems and topological dynamics of mappings defined on N-dimensional cells*, Adv. Nonlinear Stud. **5** (2005), 411–440.
[125] M. Pireddu and F. Zanolin, *Cutting surfaces and applications to periodic points and chaotic-like dynamics*, Topol. Methods Nonlinear Anal. **30** (2007), 279–320. Errata: Correction to "Cutting surfaces and applications to periodic points and chaotic-like dynamics" (Topol. Methods Nonlinear Anal. **30** (2007), 271–320), Topol. Methods Nonlinear Anal. **33** (2009), 395.
[126] M. Pireddu and F. Zanolin, *Some remarks on fixed points for maps which are expansive along one direction*, Rend. Istit. Mat. Univ. Trieste **39** (2007), 245–274.
[127] M. Pireddu and F. Zanolin, *Chaotic dynamics in the Volterra predator–prey model via linked twist maps*, Opuscula Math. (Special issue dedicated to the memory of Professor Andrzej Lasota) **28** (2008), 567–592.
[128] H. Poincaré, *Sur certaines solutions particulières du problème des trois corps*, C. R. Acad. Sci. Paris **97** (1883), 251–252.
[129] H. Poincaré, *Sur certaines solutions particulières du problème des trois corps*, Bulletin Astronomique **91** (1884), 65–74.
[130] A. Pokrovskii, O. Rasskazov and D. Visetti, *Homoclinic trajectories and chaotic behavior in a piecewise linear oscillator*, Discrete Contin. Dyn. Syst. Ser. B **8** (2007), 943–970.
[131] A. V. Pokrovskii, S. J. Szybka and J. G. McInerney, *Topological degree in locating homoclinic structures for discrete dynamical systems*, Izvestiya of RAEN, Series MMMIU **5** (2001), 152–183.
[132] F. Przytycki, *Ergodicity of toral linked twist mappings*, Ann. Sci. École Norm. Sup. **16** (1983), 345–354.
[133] F. Przytycki, *Periodic points of linked twist mappings*, Studia Math. **83** (1986), 1–18.
[134] P. H. Rabinowitz, *Nonlinear Sturm–Liouville problems for second order ordinary differential equations*, Comm. Pure Appl. Math. **23** (1970), 939–961.
[135] C. Rebelo and F. Zanolin, *On the existence and multiplicity of branches of nodal solutions for a class of parameter-dependent Sturm-Liouville problems via the shooting map*, Differential Integral Equations **13** (2000), 1473–1502.
[136] P. Reichlin, *Equilibrium cycles in an overlapping generations economy with production*, J. Econom. Theory **40** (1986), 89–102.
[137] S. Rinaldi and S. Muratori, *Conditioned chaos in seasonally perturbed predator–prey models*, Ecological Modelling **69** (1993), 79–97.
[138] C. Robinson, *Dynamical systems. Stability, symbolic dynamics, and chaos. Second edition*, Studies in Advanced Mathematics, CRC Press, Boca Raton, FL, 1999.
[139] C. Robinson, *Symbolic dynamics for transition tori*, in: Celestial mechanics (Evanston, IL, 1999), pp. 199–208, Contemp. Math. 292, Amer. Math. Soc., Providence, RI, 2002.

[140] C. Robinson, *An introduction to dynamical systems: continuous and discrete*, Pearson Prentice Hall, Upper Saddle River, NJ, 2004.
[141] F. Rothe, *The periods of the Volterra-Lotka system*, J. Reine Angew. Math. **355** (1985), 129–138.
[142] R. Rudnicki, *Ergodic measures on topological spaces*, Univ. Iagel. Acta Math. **26** (1987), 231–237.
[143] A. Ruiz-Herrera, *Chaos in predator–prey systems with/without impulsive effect*, Nonlinear Anal. Real World Appl. **13** (2012), 977–986.
[144] A. Ruiz-Herrera and F. Zanolin, *An example of chaotic dynamics in 3D systems via stretching along paths*, Ann. Mat. Pura Appl. (2012) DOI: 10.1007/s10231-012-0271-0
[145] A. Ruiz-Herrera and F. Zanolin, *Periodic solutions and chaotic dynamics in 3D equations with applications to Lotka–Volterra systems*, 2012, submitted.
[146] D. E. Sanderson, *Advanced plane topology from an elementary standpoint*, Math. Mag. **53** (1980), 81–89.
[147] A. N. Sharkowskiĭ, *Co-existence of cycles of a continuous map of the line into itself*, Ukrainskiĭ Matematicheskiĭ Zhurnal **16** (1964), 61–71; English translation: Internat. J. Bifur. Chaos Appl. Sci. Engrg. **5** (1995), 1263–1273.
[148] S. Silverman, *On maps with dense orbits and the definition of chaos*, Rocky Mountain J. Math. **22** (1992), 353–375.
[149] S. Smale, *Diffeomorphism with many periodic points*, in: Differential and Combinatorial Topology (A Symposium in Honor of Marston Morse), pp. 63–80, Princeton Univ. Press, Princeton, N. J., 1965.
[150] S. Smale, *Differentiable dynamical systems*, Bull. Amer. Math. Soc. **73** (1967), 747–817.
[151] S. Smale, *Finding a horseshoe on the beaches of Rio*, Math. Intelligencer **20** (1998), 39–44.
[152] J. Smítal, *Chaotic functions with zero topological entropy*, Trans. Amer. Math. Soc. **297** (1986), 269–282.
[153] J. Springham, *Ergodic properties of linked-twist maps*, PhD thesis, University of Bristol, available at ArXiv:0812.0899v1.
[154] J. Springham and S. Wiggins, *Bernoulli linked-twist maps in the plane*, Dyn. Syst. **25** (2010), 483–499.
[155] R. Srzednicki, *Generalized Lefschetz theorem and a fixed point index formula*, Topology Appl. **81** (1997), 207–224.
[156] R. Srzednicki, *A generalization of the Lefschetz fixed point theorem and detection of chaos*, Proc. Amer. Math. Soc. **128** (2000), 1231–1239.
[157] R. Srzednicki, *Ważewski method and Conley index*, Handbook of differential equations, pp. 591–684, Elsevier/North-Holland, Amsterdam, 2004.
[158] R. Srzednicki and K. Wójcik, *A geometric method for detecting chaotic dynamics*, J. Differential Equations **135** (1997), 66–82.
[159] R. Srzednicki, K. Wójcik and P. Zgliczyński, *Fixed point results based on the Ważewski method*, Handbook of topological fixed point theory, pp. 905–943, Springer, Dordrecht, 2005.
[160] P. Štefan, *A theorem of Šarkovskii on the existence of periodic orbits of continuous endomorphisms of the real line*, Comm. Math. Phys. **54** (1977), 237–248.
[161] R. Sturman, *The linked twist map approach to fluid mixing*, in: Dynamical Systems and Statistical Mechanics, London Mathematical Society Durham Symposium 2006, available online at: `http://www.maths.dur.ac.uk/events/Meetings/LMS/2006/DSSM/talks.html`.
[162] R. Sturman, J. M. Ottino and S. Wiggins, *The mathematical foundations of mixing*, Cambridge Monographs on Applied and Computational Mathematics, vol. 22, Cambridge University Press, Cambridge, 2006, The linked twist map as a paradigm in applications: micro to macro, fluids to solids.
[163] A. Szymczak, *The Conley index and symbolic dynamics*, Topology **35** (1996), 287–299.

[164] P. Táboas, *Periodic solutions of a forced Lotka-Volterra equation*, J. Math. Anal. Appl. **124** (1987), 82–97.

[165] M. Vellekoop and R. Berglund, *On intervals, transitivity = chaos*, Amer. Math. Monthly **101** (1994), 353–355.

[166] J. Waldvogel, The period in the Lotka-Volterra system is monotonic, *J. Math. Anal. Appl.* **114** (1986), 178–184.

[167] P. Walters, *An introduction to ergodic theory*, Graduate Texts in Mathematics, vol. 79, Springer-Verlag, New York, 1982.

[168] G. T. Whyburn, *Quasi-closed sets and fixed points*, Proc. Nat. Acad. Sci. U.S.A. **57** (1967), 201–205.

[169] S. Wiggins, *Global Bifurcations and Chaos: Analytical Methods*, Appl. Math. Sciences, vol. 73, Springer-Verlag, New York, 1988.

[170] S. Wiggins, *Chaos in the dynamics generated by sequence of maps, with application to chaotic advection in flows with aperiodic time dependence*, Z. angew. Math. Phys. **50** (1999), 585–616.

[171] S. Wiggins and J. M. Ottino, *Foundations of chaotic mixing*, Philos. Trans. R. Soc. Lond. Ser. A Math. Phys. Eng. Sci. **362** (2004), 937–970.

[172] D. Wilczak, *Chaos in the Kuramoto–Sivashinsky equations–a computer-assisted proof*, J. Differential Equations **194** (2003), 433–459.

[173] K. Wójcik, *Lefschetz zeta function and the periodic points*, Bull. Polish Acad. Sci. Math. **43** (1995), 59–64.

[174] K. Wójcik and P. Zgliczyński, *Isolating segments, fixed point index, and symbolic dynamics*, J. Differential Equations **161** (2000), 245–288.

[175] M. Wojtkowski, *Linked twist mappings have the K-property*, in: Nonlinear dynamics (Internat. Conf., New York, 1979), pp. 65–76, Ann. New York Acad. Sci., vol. 357, New York Acad. Sci., New York, 1980.

[176] X-S Yang, H. Li and Y. Huang, *A planar topological horseshoe theory with applications to computer verifications of chaos*, J. Phys. A **38** (2005), 4175–4185.

[177] X-S Yang and Q. Li, *A computer-assisted proof of chaos in Josephson junctions*, Chaos Solitons Fractals **27** (2006), 25–30.

[178] C. Zanini and F. Zanolin, *Complex dynamics in a nerve fiber model with periodic coefficients*, Nonlinear Anal. Real World Appl. **10** (2009), 1381–1400.

[179] C. Zanini and F. Zanolin, *Periodic solutions for a class of second order ODEs with a Nagumo cubic type nonlinearity*, Discrete Contin. Dynam. Systems **32** (2012), 4045–4067.

[180] C. Zanini and F. Zanolin, *An example of chaos for a cubic nonlinear Schrödinger equation with periodic inhomogeneous nonlinearity*, Adv. Nonlinear Stud. **12** (2012), 481–499.

[181] E. Zeidler, *Nonlinear functional analysis and its applications. I. Fixed-point theorems*, Springer-Verlag, New York, 1986.

[182] P. Zgliczyński, *Fixed point index for iterations of maps, topological horseshoe and chaos*, Topol. Methods Nonlinear Anal. **8** (1996), 169–177.

[183] P. Zgliczyński, *Computer assisted proof of chaos in the Rössler equations and in the Hénon map*, Nonlinearity **10** (1997), 243–252.

[184] P. Zgliczyński, *On periodic points for systems of weakly coupled 1-dim maps*, Nonlinear Anal., Ser. A: Theory Methods **46** (2001), 1039–1062.

[185] P. Zgliczyński, *Covering relations, cone conditions and stable manifold theorem*, J. Differential Equations **246** (2009), 1774–1819.

[186] P. Zgliczyński and M. Gidea, *Covering relations for multidimensional dynamical systems*, J. Differential Equations **202** (2004), 32–58.

Index

A
affine map 160
almost periodic 6
arcwise connected 153
arcwise connectedness 152
Arzela–Ascoli 5
Arzela–Ascoli Theorem 13
autonomous 5
autonomous Kolmogorov systems 39
auxiliary functions 153
average conditions 82–84
average inequalities 76

B
Bernoulli shift 175, 176, 179–181, 191
bifurcation theory 145
Block–Coppel chaos 197
Bolzano Theorem 152, 153, 165, 218
Brouwer degree 148
Brouwer fixed point 69, 97, 127, 133, 145
Brouwer Theorem 147
bulging horseshoe 142

C
chaos 124, 172
Chaos Lemma 180
chaotic dynamics 130, 137, 168–170, 174–176, 183, 190, 206, 212, 223
compact uniform attractor 4, 6–8, 15, 16
competitive exclusion 77
competitive system 82, 119
complex dynamics 172, 183, 207
conjugacy 183
conjugate 183
Conley index 123, 172
Conley–Ważewski theory 129
continuum 133, 147, 148, 150, 168
counterexample 142
covering 183
crossing 136, 171, 183
Crossing Lemma 145, 151
crossing number 131
cutting set 152
cutting surface 133, 166
cutting surfaces 128

D
deformation relation 168
degree theory 147
Devaney chaos 192, 197
Devaney chaotic maps 192
diagonal dominance 69

E
edge subshift 186, 187
equicontinuous 4, 5, 13
existence and uniqueness 116
expansive 123
extinction 65, 109, 114, 119
extinction of species 84

F
family of expanders 141
fixed point 47, 130, 132, 133, 135, 142, 145, 168
fixed point theorems 136

G
global repellor 53
globally asymptotically stable 3, 40, 65, 100, 109
globally attractive 39
Gronwall's inequality 16

H
homeomorphism 132, 162, 164, 190
homoclinic point 220
horseshoe 169
Hurewicz–Wallman Lemma 156

I
impulses 91, 92
impulsive differential inequality 116
impulsive equation 95
impulsive linear equation 113
impulsive logistic equation 96, 102, 111
impulsive logistic equations 94
impulsive Lotka–Volterra system 95
impulsive perturbations 92
impulsive system 104
invariant 175
invariant set 179
invariant under homeomorphisms 161

J
J-permanent 2

K

Kolmogorov systems 1, 5, 39, 207

L

Lefschetz fixed point theory 172
Lefschetz number 123, 127
Leray–Schauder 127
Leray–Schauder Continuation Principle 145
Leray–Schauder Continuation Theorem 148, 160
Li–Yorke chaos 198
linked twist maps 124, 174
Lipschitzian 1
locally arcwise connected 153, 154
logistic equation 65, 67, 79
Lotka–Volterra system 6, 20, 26, 76, 215
Lyapunov function 40, 81, 90, 94, 98, 100, 105, 107, 115, 120
Lyapunov method 67

M

Markov partitions 129

N

nonautonomous 68, 69
nonincreasing 9, 10
nullcline planes 59

O

oriented N-dimensional rectangle 164, 166
oriented N-dimensional rectangles 167

P

partially permanent 2, 3, 19
periodic 6
periodic points 123, 172, 176
periodic solutions 206, 209
permanent 2, 16, 38, 97, 108
Perron–Frobenius Theorem 49
piecewise continuous 91, 92, 101
planar map 142, 144, 163
planar stretching property 168
Poincaré map 124, 206, 207, 211, 222, 223
Poincaré–Miranda Theorem 127, 151, 156, 221
population dynamics 124
positive solution 77–79, 81, 83, 110
positive T-periodic solution 97
predator–prey model 124
predator–prey systems 209

R

repellor 53

S

SAP method 124
scrambled set 191
semiconjugacy 183, 185
slice 135
Smale horseshoes 199, 213
split Lyapunov function 42
starting set 133, 135
stretching 137
stretching along the continua 141
Stretching Along the Paths 207
stretching along the paths 130, 164, 168, 198
Stretching Along the Paths(SAP) 123
stretching properties 220
strongly competitive 53
subshift 185, 189
subsystem 21
superlinear 123
survival of species 82
symbolic dynamics 181

T

T-periodic 113
T-periodic solution 65, 68
target set 133, 135
Tietze–Urysohn Theorem 165
topological degree 158, 172
topological horseshoes 164, 170, 172
topological method 123
topologically conjugate 183
topologically semiconjugate 183
two-sided Bernoulli shift 207

U

uniformly bounded 2, 5, 38
uniformly bounded away 2, 3

V

vertex subshift 186
vertex subshifts 187
Volterra predator–prey model 206, 207
Volterra's principle 208

W

Whyburn Lemma 145–147, 150
winding number 145

Bei Fragen zur Produktsicherheit wenden Sie sich bitte an:
If you have any questions regarding product safety,
please contact:

Walter de Gruyter GmbH
Genthiner Straße 13
10785 Berlin
productsafety@degruyterbrill.com